Nuclear Reactor Safety

Nuclear Reactor Safety
On the History of the Regulatory Process

David Okrent

The University of Wisconsin Press

Published 1981

The University of Wisconsin Press
114 North Murray Street
Madison, Wisconsin 53715

The University of Wisconsin Press, Ltd.
1 Gower Street
London WC1E 6HA, England

First printing

Printed in the United States of America

For LC CIP information see the colophon

ISBN 0-299-08350-0

Contents

Preface

This manuscript is derived from a long report which examines the history of the evolution of light water reactor safety. In casting about for an approach to writing on this subject, I inevitably relied on the perspective I knew best, namely, that of a member of the Advisory Committee on Reactor Safeguards (ACRS). I was first appointed to the ACRS as of November 1, 1963, and at its fifty-first meeting, November 7–8, 1963, as a brand new member, I listened to committee discussion of the proposed Ravenswood reactor, among other things.

Partly because of ACRS self-imposed restrictions on member interpretation of past ACRS actions, and partly because I decided not to act as a major filter on what information and opinion was available, I decided in writing the long report, which was about 1,200 pages long, to rely heavily on extracts and even complete reproductions of large portions of ACRS minutes, letters, and documents. The present book represents a much condensed version of the earlier long report; most of the quoted source material has been paraphrased in abbreviated fashion.

I did not have access to the files of the regulatory staff; hence, this history will suffer from a lack of the insight which could have been obtained from staff documents. Having even less access to the files of regulatory groups in other countries, I have avoided trying to present any history of foreign safety requirements for light water reactors.

Chapters 3–14 form the central portion of the document. They are intended to provide a historical view of the development of siting policy and the major safety issues which interacted strongly with siting policy. The other chapters represent a very incomplete selection of the very many important issues and developments in light water reactor safety, and they are presented in much less depth. Coverage of the loss-of-coolant accident (LOCA) has deliberately been abbreviated to include only a few selected aspects. It would require a book at least as long as this one to do justice to this one topic.

The original report was completed in the spring of 1978, and did not comment on the implications of the Three Mile Island accident, which occurred in March 1979. To a limited extent, this manuscript provides some personal opinions. However, it is intended to look at history, not to examine and comment on the question, "How safe is safe enough?"

Acknowledgments

The long report on which this manuscript is based was prepared while I had the benefit of a sabbatical leave from the University of California, Los Angeles, and a (second) fellowship from the John Simon Guggenheim Memorial Foundation. During this period (July 1977–June 1978) I was a visitor with the Département de Sureté Nucléaire, Commissariat à l'Energie Atomique, at Saclay and Fontenay-aux-Roses, France; at the International Atomic Energy Agency, Vienna, Austria; and at the Department of Nuclear Engineering, Technion–Israel Institute of Technology, Haifa, Israel, where I held the Isaac Taylor Chair.

Various individuals and groups have contributed in one fashion or another and I wish to express my appreciation, I hope without too many oversights: Donald G. Browne and Thomas E. McKone for editorial assistance; Jolie McNulty for typing; the Staff of the Nuclear Engineering Department at the Technion for typing, Xeroxing, and a pleasant environment; the staff of the Advisory Committee on Reactor Safeguards (ACRS) for finding various documents; Milton Plesset and Stephen Lawroski for strong support of the original idea; Morton Libarkin, Raymond F. Fraley and Dade W. Moeller for struggling through an early, partial draft of the report and supplying detailed suggestions and comments; Irving Rockwood for making many invaluable suggestions for the manuscript of the book; and the author's family for living for months with boxes of documents in our home while I was on sabbatical leave and for accepting the reductions in leisure time introduced by the writing of this manuscript.

Finally, I would like to express my appreciation for having had the opportunity to work with the members of the ACRS for many years.

Glossary of Terms

ACRS	Advisory Committee on Reactor Safeguards
AEC	Atomic Energy Commission
AIF	Atomic Industrial Forum
ASLB	Atomic Safety and Licensing Board
ASME	American Society of Mechanical Engineers
ATWS	Anticipated Transients Without Scram
BASF	Badische Anilin-und Soda-Fabrik AG
BWR	Boiling Water Reactor
Class 9	Accident which goes beyond the design basis for the reactor and causes a radioactive release in which consequences exceed 10 CFR Part 100
DRL	Division of Reactor Licensing
EBR	Experimental Breeder Reactor
EPRI	Electric Power Research Institute
ECCS	Emergency Core-Cooling System
FSAR	Final Safety Analysis Report
IAEA	International Atomic Energy Agency
JCAE	Joint (congressional) Committee on Atomic Energy
LOCA	Loss-of-Coolant Accident
LOFT	Loss-of-Fluid Test
LWR	Light Water Reactor
MCA	Maximum Credible Accident
MTR	Materials Testing Reactor
MWe	Megawatt Electric
MWt	Megawatt Thermal
NACA	National Advisory Committee for Aeronautics
NEPA	National Environmental Policy Act
NRC	Nuclear Regulatory Commission
ORNL	Oak Ridge National Laboratory
PDF	Population Density Factor
PSAR	Preliminary Safety Analysis Report
PWR	Pressurized Water Reactor
RSI	Reactor Site Index

Scram	Rapid insertion of all control rods into core as fast as the mechanisms permit in order to terminate the chain reaction
SEFOR	Southwest Experimental Fast Oxide Reactor
SPF	Site-Population Factor
SRE	Sodium Reactor Experiment
SSE	Safe Shutdown Earthquake

Chronology of Some Important Events in the History of Light Water Reactor Safety

1947
  Establishment of first Reactor Safeguards Committee *Chapter 1*
1950
  Issuance of WASH-3, including "rule of thumb" siting criterion
    Chapter 3
1952
  Evolution of concept of containment *Chapter 3*
1953
  Expression of safety philosophy by Edward Teller *Chapter 1*
1953-1954
  Announcement and approval of Shippingport PWR *Chapter 3*
1956
  Approval of construction of Indian Point 1 reactor *Chapter 3*
  Exchange of letters between Senator Hickenlooper and AEC Chairman
    Libby *Chapter 3*
1957
  Issuance of WASH-740 concerning effects of large reactor accidents
    Chapter 1
  Passage of the Price-Anderson Act *Chapter 1*
  Establishment of statutory ACRS *Chapter 1*
  First dissent by ACRS member (Abel Wolman) on Plum Brook reactor
    Chapter 3
1958
  AEC Chairman McCone considers the ACRS overconservative *Chapter 3*
  Discussion of "maximum credible accident" concept by Clifford Beck
    Chapter 3
  ACRS reports unfavorably on first proposed site (and reactor concept)
    for Piqua, Ohio *Chapter 3*
1960
  ACRS comments on nuclear power plants in California *Chapter 3*

ACRS decision not to issue report on problems arising from primary system rupture when AEC proposed establishment of task force to study the matter *Chapter 8*

The ACRS safety research letter of October 12, 1966, on matters related to large-scale core melt, LOCA–ECCS, and other concerns *Chapter 11*

Initiation of heavy section steel test program of safety research for reactor pressure vessels *Chapter 15*

1966–1967

Evolution of requirements for improved integrity of pressure vessels *Chapter 15*

1967

Browns Ferry review and the asterisked items *Chapters 9, 18*

Task force report on power reactor emergency cooling *Chapter 11*

Proposal for probabilistic approach to safety evaluation by F. R. Farmer *Chapter 12*

Finding against the Burlington site by the ACRS and the regulatory staff *Chapter 13*

AEC rejection of proposed limit on population density around reactor site *Chapter 13*

Development of AEC requirements for pre-operational and in-service inspection of reactor pressure vessels *Chapter 15*

AEC publication of second version of general design criteria for comment *Chapter 15*

Revision of safety research program in LOFT as part of new emphasis on ECCS *Chapter 18*

1967–1968

Separation of protection and control *Chapter 16*

1968

Development of ACRS site-population index *Chapters 9, 13*

Zion review *Chapter 10*

Removal of core catcher from Indian Point 2 reactor *Chapters 10, 11*

ACRS letter of February 26, 1968, on task force report *Chapter 11*

Bolsa Island, Trap Rock, Bowline and Montrose sites *Chapter 13*

Oyster Creek review and the development of more stringent quality assurance requirements *Chapter 15*

Steam-line break accident *Chapter 15*

1969

ACRS draft report of July, 1969, on location of power reactors at sites of population density greater than Indian Point and Zion *Chapter 13*

Birth of ATWS *Chapter 16*

1969–1973
  Newbold Island review *Chapter 14*
1970
  Meetings of ACRS and regulatory staff with representatives of nuclear
    industry concerning metropolitan siting *Chapter 13*
  ACRS report on use of water-cooled power reactors at sites more
    densely populated than those employed to date (never issued)
    Chapter 13
  Initiation of safety guides (later renamed regulatory guides) as means of
    defining technical positions acceptable to regulatory staff and ACRS
    Chapter 15
  Adoption of AEC regulation 10 CFR 50.109 Backfitting *Chapters 15, 17*
1970–1971
  Study by duPont on molten core retention and ACRS letter of January
    11, 1971, to Milton Shaw on safety research program *Chapter 11*
1971
  Adoption of general design criteria as Appendix A to 10 CFR Part 50
    Chapter 15
  Interim acceptance criteria for ECCS *Chapter 18*
  Issuance of 10 CFR Part 51 on environmental protection *Chapter 19*
1971–1972
  Revision of ACRS report on acceptance criteria for ECCS *Chapter 18*
1972
  Federal Advisory Committee Act *Chapter 1*
  Letter of February 3, 1972, from Milton Shaw to ACRS opposing
    research and development on molten core retention *Chapter 11*
  ACRS letter of February 10, 1972, on water reactor safety research
    Chapter 18
  First ACRS letter on generic items *Chapter 18*
1973
  WASH-1270, A staff position on ATWS *Chapter 16*
  AEC publication of seismic and geologic siting criteria *Chapter 17*
  Reorganization of safety research in AEC, placing LWR safety research
    under H. Kouts *Chapter 18*
  Final acceptance criteria for ECCS *Chapter 18*
1974
  Issuance of WASH-1285, a report by the ACRS on pressure vessel
    integrity *Chapter 15*
  Energy Reorganization Act of 1974, establishing the NRC, including its
    own safety research program *Chapter 18*
  Issuance of draft WASH-1400 *Chapter 19*

1975

Birth of NRC *Chapter 1*

Site-population density guidelines adopted in NRC standard review plan *Chapter 14*

Browns Ferry fire *Chapter 18*

Final version of WASH-1400 *Chapter 19*

1976

The Government in the Sunshine Act *Chapter 1*

1976–1978

Consideration of class 9 accidents in review of floating nuclear power plants *Chapter 19*

1977

ASLB appeal board ruling concerning emergency evacuation and definition of exclusion and low-population zone boundaries *Chapter 19*

ASLB hearing board examination of class 9 accidents in connection with Black Fox reactor *Chapter 19*

1978

First annual report by ACRS to Congress on NRC Safety Research Program *Chapter 18*

NUREG-0438, report to Congress by NRC on plan for research to improve the safety of light water nuclear power plants *Chapter 18*

1979

Three Mile Island accident *Chapter 19*

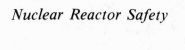
Nuclear Reactor Safety

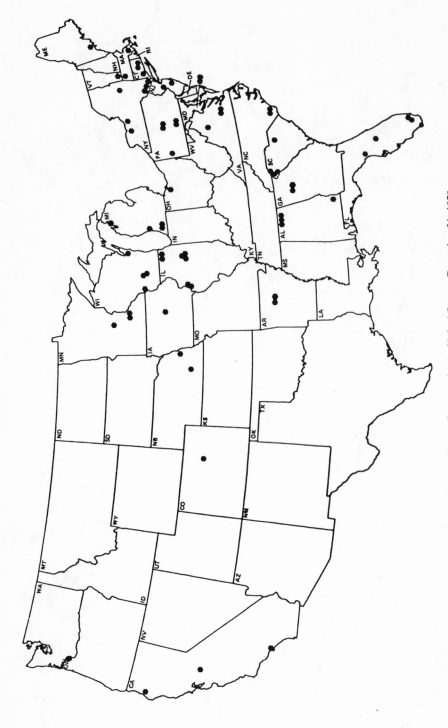

Licensed Nuclear Power Reactors in the United States, December 31, 1978

1

The ACRS
and the Regulatory Process

The Atomic Energy Act of 1946, establishing tight Federal government control, essentially was designed to safeguard the United States monopoly of atomic bombs and to expand the military applications of the new technology. The Congress decided on civilian control of atomic energy, embodied in a five-member Atomic Energy Commission (AEC). The new law did provide a statutory basis for peaceful applications under a system of government licensing, and many of those who participated in the Congressional hearings and debates in 1946 anticipated the production of electrical power. But not until passage of the Atomic Energy Act of 1954 did the AEC begin to move in a major way toward the licensing of civilian nuclear power facilities.

During the period between 1946 and 1954, the AEC was responsible for the operation and safety of the military reactors employed for production of plutonium and tritium. And the AEC initiated a reactor program which included the Materials Testing Reactor and the Experimental Breeder Reactor (EBR-1). Both were built at the National Reactor Testing Site in Idaho, and EBR-1, a small, one megawatt thermal (1 MWt) fast breeder reactor, was the first nuclear reactor in the world to generate electricity.

The regulation of nuclear power reactors had its inception in the United States in late 1947. In June of that year, the AEC discussed with its General Advisory Committee the problem of evaluating the safety of nuclear

3

reactors. It was concluded that a panel should be established to advise the commission on reactor safety matters. This was done by inviting highly qualified individuals with backgrounds in appropriate scientific disciplines to serve on a Reactor Safeguards Committee reporting to the AEC general manager. Beginning in the fall of 1947, meetings of this committee were held to consider and advise upon matters referred to it. In 1951 the commission also established an Industrial Committee on Reactor Location Problems charged with the responsibility of advising upon the siting of nuclear reactors.* The Reactor Safeguards Committee and the Industrial Committee on Reactor Location Problems were combined by the AEC in July of 1953 and renamed the Advisory Committee on Reactor Safeguards, or ACRS.

The ACRS was assigned the following functions: (1) reviewing hazards summary reports prepared by organizations planning to build or operate reactor facilities. (2) Advising the AEC regarding the consistency of proposed reactor locations with accepted industrial safety standards, taking into account the proximity of surrounding population and property.

The first members of the ACRS were as follows: C. Rogers McCullough, Monsanto Chemical Co., chairman; Manson Benedict, Massachusetts Institute of Technology; Willard P. Conner, Jr., Hercules Powder Co.; R. L. Doan, Phillips Petroleum Co.; Hymer Friedell, Western Reserve University; I. B. Johns, Monsanto Chemical Co.; Mark M. Mills, North American Aviation, Inc.; Kenneth R. Osborn, Allied Chemical and Dye Corp.; Donald A. Rogers, Allied Chemical and Dye Corp.; Reuel C. Stratton, Travelers Insurance Co.; Edward Teller, University of California; Abel Wolman, Johns Hopkins University; Harry Wexler, U.S. Weather Bureau, Department of Commerce; and C. R. Russell, secretary, AEC.

AEC Safety Philosophy

Very little had been said publicly prior to 1953 on AEC philosophy regarding reactor safety. Therefore, the following statement, made in 1953 to the Joint Committee on Atomic Energy by Edward Teller, formerly chairman, Reactor Safeguards Committee, was of particular interest:

Up to the present time we have been extremely fortunate in that accidents in nuclear reactors have not caused any fatalities. With expanding applications of

*Reminiscing about 25 years later, Richard Doan recalled that the AEC felt that the Reactor Safeguards Committee, which consisted largely of academics, was too conservative. The AEC established a new committee comprised of members from industry to advise it on reactor siting. The "Industrial Committee" independently provided the same "conservative" advice as had the first committee, so the two were combined by the AEC.

nuclear reactors and nuclear power, it cannot be expected that this unbroken record will be maintained. It must be realized that this good record was achieved to a considerable extent because of safety measures which have necessarily retarded development.

The main factors which influence reactor safety are, in my opinion, reasonably well understood. There have been in the past years a few minor incidents, all of which have been caused by neglect of clearly formulated safety rules. Such occasional accidents cannot be avoided. It is rather remarkable that they have occurred in such a small number of instances. I want to emphasize in particular that the operation of nuclear reactors is not mysterious and that the irregularities are no more unexpected than accidents which happen on account of disregard of traffic regulations.

In the popular opinion, the main danger of a nuclear pile is due to the possibility that it may explode. It should be pointed out, however, that such an explosion, although possible, is likely to be harmful only in the immediate surroundings and will probably be limited in its destructive effects to the operators. A much greater public hazard is due to the fact that nuclear plants contain radioactive poisons. In a nuclear accident, these poisons may be liberated into the atmosphere or into the water supply. In fact, the radioactive poisons produced in a powerful nuclear reactor will retain a dangerous concentration even after they have been carried downwind to a distance of ten miles. Some danger might possibly persist to distances as great as one hundred miles.

It would seem appropriate that Federal regulations should apply to a hazard which is not confined by state boundaries. The various committees dealing with reactor safety have come to the conclusion that none of the more powerful reactors built or suggested up to the present time are absolutely safe. Though the possibility of an accident seems small, a release of the active products in a city or densely populated area would lead to disastrous results.

It has been therefore the practice of these committees to recommend the observance of exclusion distances, that is, to exclude the public from areas around reactors, the size of the area varying in appropriate manner with the amount of radioactive poison that the reactor might release. Rigid enforcement of such exclusion distances might hamper future development of reactors to an unreasonable extent. In particular, the danger that a reactor might malfunction and release its radioactive poison differs for different kinds of reactors.

It is my opinion that reactors of sufficiently safe types might be developed in the near future. Apart from the basic construction of the reactor, underground location or particularly thoughtfully constructed safety devices might be considered.

It is clear that no legislation will be able to stop future accidents and avoid completely occasional loss of life. It is my opinion that the unavoidable danger which will remain after all reasonable controls have been employed must not stand in the way of rapid development of nuclear power. It also would seem that proper legislation at the present time might make provisions for safe construction and safe operation of nuclear reactors.

It would seem reasonable to extend the AEC procedures on reviewing planned reactors and supervising functioning reactors to nuclear plants under the control of private enterprise. To what extent these functions should be advisory or regulatory is a difficult question. I feel that ultimate responsibility for safe operation will have to be placed on the shoulders of the men and the organizations most closely connected with the construction and the operation of the reactor.

The Beginning of the Use of Nuclear Power
for the Generation of Electricity

In the years from 1946 to 1954 there was considerable effort on the development of the nuclear submarine, as well as the initiation of research programs on a variety of possible reactor designs suitable for the production of electricity. There was substantial agreement that the technical questions about whether a reactor could produce electricity on a large scale had been essentially solved, but economic viability was another question.

One development of especial significance during this period was the AEC's decision to support the building of a demonstration pressurized water reactor (PWR). A small PWR had been developed in the submarine propulsion program, and a larger unit, employing slightly enriched uranium, had been proposed for a nuclear-powered aircraft carrier, but President Eisenhower decided to cancel this project in 1953. However, in his famous "Atoms for Peace" speech at the United Nations on December 8, 1953, Eisenhower proposed the establishment of an International Atomic Energy Agency (IAEA) to manage the use of atomic energy "to serve the peaceful pursuits of mankind" and generally endorsed the development of nuclear energy. AEC support of a civilian PWR both helped launch the expeditious development of electricity generation from nuclear power in the United States, and enabled the development of a power plant having characteristics suitable for an aircraft carrier propulsion unit.

In December 1953 the AEC invited private industry to submit proposals for the PWR project. Three months later, AEC Chairman, Lewis L. Strauss, announced the selection of the Duquesne Light Company which had offered to furnish a site at Shippingport, Pennsylvania, and would operate the plant. It would be designed and constructed by Westinghouse under close supervision by the AEC Naval Reactors Division. Ground was broken for the Shippingport plant in September 1954, and it began to operate in 1957.

The Atomic Energy Act of 1954 was passed after much debate concerning the merits of public versus private power. Among other things the act provided for "a program to encourage widespread participation in the development and utilization of atomic energy for peaceful purposes to the maximum extent consistent with the common defense and security and with the health and safety of the public." The act provided for the issuance of commercial licenses for power reactors, and gave the AEC the responsibility for protecting adequately the public health, safety, life, and property. The AEC could issue a construction permit once an application was received in writing. Then, as long as the facility met the provisions of the act and the rules and regulations of the commission, an operating license could be issued. The act allowed a public hearing "upon the request of any person whose interest may be affected by the proceeding."

Thus the AEC was given the general responsibility both for encouraging nuclear power and for licensing it. The Congress left it to the AEC to determine what constituted "adequate" protection of the public health and safety.

The AEC formed a Division of Civilian Application in 1955 to license reactors and established within this division a Hazards Evaluation Branch. The Hazards Evaluation Branch initially relied heavily on the opinions of the Advisory Committee on Reactor Safeguards (ACRS), which had been formed in 1953 and up to 1955 had functioned as the major advisory group on safety issues of the heretofore Government-owned atomic projects.

Under the initial licensing regulations, reviews of applications for construction permits were evaluated by the Hazards Evaluation Branch, which next (or concurrently) sent the application to the ACRS for review and then forwarded it, through the AEC general manager, to the commissioners. The commission did not publish a document setting out the safety problems, if any, raised by an application, nor did it make publicly available the reports it received from the ACRS. Also, public notice of commission action on an application represented a *fait accompli*.

The Price-Anderson Act

In 1956, the commissioners, in a split vote, authorized a construction permit to a utility consortium headed by Detroit Edison to build a 100 MWe developmental fast breeder reactor in Lagoona Beach, Michigan (known as the Fermi reactor). The AEC did so despite both an ACRS report and a staff analysis questioning whether the consortium could provide adequate information assuring safe operation prior to the time for issuance of an operating license.

The Joint Congressional Committiee on Atomic Energy (JCAE) learned of this from the only dissenting commissioner, Thomas Murray, and the incident led to a series of Congressional hearings and new legislation in 1957, which, among other things, established the ACRS as a statutory body, and required it to review all applications for construction and operation of nuclear facilities and to make a public report on its review. The new law also made public hearings on all applications for licenses for nuclear reactors mandatory.

The Price-Anderson act in 1957 also established the liability limits and no-fault provisions for insurance on nuclear reactor accidents. One of the important technical inputs into establishing this aspect of the law was a report given the number WASH-740 and entitled, "Theoretical Possibilities and Consequences of Major Accidents in Large Nuclear Power Plants," which was prepared by Brookhaven National Laboratory and published by the AEC in March, 1957 (AEC 1957).

In 1961 the structure of the AEC was modified so that, while the AEC maintained both promotional and regulatory functions, the regulatory staff no longer reported to the commission via the general manager. Instead, there was established a Director of Regulation who reported directly to the commission. The AEC safety research program, however, was left in the Division of Reactor Development on the promotional side of the commission under the general manager, and hence was not under the direct control of the regulatory staff.

Public hearings were initially conducted by a Hearing Examiner who was a lawyer. In 1962, Congress amended the law and authorized the AEC to appoint Atomic Safety and Licensing Boards (ASLB) composed of two technical members and a lawyer. In 1969, the AEC created Atomic Safety Licensing Appeal Boards, which took over from the commission itself the function of ruling on an appeal from an ASLB hearing. The ASLB hearing now provides a direct opportunity for legal intervention by the public, as well as a legal opportunity for an applicant for a construction permit to disagree with the regulatory staff. If not overruled following an appeal to the commission, the ASLB ruling is binding on the regulatory staff.

The ACRS Role in the Licensing Process

With the institution of the Atomic Safety Licensing Appeal Boards in 1969, the regulatory process took the form which it has retained up through the present, except that the entire AEC regulatory apparatus was divorced from the AEC and a new Nuclear Regulatory Commission (NRC) with five commissioners was established and began work in early 1975. The regulatory process would proceed as follows: When a utility had purchased a nuclear steam supply system and had arranged for the engineering services required for the balance of the nuclear power plant, which costs far more than the nuclear reactor itself, the utility would have a preliminary safety analysis report (PSAR) prepared and would submit this report as part of its application for a construction permit. The regulatory staff would undertake a review of the specific application, including safety-related aspects of both the proposed site and the plant design. The regulatory staff would send out a large number of written questions to the utility concerning the

PSAR. These would be answered in writing and would usually be followed by two more rounds of questions.

When the review of the regulatory staff was nearing completion, it would prepare a safety evaluation report which would summarize the findings of the regulatory staff, including any reservations or unresolved issues.

The ACRS would then complete its review of the application for the construction permit and submit a public letter to the AEC (or NRC) giving its advice and recommendations concerning the application. After the ACRS letter had been submitted, an ASLB Hearing could proceed and be completed. The ACRS letter would not be a part of the ASLB record and the ACRS recommendations need not be followed by the regulatory staff or even discussed at the ASLB Hearing.

An ASLB Hearing is mandatory at the construction permit stage, even if there is no intervention concerning the case.

When the construction of a nuclear power plant was well along, the utility would submit a final safety analysis report (FSAR) as part of its application for an operating license. This report would provide more detail on the actual design of the plant and would reflect changes made since the PSAR. The regulatory staff would follow a similar review procedure as it did earlier for the construction permit. When the staff believed it was ready to proceed, it would again issue a safety evaluation report. Review by the ACRS and a public ACRS report on the application for an operating license would again be mandatory. However, a public ASLB Hearing would only be required if there were intervention into the proceedings.

The 1957 legislation, which provided for a statutory ACRS, stated that the members of the ACRS were to be appointed by the AEC commissioners for a term of four years each. The ACRS itself was to designate one of its members to act as committee chairman. In practice this election has been for a calendar year, and since 1958, the committee bylaws have provided that the chairman could not succeed himself. The law allows a maximum of 15 members. The members, who serve part-time, were not to be AEC employees, and since they would not be dependent upon the AEC for their livelihood, would be free of economic pressure from the AEC.

The manner in which the ACRS functions is of some interest. Although changes have occurred through the years, much has remained the same. The regulatory staff constitutes the group legally responsible for reactor licensing, and, as such, receives both applications for construction permits and also the various safety-related documents, such as preliminary safety analysis reports (PSAR) and final safety analysis reports (FSAR). Copies of these are sent to the ACRS. The written questions prepared by the regulatory staff and the responses by the applicant are sent to the ACRS, when they are ready. These and other relevant documents are made available to each ACRS member in advance of the committee meeting at which the project is to be considered. The ACRS members are usually experts in specific areas of importance in reactor safety and will pay special attention to these specific areas in their individual review of documents. The ACRS appoints a subcommittee for each project. There are also subcommittees appointed to consider general questions such as fires, reactor pressure vessels, or earthquakes. It is the function of an ACRS subcommittee to develop the information it believes the full committee needs for review of the project or general question, and the subcommittee may hold one or several meetings performing its function. It is not the function of the subcommittee to provide any opinion to the applicant or the staff, or necessarily, even to the full committee. Any ACRS member can attend any subcommittee meeting.

Prior to review of a project by the full committee, the safety evaluation report prepared by the regulatory staff is also furnished to the ACRS members.

At subcommittee and full committee meetings, representatives of the regulatory staff and of the reactor designer and reactor operator appear before the assembled committee members to describe the proposed reactor and to discuss whichever safety aspects the committee wishes to emphasize. There is a free technical exchange between the committee and the staff and between the committee and the representatives of the applicant. The ACRS can send written questions and request written answers from an applicant, usually under the auspices of the staff, but this is not commonly done.

Prior to the early 1970s, with the advent of the Government in the Sunshine Act and the Federal Advisory Committee Act, ACRS meetings were closed to the public.* In those days the ACRS could meet separately with the staff, with the staff and applicant, or in closed executive session.

When the ACRS decided it had sufficient information on which to reach a decision, it met in executive session and prepared a letter report to the chairman of the Atomic Energy Commission. The report is signed by the chairman of the ACRS but prepared by the committee as a whole. The ACRS report is almost always a consensus. An individual member has the right to issue his own separate opinion as part of the report.

As noted earlier, the ACRS letter on a construction permit or operating license is only advisory, and is not a formal part of the ASLB hearing. However, the ASLB must have a report by the ACRS before the Board can complete its action. Hence, a potential power of the ACRS is that of delay.

*The Federal Advisory Committee Act, Public Law 92-463, dated October 6, 1972, required that meetings of most Federal advisory committees, including the ACRS, be held in the open. The Government in the Sunshine Act, which was passed in 1976, required that government agencies conduct their activities in the open.

If it lacked information sufficient to reach a decision, it could not issue its own report on a specific case, thereby holding that case up.

Prior to 1965, the ACRS held full committee meetings every two months, and these lasted for two or three days. Special meetings were called as needed. Since 1965, the ACRS has generally held three-day-long, full committee meetings each month, with occasional special meetings.

There is no such thing as a standard ACRS meeting agenda. The agenda, which is established by the ACRS itself, depends on the status of upcoming construction permit and operating license reviews, on the existence of any recent unusual operating experience, on the need to discuss the safety research program or some safety issue of general interest, etc.

ACRS members are appointed by the AEC (or NRC) commissioners. Most but not all appointments have been made based on recommendations by the ACRS itself. These recommendations are for individuals who are expert in a new area of growing importance or an area in which a gap has been created due to departure of a previous member. The commissioners could and sometimes did try to influence the ACRS by choosing an appointee who they anticipated would not be overly stringent with regard to safety requirements.

It is important to emphasize that the statutory ACRS elects its own chairman, establishes its own agenda, selects its own consultants, and decides for itself when it is ready to write a report. And, while the ACRS has obligations to respond to specific requests for advice from the commission, it can on its own initiative also take up matters falling in its general province of responsibility.

That this very considerable independence of function should exist is, in a sense, remarkable. Such independence did not occur automatically. There were efforts from within the AEC to limit the freedom and scope of the prestatutory Reactor Safeguards Committee, and to restrict the scope and operations of the statutory ACRS during the first years of its existence. For example, the regulatory staff recommended that the ACRS not be allowed to meet in executive session or to have its own consultants. Again, during the latter half of the 1960s, various voices were raised from among the Congress, the nuclear industry, and within the AEC itself in favor of curtailing the ACRS responsibilities and scope of action. However, such curtailment of ACRS independence has not occurred.

2

A Brief Overview
on Siting of Nuclear Reactors

The complexity and great breadth of the field of nuclear reactor safety make it impossible even to attempt to review all of its history. Contributions to the evolutionary process have come from many countries and from many individuals and institutions within a country. Rather than try to do justice to all the participants in this process, we shall try in this semi-historical review to look at several aspects of reactor safety from the particular vantage point of the ACRS.

The siting of reactors has an important influence on the hazards and risks posed by the reactor. For a large uncontrolled air-borne release of radioactive material arising from some unforeseen accident, distance to surrounding population is important. Not only does it provide more time for warning and evacuation, but also the concentration of radioactivity in the air will generally drop with distance. Thus, even if a major release of radioactivity has occurred, the likelihood of a lethal dose falls off very rapidly and becomes very small at 10 or 15 miles. On the other hand, the total societal dose, which results from smaller exposures of radioactivity to very many people, is less sensitive to population distribution, since many sites have similar average population densities at distances greater than 25 miles.

Of course, the siting of reactors affects public safety in other ways including the effects of small routine releases of radioactivity, the effects of smaller, controlled accidental releases, and the potential effects from radioactivity which gets into the ground and may later appear in drinking water.

Reactor siting has been the subject of several International Atomic Energy Agency Conferences (IAEA 1962, 1963, 1967, 1973, 1974) and has been featured in the reactor safety discussions at the four United Nations "Atoms for Peace" Conferences held in Geneva, Switzerland (United Nations 1955, 1958, 1964, 1971). Many countries have developed their own approaches to dealing with the siting of power reactors. However, site criteria, in general, are far from definitive, tend to be comparative or relative, and are built around precedent, once some history of siting has occurred.

This examination of reactor siting will deal primarily with demographic (population) considerations. We will find it convenient to divide the discussion loosely into several time-periods: pre–1961; 1961–1965; late 1965–1966; 1966–1968; 1968–1970; and 1970–present.

We shall see that in 1950 the first AEC Reactor Safeguards Committee produced a very restrictive rule of thumb site criterion which related the reactor power to the required exclusion distance (or radius of the very lightly populated region surrounding the reactor that was owned by, or under direct control of the reactor operator). In the rule of thumb, the occurrence of a gross release of radioactivity from an uncontained reactor was assumed, and the site criterion required a large exclusion radius to meet its requirements. However, pressures were already building in 1950 to construct reactors at sites which strongly violated the rule of thumb, and the concept of reactor containment emerged quickly.

In the mid-1950s, in addition to several test and experimental reactors, three light water power reactors of substantial size that employed containment buildings to hold the radioactivity in case of an accident were approved for construction. This included the one at the Indian Point, New York, site, which remains to this day the most populated site used in the United States for a light water reactor (LWR). In the late 1950s several other smaller reactors, all with containment buildings and at rural sites, were approved. During the same period, a few small power reactors (60 MWt) were proposed for sites within or at the very edge of small- or medium-size cities; these were rejected or forced to move to somewhat less populated sites, which were grudgingly accepted by the ACRS.

The review of these reactors was on a case-by-case basis, a judgment being reached concerning acceptability of each specific reactor-site combination. No AEC reactor site criteria had ueen published; and, from a review of the ACRS minutes and other documents, there does not appear to have been any single guiding philosophy.

The AEC commissioners themselves seem not to have exercised strong direction on the development of siting criteria in the period prior to 1960. They may, however, have exercised strong or decisive influence on the favorable decisions for construction of the Shippingport, Dresden, and Indian Point reactors (the detailed history of these early, precedent-setting reactor reviews is not available to the author). In the late 1950s, the commissioners appeared particularly interested in the development of written criteria that were as specific as possible. This was to help to avert the awkward situations wherein considerable work was expended on reactor proposals involving sites that were later evaluated unfavorably by the ACRS and/or the AEC regulatory staff.

The first rather detailed expression of ACRS philosophy on power reactor siting came in its reports to the AEC on this subject in October and December of 1960. Independently, although in close association with the ACRS, the AEC regulatory staff (then called the Hazards Evaluation Branch) under the leadership of Dr. Clifford Beck, were developing site criteria. In 1961, the AEC published for comment in the Federal Register proposed site criteria which included the concepts of an exclusion area, a low-population (or evacuation) zone, and a population center distance. Also for purposes of site evaluation, an accident was postulated in which the noble gases and half the radioiodine were released to a containment building that was assumed to maintain its integrity, and in which guideline doses of 25 rem whole body and 300 rem to the thyroid were not to be exceeded under the specified conditions. This postulated accident (the maximum credible accident or MCA), whose consequences were not to be exceeded by any credible accident, became the focus of siting evaluation.

Until 1966, the 10 CFR Part 100 reactor site criteria,* which were adopted by the AEC in 1962, led to a somewhat stylized reactor safety review, in which primary emphasis was placed on containment design and certain engineered safety features that could enable the increasingly higher power reactors to meet the guidelines doses. Reactors were receiving construction permits based on rather general safety criteria and sketchy design information. Most safety improvements which developed were related to meeting the requirements of the postulated MCA.

We shall see that the manner in which the Part 100 site criteria of 1962 were developed and implemented made them lose the basic impact of two major philosophic recommendations made by the ACRS in its October, 1960, report. These were the concept of a limit on the integrated man-rem dose from a credible accident and, more importantly, the recommendation

*10 CFR Part 100 refers to Title 10, Code of Federal Regulations, Part 100. Title 10 is used for nuclear power. The AEC and NRC place regulations in Part 20, 50, 100, etc., as they think appropriate for categorization.

that there be no catastrophic effects to a population center from an uncontained accident much worse than the maximum credible accident (which was assumed to be contained). And we shall find the regulatory staff and the ACRS approving new reactors without direct consideration of these concepts during the period 1963–1965.

Beginning in late 1965, the ACRS began to give increased emphasis to events that mght lead to consequences far worse than the stylized MCA. And in 1966, there occurred a revolution in LWR reactor licensing which for the next 12 years led both to greatly increased measures to reduce the probability of occurrence of serious accidents and also to greatly increased attention to safety features needed to prevent core melt in the event of accidents.

The ACRS and the staff, in general, opposed the metropolitan or the near-metropolitan siting of power reactors, even if Part 100 could be met. The staff took the position in the 1960s that more experience was needed before metropolitan siting could be approved. The ACRS generally favored improvements in safety design in order for metropolitan siting to be considered. For those reactors it approved, the regulatory staff stayed rather stringently within the prescriptions of Part 100 and its maximum credible accident (later designated the design basis accident), and declined to publicly discuss or examine safety aspects related to accidents which went beyond the MCA (e.g., in which containment integrity was lost). This trend of not examining Class 9 accidents (accidents far exceeding Part 100 guidelines) and not evaluating measures which might ameliorate their impact remained unchanged by the regulatory staff essentially until 1976, when limited formal consideration of such accidents was first introduced, under ACRS pressure, into the review of the floating nuclear power plant.

The regulatory stance of the AEC commissioners themselves during the period 1960–1974 seems to have depended on the role they were playing. The commissioners did not favor adoption of any formal moratorium on metropolitan siting. When the highly populated Bolsa Island reactor site was under active consideration in the mid-1960s, the commissioners resisted the development of detailed siting criteria which might pose a hurdle to this project. Similarly, when in 1966 the ACRS was about to formally recommend the development and implementation of measures to deal with safety concerns arising from large-scale core melt, the commissioners instead proposed a task force to "study" the matter, thereby delaying (indefinitely) such a recommendation.

On the other hand, when required to accept a judicial role, such as that in connection with their review of the ASLB hearing on the Malibu site, which is discussed in Chapter 17, the commissioners supported the more conservative position of the Atomic Safety and Licensing board over that of the regulatory staff. In their decision on the Malibu site the commissioners, in a sense, provided the first quantitative guideline on acceptable risk, namely, they stated that the fact that the fault in question at Malibu had not moved in about 10,000 years did not provide adequate assurance that the plant need not be designed for surface displacement.

It must be noted that there is no definitive basis for judging that any approach or position adopted by individuals or groups during this era was right or wrong. Up to the present, at least, the regulation of nuclear power reactors has led to less adverse effect on the public health and safety than almost any other equivalent technology. It has been argued by some that there has been more protection (and expense for safety) than appropriate.

From 1966 on, while there has been no significant change in the reactor siting criteria, and while these criteria have been pursued rather legalistically by the regulatory staff (and therefore by applicants), there have been major changes in reactor design criteria and in the depth of safety review aimed at reducing the probability of an accident exceeding the guidelines of Part 100. However, the unwillingness to look at class 9 accidents or other badly degraded accident conditions by either the nuclear industry or the regulatory staff may have led to a less than optimum approach with regard to overall reduction in public risk for the same expenditure of effort and money as has been incurred. This matter is now getting active attention, as is the matter of revised site criteria.

3

Reactor Siting Before 1961

In the first few years after World War II, the relatively few reactors built in the United States for peaceful uses of atomic energy were of low power (less than 50 megawatts of thermal energy, i.e., 50 MWt), and usually sited on large government reservations away from large population centers. However, during the 1950s reactors of power up to 600 MWt were proposed for construction at a range of sites including some at the edge of small cities and one about 25 miles from New York City (Indian Point).

Siting affects safety since, in the event of a large uncontrolled release of radioactivity resulting from a serious reactor accident, the level or intensity of radiation will generally fall off with distance. Also, people living further from the reactor have more time to receive a warning and evacuate an area threatened with large radiation doses.

During the dozen years prior to 1961, the AEC did not have a formal regulation which defined the criteria by which it judged site acceptability for a test reactor or power reactor. Proposals for construction of a reactor were made to the AEC regulatory staff and the ACRS and their acceptability was decided on a case-by-case basis. Most of the reactors reviewed were AEC-owned reactors, or were reactors supported and financed in part by the reactor development side of the AEC. The regulatory staff was a relatively small organization, responsible to the general manager of the AEC, and potentially subject to considerable pressure. Prior to 1957, the

ACRS was not a statutory committee. The commissioners could and did choose not to follow ACRS advice on specific cases, such as the Fermi 1 reactor.

Major precedents on power reactor siting were set by approval of the Shippingport and Indian Point reactors without extended review of alternate siting possibilities. On the other hand, when the AEC Division of Reactor Development supported small demonstration power reactors in small or moderate-size cities at Piqua, Ohio, Jamestown, New York, and Point Loma, California, the statutory ACRS advised against acceptance of the sites, and the recommendations were accepted by the AEC commissioners.

Rule of Thumb

In 1950 the AEC Reactor Safeguards Committee prepared a report (AEC 1950) which contained the first publicly documented site guidance. In that report the Reactor Safeguards Committee assumed that each reactor might suffer an accident which led to gross overheating or melting of the fuel and rupture of the primary system which contained the coolant and the fuel. The radioactivity normally locked in the fuel was assumed to escape in an uncontrolled fashion from the relatively conventional building which housed the reactor. Allowing for meteorological effects on the transport and dispersion of the radioactivity, the Reactor Safeguards Committee recommended that beyond a radius R in miles, which depended on the reactor thermal power, P(kwt), according to the relationship

$$R = 0.01\sqrt{P}$$

the calculated radiation exposure should be less than 300 rem (which is roughly the threshold for a lethal dose), or that evacuation should be possible. For distances less than R miles from the reactor, residences would be excluded; hence, it was called an exclusion radius. For a 30 megawatt thermal (MWt) plant, this so-called rule of thumb formula led to an exclusion area of radius 1.7 miles. For a 3,000 MWt plant of the type commonly used today for electric power, the exclusion radius would be 17.3 miles according to this formula.

The Reactor Safeguards Committee appeared to be particularly concerned with accidents involving a sudden large rise of power which could cause gross fuel melting and possibly pressures large enough to disrupt the reactor core and the walls of the primary system pressure boundary, which contained the reactor coolant. (Such accidents are sometimes designated as large, uncontrolled reactivity excursions, where the term reactivity is a measure of the rate at which the power tends to rise with time. When a reactor is exactly critical, there is no excess reactivity, and the power stays constant. If, for example, a control rod is removed, there will be an increase in reactivity, followed by a resulting increase in power until control rod insertion or some inherent feature of the reactor reduces the excess in reactivity.)

The potential for core overheating and melting from inadequate removal of heat caused by radioactive decay of the fission products after shutdown of the chain reaction was not ignored. But reactivity excursions and the potential for autocatalytic effects, that is, reactor features which inherently aggravate a reactivity excursion by adding rather than losing reactivity with an increase in power, received great emphasis. Considering the weapons background of many of the committee members, and the characteristics of some of the reactors then reviewed, this emphasis is not surprising.*

The Introduction of Containment

The large exclusion distances required by the rule of thumb criterion would have allowed rather few sites in the United States to qualify for uncontained, large nuclear power plants. However, pressures built up very rapidly for the use of sites with smaller exclusion radii. Within a year or two a new concept was developed to allow for relaxation of the criteria governing an uncontained fission product release. This was to place a strong containment building around the reactor to hold in the radioactive fission products released in an acident. A relatively modest-power prototype, the Submarine Intermediate Reactor, was proposed and approved for construction within a large steel sphere at a site near West Milton, New York, which violated the rule of thumb criterion.

The first "civilian" nuclear power plant was the Shippingport Atomic Power Station which was publicly announced in October 1953. This plant, which was also called the pressurized water reactor (PWR), was designed, built, and owned by the government, but was operated by a utility (Duquesne Light Company) under the stringent guidance of the Division

---

*It is of some interest that back in 1950, the document WASH–3 (AEC 1950) called special attention to sabotage as a potentially important source of serious reactor accidents. WASH–3 was particularly concerned with sabotage by a knowledgeable person having access to the plant. About 16 years later, during the construction permit review of the Turkey Point 3 and 4 site in southern Florida, the question of attack on a reactor from an unfriendly country was raised by an intervenor. The AEC held that the Department of Defense and not the AEC was responsible for defense against any such aggression. During the 1970s, the matter of sabotage, both by an insider and by armed intruders, was given much increased emphasis by the ACRS and the regulatory staff, and a considerable augmentation of protective requirements was initiated.

of Naval Reactors of the AEC. The PWR clearly would not have met the 1950 rule of thumb criterion at its Shippingport, Pennsylvania, site which was about 420 acres in area and about 20 miles from Pittsburgh, and was in a region with more population than was characteristic of the remote sites employed for weapons material production and for engineering test reactors. However, a containment building was provided around this reactor.

The economic acceptability of the rule of thumb criterion for power reactors had already been questioned because of the high cost of land for such large exclusion areas, and because of the desire or need of utilities to have their power plants near load centers. However, the decision leading to the selection of the Duquesne Light Company proposal appears to have been based, in part, on the interest by the government in maintaining an AEC technological program with a potential for application to future naval reactors projects and, in part, on the need to find a willing and able utility. It is not clear how matters such as surrounding population density entered into the decision-making process on Shippingport.

Soon after the Shippingport decision, proposals were made by utility companies for several privately owned, large power reactors, and the AEC had to consider applications for construction permits. All were to have containment vessels, and three of them were in the general vicinity of large cities: Commonwealth Edison proposed the Dresden 1 boiling water reactor (BWR), about 35 miles southwest of Chicago, Illinois; Consolidated Edison proposed the Indian Point 1 PWR, 24 miles north of New York City; and Detroit Edison proposed the Enrico Fermi fast reactor, 25 miles south of Detroit.

The Indian Point 1 Reactor

Indian Point 1 reactor with a power of 585 MWt was at the most highly populated site approved for a reactor of that size, a site which ultimately has represented a landmark case. No large power reactor has been built in the United States in the ensuing quarter century at a site having a greater surrounding population density than Indian Point.

A letter from Harold Price, Director of the AEC regulatory staff, to Vice President J. F. Fairman of Consolidated Edison, dated May 16, 1955, states:

at the meeting in New York on May 2, 1955, ... it was agreed among the technical people present that there had not been assembled sufficient information on which to base an evaluation of the actual and potential hazard of operating a reactor of the type contemplated at the site you have chosen. Thus while we have no reason to believe that such a reactor could not be operated safely at the chosen site, neither do we have sufficient information to allow us to state with reasonable assurance that it can be.

Actually no formal design criteria or siting criteria existed in 1955, and rather little preliminary design information was available in 1955–56 when the ACRS reviewed the construction permit request. The applicant did submit five "hazard reports" which dealt with the reactor site and, in a very conceptual fashion, with various design aspects. There is no indication in the ACRS records of any emphasis on demographic considerations during the review of the Indian Point site, nor do these records provide any insight concerning the basis on which the ACRS judged this reactor-site combination to be acceptable.

On January 6, 1956, the ACRS wrote a very brief report to the AEC in favor of the construction of the Indian Point 1 reactor. In the report the committee noted that considerable work remained to be done to verify the design calculations and it recommended that the final design and hazard summary report should be reviewed before the reactor went into operation. Approval of a reactor for this site represented a major step in the commercialization of nuclear power, one which had strong support from the development (or promotional) side of the AEC. The site was approved prior to the existence of a statutory ACRS, at a time when ACRS reports did not have to be made public and when the commissioners themselves took primary responsibility in deciding on the acceptablity of some reactors, for example, the Fermi 1 fast reactor.

A Letter from Senator Hickenlooper

On February 15, 1956, while construction of the Shippingport reactor was being completed, Senator Bourke Hickenlooper wrote to the ACRS asking questions concerning the siting of power reactors in populated areas. His letter is reproduced below.

As a member of the Joint Committee on Atomic Energy, also as a member of the Senate and as a citizen, I am, and have been, concerned about the question of safety of the public involved in the location and operation of atomic reactors. I would, therefore, like to ask the following questions, which to my knowledge have not been answered heretofore, and I shall appreciate categoric concise answers.

1. Will the operation of the Reactor presently under construction at Shippingport, Pennsylvania, create any potential dangers or the possibility of danger to the citizens in that densely populated area? (This question includes not only the possibility of hazards under so-called normal operation, but hazards or injury which could come from human error in operating the Reactor or from natural catastrophe, sabotage or enemy attack.)

2. I ask the same question about the operation of sizeable commercial type reactors so far as they might be located in any other area of average or concentrated density in the United States.

3. Is it and will it be, in the light of all present knowledge and that anticipated in the foreseeable future, more desirable, from the standpoint of safety to the public against actual or possible danger or injury, to locate such reactors in areas of practically no density of population, such as, for instance, the Arco area, rather than in areas of populations of normal to concentrated density?

The response was made, not by ACRS Chairman McCullough, but by Dr. Willard F. Libby, Acting Chairman of the Atomic Energy Commission, on March 14, 1956:

In your letter of February 15, 1956, to the Commission's Advisory Committee on Reactor Safeguards, you asked for categoric, concise answers to three questions dealing with the relationship between reactor isolation and possible hazards to the public.

Regardless of location or isolation, there is no such thing as an absolutely safe nuclear reactor—just as there is no such thing as an absolutely safe chemical plant or oil refinery. There is always present, regardless of the remoteness of its probability, a finite possibility of the occurrence of an event, or series of events, the result of which is the release of unsafe quantities of radioactive material to the surrounding area. Should such a release occur, the number of persons receiving excessive radioactive exposure would, of course, be directly related to the density and distribution of population in the area surrounding the reactor.

It follows that the answer to each of the three questions you raise is "yes." Thus, there is a remote *possibility* of danger to citizens in the vicinity of the Shippingport reactor or any other sizeable reactor, and, it is, therefore, more desirable from the standpoint of safety alone to locate reactors in areas of low, rather than normal, population density.

Such answers, standing by themselves can, however, be misleading and could result in misunderstanding and misconception. For this reason, I should like to expand my answers to include a discussion of some of the more important factors which we must consider in determining the location of any given reactor.

As previously stated, if considerations were given to safety factors alone, atomic reactors should be located in areas of lowest possible population density. However, the growth and development of an atomic energy industry cannot proceed under conditions of isolation which are significantly different from those which have been found to be applicable to most other industries.

While it is true that the potential danger to the public from a nuclear accident is only that of the release of radioactive materials and not that of an atomic explosion, still the maximum conceivable damage which can be caused by such an accident is far greater than that which can result from normal industrial accidents. Therefore, it is incumbent upon the new industry and the Government to make every effort to recognize every possible event or series of events which could result in the release of unsafe amounts of radioactive material to the surroundings and to take all steps necessary to reduce to a reasonable minimum the probability that such events will occur in a manner causing serious overexposure to the public.

The licensing provisions of the Atomic Energy Act of 1954 have made it possible for the Commission to establish a regulatory program designed to assure that these objectives will be achieved. Under our regulations no license will be issued for the operation of any reactor, regardless of size or intended use, until the scientists and engineers who conceived and designed the reactor have made a complete evaluation of all potential hazards of their particular reactor, and of the adequacy of the steps they have taken in design and operating procedures to minimize the probability of occurrence of an accident which would result in the release of unsafe quantities of radioactive materials to the surroundings. This evaluation, which is reported in a detailed "Hazard Summary Report" to the Commission, is used by the technical experts on the Commission's staff working in close collaboration with the experts of the Commission's Advisory Committee on Reactor Safeguards to determine whether or not the operation of the reactor can be carried out in a manner that gives reasonable assurance that the health and safety of the public will not be endangered.

The financial incentive of the owners of the reactor to take all steps necessary to protect their investment, as well as to decrease their potential public liability, and the legal and moral responsibilities of the Commission to protect the public from overexposure to radioactivity, are resulting in a system which is characterized by an attitude of caution and thoroughness of evaluation unique in industrial history. Every phase of the reactor design and operating procedure is reviewed separately and as a part of the whole. The inherent nuclear, chemical, metallurgical, physical and mechanical characteristics of the fuel, moderator, coolant, neutron absorbers and structural materials are carefully considered in connection with the electrical, mechanical, chemical, physical, metallurgical and nuclear characteristics of the control and safety systems, the heat removal systems, the pressure systems, and so on, to assure that the probability of an operating mishap has by adequate design and operating precautions been brought to an acceptably low level.

Not only must the evaluation show that the designers have taken all reasonable precautions to assure that the probability of a mishap is satisfactorily low, it must further show what steps have been taken to protect the public in the event the highly improbable incident did occur and unsafe quantities of radioactive materials were released from the reactor itself. It is in this evaluation of what is essentially a vital second line of defense for the public that the relationship of the characteristics of the location of the reactor to the ability of the building to contain radioactive materials which might be released becomes an important factor. It is during this phase of the study that the hydrology, meteorology, geology and seismology of the area; the existing and potential population density and distribution; the type of existing and potential activity in the area (i.e., agricultural, commercial, industrial, residential, etc.); the use of the surface and surface waters for industrial or personal consumption; and other factors pertinent to the specific location, are considered in order to be sure that the degree of containment is adequate for the location chosen.

If, for example, it is possible to show that under the most adverse set of circumstances which might occur, the structure of the building containing the reactor would not be expected to allow the release of any significant amount of radioactive materials into the surrounding area, such factors as the proximity of the reactor to densely populated areas would be less important than otherwise. Likewise, if the distance from densely populated areas were so great that under the most adverse conditions it would be reasonable to expect that there would be little exposure of the public, the degree of containment would not be so important.

It is expected that power reactors, such as that now under construction at Shippingport, Pennsylvania, will rely more upon the philosophy of containment than isolation as a means of protecting the public against the consequences of an improbable accident, but in each case there will be a reasonable distance between the reactor and major centers of population.

In summary then, our safety philosophy assumes that the potential danger from an operating atomic reactor is very great and that the ultimate safety of the public is dependent upon three factors:

1. Recognizing all possible accidents which could release unsafe amounts of radioactive materials;

2. Designing and operating the reactor in such a way that the probability of such accidents is reduced to an acceptable minimum;

3. By appropriate combination of containment and isolation, protecting the public from the consequences of such an accident, should it occur.

It is interesting that in 1956, Senator Hickenlooper raised questions which once again were to be found in the forefront of discussion twenty years later. AEC Chairman Libby pointed out that there is no absolutely safe reactor, just as there is no absolutely safe chemical plant. Libby discussed the principle of defense in depth and stated: "It is expected that power reactors such as that now under construction at Shippingport, Pennsylvania, will rely more upon the philosophy of containment than isolation as a means of protecting the public against the consequence of an improbable accident, but in each case there will be a reasonable distance between the reactor and major centers of population."

Libby suggested that under the most adverse circumstances, containment might be expected not to allow the release of any significant amount of radioactivity into the surrounding area. But he did not guarantee its infallibility.

In addition to the precedent setting Shippingport, Dresden, and Indian Point reactors, several other commercial and public power reactors received approval for construction in the time-period up through 1960, as shown in table 1. And several engineering test reactors were approved.

However, not all reactor proposals received approval during this time-period. With support from the development side of the AEC, a few, relatively low power reactors were proposed for sites right at the edge of small- or medium-size cities. These proposals were rejected or a shift to a somewhat less-populated site was required. The ensuing disruption in the projects led

TABLE 1
Power reactors approved before 1960

Name	State	Reactor Type	Power (MWt)	Exclusion Distance (mi.)	Distance from City (mi.)	Date Site Approved
Shippingport	Pa.	Pressurized Water	231	.4	7.5	Before 1955
Indian Point	N.Y.	Pressurized Water	585	.3	17	May 1955
Dresden 1	Ill.	Boiling Water	630	.5	14	July 1955
Fermi 1	Mich.	Fast Reactor	300	.75	7.5	July 1955
Yankee	Mass.	Pressurized Water	485	.5	21	July 1957
Elk River	Minn.	Boiling Water	58	.23	220	August 1958
Piqua	Ohio	Organic	48	.14	27	January 1959
Carolinas-Va.	S.C.	D_2O; Tube type	63	.50	25	January 1959
Hallam	Nebr.	Sodium Graphite	240	.25	17	July 1959
Saxton	Pa.	Pressurized Water	20	.17	20	September 1959
Pathfinder	S.D.	Boiling Water	203	.5	3.5	December 1959
Big Rock	Mich.	Boiling Water	240	.5	13.5	March 1960
Humboldt Bay	Calif.	Boiling Water	202	.25	3.5	March 1960
Bonus	P.R.	Boiling Water	50	.25	12	March 1960
Peach Bottom	Pa.	Gas-Cooled	115	.57	21	March 1960

to a demand for the development of site criteria. We shall now review briefly several of the case-by-case site reviews in the period 1957–1960.

The First ACRS Dissent

At its second meeting, November 1–3, 1957, the statutory ACRS wrote reports favorable to operation of the Shippingport PWR, to construction of the 30 MWt General Electric Test Reactor in Pleasanton, California, and to construction of the 60 MWt National Advisory Committee for Aeronautics (NACA) Test Reactor (Plum Brook) three miles from Sandusky, Ohio. It is of some interest to note that the Plum Brook report was the first to contain a dissent in the form of additional remarks by an ACRS member. The ACRS report itself noted the potential risks imposed by the proposed experiments for defense purposes which would cause the test fuel elements to fail; and, while accepting the proposed site, the ACRS indicated that a less-populated site would be preferable. In his additional remarks member Abel Wolman stated:

While I agree with all that the Committee has stated, I feel that I must add some remarks for purposes of clarifying my own position. In view of the prospect of future continuing debates as to the safety of conducting essential experiments at this site, I would recommend against the site on the information presently available. I believe that the applicant should be required to consider the availability of other sites at which operation of the reactor would be feasible and which would afford a higher degree of protection to the health and safety of the public.

It is unrealistic to permit operation at this site if experiments of importance to the national defense are likely to have to be curtailed because of the site. The realities of human behavior are such that operation of experiments, the hazards of which may be uncertain, are likely to be permitted if they are important to the national defense.

I do not believe that we should freeze on a site in a situation like this merely because an applicant has chosen it.

During its first decade as a statutory committee, the ACRS made it a matter of policy and practice to try to present a single unanimous report, and sometimes several extra committee meetings were held in order to work out resolution of a difficult issue in a manner acceptable to all members. The next written dissent was not to occur until 1967, in the ACRS report concerning construction of the Browns Ferry 1 and 2 reactors.

Reactor Site Reviews in 1958

At its ninth meeting, August 4–5, 1958, the ACRS reviewed the proposed construction of the 58 MWt Elk River boiling water reactor. In its report the committee noted that, as a matter of policy, they considered it undesirable to locate a nuclear reactor of this power close to a growing community, and that, in the event of a major accident, a few people might be exposed to higher radiation dosage than was considered acceptable. (A dose of 4,800 rem to the thyroid is discussed in the meeting minutes.) The ACRS concluded, however, that, subject to the containment vessel meeting its specified leak rate, the reactor would not represent an undue hazard to the public.

At the same meeting the ACRS heard a presentation concerning a proposed organic-cooled power reactor for the city of Piqua, Ohio. The reactor thermal power was only 45 MWt. However, the site was just outside the city limits and only a few hundred feet from a temporary residential area. In its report of August 5, 1958, the ACRS concluded tentatively that the proposed site was not a suitable one.

At the tenth ACRS meeting, October 15–17, 1955, there was further discussion of the proposed Piqua reactor. The reactor designer, Atomics International, stated that its analyses had not found an accident which could release significant quantities of radioactive materials and that it was not proposing containment. The AEC regulatory staff, on the other hand, said experience to date was inadequate to support this position, even if it might be true. The regulatory staff felt that containment was mandatory in a moderately populated region. Dr. F. Pittman, Acting Director of the AEC Division of Reactor Development, said that, in the future, sites for AEC-supported reactors would receive early review to help preclude situations like that of the Piqua reactor.

At its eleventh and twelfth meetings, the ACRS continued discussion of the Piqua reactor. A new site was proposed; however, 700 people would still live within a mile of the reactor.

The ACRS wrote a report in which it stated that the newer site was more suitable than the original one. However, it also stated that: "the Committee does not consider the installation at this site of a nuclear power plant of this capacity of a relatively untried type to be without undue public hazard until the proposed unconventional type of containment is replaced by a more substantial and dependable system."

At the thirteenth ACRS meeting, January 8–10, 1959, a more conventional containment building was proposed for the Piqua reactor, and the committee wrote a report favorable to the new site.

The minutes of the tenth ACRS meeting note that the ACRS Chairman "reported a meeting with Mr. McCone, the new Chairman of the AEC, who stated that he considered the ACRS overconservative and contributing to the slowdown of nuclear power development due to financial and safety restrictions." This was not to be the last time that the ACRS would be so labeled.

The minutes of the tenth meeting also report considerable discussion about site-selection criteria. It was noted that the probability of the maximum credible accident had not been analyzed or discussed in detail. The question was raised whether radiation doses like 4,800 rem to the thyroid might be acceptable if a very limited number of people (say 25) were involved. And there was a search for analogies in other aspects of industrial safety, such as release of noxious chemicals. At this meeting the ACRS passed a resolution which would require an exclusion radius of 1/4 mile or more for reactors of power equal to 10 MWt or greater.

Presumably because of the difficulties encountered in obtaining approval for the Piqua site, the AEC requested the ACRS to provide a comparison of the standards applied in evaluating sites for the Elk River reactor; the VBWR at Vallecitos, California, the Sodium Reactor Experiment (SRE) at Santa Susana, California, and the Shippingport PWR. Excerpts from the ACRS response of December 15, 1958, follow below:

It need scarcely be emphasized that the question of site evaluation is complex. A large number of variable factors, many not strictly comparable from site to site, must be considered. Exact, completely objective, numerical site criteria are difficult to formulate, however convenient and desirable these might be. But the Committee attempts to bring a consistent philosophy to the reactor hazards problem and to provide a common basis for site judgements.

Three distinct types of reactor are involved in the group in question. These are of the sodium graphite, pressurized water, and boiling water types.

SRE, a low power (5 Mw thermal) reactor of the sodium graphite type, operates at atmospheric pressure in an underground location. The primary coolant is contained in a stainless steel shell which is in turn contained in a sealed concrete structure. Secondary coolant from primary heat exchangers located within the containment structures gives up its heat in external steam boilers. A rupture of the primary system will not cause melting of fuel or release of fission products therefrom. For these reasons, and because the SRE is located in a relatively large exclusion area (1.4 miles minimum radius), immediately surrounded by a sparsely populated district, no containment vessel of the type used for pressurized reactors is employed.

The PWR reactor of the pressurized water type is provided with an exclusion distance of approximately 0.5 miles. It is fully contained and provided with biological shielding of the containment structures. It is designed to contain the vapor and energy released in the event of a rupture of the primary water system and one steam generator. In addition, the interconnected containment vessels are designed to contain the energy resulting from significant metal–water reactions.

The VBWR and Elk River reactors are of the boiling water type. In the VBWR, the coolant is vaporized and is used for the direct drive of turbogenerators. In the Elk River reactor, radioactive steam is taken to a heat exchanger, providing a barrier. Both are provided with containers designed to prevent release of vapors resulting from a break in the cooling system.

In attempting to decide for a particular reactor whether a given exclusion distance provides adequate protection for public safety, the Committee evaluates design features such as containment vessels, missile shields, biological shields, hydrology, meteorology, and geology, all of which affect reactor safety, particularly when a reactor is located near a populous area. Thus it was felt that the Elk River site would provide an acceptable degree of protection to the public, in view of the isolated primary system and the vapor containment provided. Like considerations were applied in the case of the PWR reactor. The SRE has somewhat less containment, but has a greater exclusion radius than the others mentioned.

At a previous meeting of the ACRS subcommittee on site-selection criteria, November 16, 1958, ACRS member Willard Conner had pointed out that to expect significant consistency among approved sites would be unwarranted, since the technique used was to ask the ACRS to review a proposal and comment on the safety of that proposal. It would be unwise for the ACRS to assert in any given case that too large an exclusion area had been selected.

Dr. Conner went on to say that the real guide should be that almost everybody in the vicinity should have a reasonable chance of escaping serious injury in the event of a reactor accident. To assume that this is the case, one must, aside from estimates of possible accidents, be provided with radiation dosage criteria which represent acceptable emergency doses below the threshold of serious injury.

During an ensuing discussion on site criteria, ACRS member Henry Newson presented an approach in which site undesirability was proportional to an integration over reactor power and fission product escape fraction, population density, and probability of fission product escape. ACRS member Conner suggested that a numerical evaluation would not occur during the generation of the ACRS members present.

ACRS member (and then chairman) Rogers McCullough hypothesized a possible approach to reactor safety and siting in which a total societal risk in terms of a total damage dose (20,000 roentgens/year) would be accepted for the nuclear industry. It was estimated that this might mean 20 statistical deaths per year attributable to the nuclear industry, whereas the electric industry was estimated as experiencing a fatal accident rate of 150/year.

At the fifteenth ACRS meeting, April 16–18, 1959, it was reported that, at a recent symposium, members of the AEC Division of Biology and Medicine indicated that, contrary to earlier hypotheses, no threshold existed for biological damage from radiation. This tended to confirm the validity of a previous ACRS interest in limiting the total population dose from accidents (or routine emissions).

In a memorandum of April 23, 1959, from McCullough to H. L. Price, Director, Division of Licensing and Regulation, the ACRS suggested a regulation for the AEC, the impact of which would have been to require formal site review as the first step in requiring a reactor license. This was not adopted.

The First Proposed AEC Site Criteria

On May 23, 1959, the AEC published in the Federal Register for public comment prior to AEC formulation the following notice of proposed rule making concerning site criteria.

Factors considered in site evaluation for power and test reactors:

a. *General.* The construction of a proposed power or test reactor facility at a proposed site will be approved if analysis of the site in relation to the hazards associated with the facility gives reasonable assurance that the potential radioactive effluents therefrom, as a result of normal operation or the occurrence of any credible accident, will not create undue hazard to the health and safety of the public.

b. *Exclusion distance around power and test reactors.* Each power and test reactor should be surrounded by an exclusion area under the complete control of the licensee. The size of this exclusion area will depend upon many factors including among other things reactor power level, design features and containment, and site characteristics. The power level of the reactor alone does not determine the size of the exclusion area. For any power or test reactor, a minimum radius on the order of one-quarter mile will usually be found necessary. For large power reactors a minimum exclusion radius on the order of one-half to three-quarter miles may be required. Test reactors may require a larger exclusion area than power reactors of the same power.

c. *Population density in surrounding areas.* Power and test reactors should be so located that the population density in surrounding areas, outside the exclusion zone, is small. It is usually desirable that the reactor should be several miles distant from the nearest town or city and for large reactors a distance of 10 to 20 miles from large cities. Where there is a prevailing wind direction it is usually desirable to avoid locating a power or test reactor within several miles upwind from centers of population. Nearness of the reactor to air fields, arterial highways and factories is discouraged.

d. *Meteorological consideration.* The site meteorology is important in evaluating the degree of vulnerability of surrounding areas to the release of air-borne radioactivity to the environment. Capabilities of the atmosphere for diffusion and dispersion of air-borne release are considered in assessing the vulnerability to risk of the area surrounding the site. Thus a high probability of good diffusion conditions and a wind direction away from vulnerability areas during periods of slow diffusion would enhance the suitability of the site. If the site is in a region noted for hurricanes or tornadoes, the design of the facility must include safeguards which would prevent significant radioactivity releases should these events occur.

e. *Seismological considerations.* The earthquake history of the area in which the reactor is to be located is important. The magnitude and frequency of seismic disturbances to be expected determine the specifications which must be met in design and construction of the facility and its protective components. A site should not be located on a fault.

f. *Hydrology and geology.* The hydrology and geology of a site should be favorable for the management of the liquid and solid effluents (including possible leaks from the process equipment). Deposits of relatively impermeable soils over groundwater courses are desirable because they offer varying degrees of protection to the groundwaters depending on the depth of the soils, their permeability, and their capacities for removing and retaining the noxious components of the effluents. The hydrology of the groundwaters is important in assessing the effect that travel time may have on the contaminants which might accidentally reach them to the point of their nearest usage. Site drainage and surface water hydrology is important in determining the vulnerability of surface water courses to radioactive contamination. The characteristics and usage of the water courses indicate the degree of risk involved and determine safety precautions that must be observed at the facility in effluent control and management. The hydrology of the surface water course and its physical, chemical and biological characteristics are important factors in evaluating the degree of risk involved.

g. *Interrelation of factors.* All of the factors described in paragraphs a through f of the section are interrelated and dictate in varying degrees the engineered protective devices for the particular nuclear facility under consideration, and dependence which can be placed on such devices. It is necessary to analyze each of the environmental factors to ascertain the character of protection it might afford for operation of the proposed facility and of the kind of restrictions it might impose on the proposed design and operation.

There does not appear to have been prior ACRS approval of the draft site criteria. And the public comments, which were received largely from the nuclear industry, were highly unfavorable. A few examples follow below.

(1) *Philip Sporn*, President, American Electric Power Service Corp., stated:

Any standard set-up today, no matter how unreasonable and unnecessarily broad and supersafe, is going to be hard to re-do in the years to come.

Whatever finally comes out in lieu of this particular rule should be clearly marked as not being a rule or regulation. It should be broad and not get into cold statements such as setting distances from large cities. Regulations will be millstones around the neck of an industry which is just starting. This particular set of rules should be suspended in the interim. It has already been a real service by bringing out the things it was designed to do.

(2) *Leonard F. C. Reichle*, Nuclear Engineering Director, Ebasco Services, Inc., said:

The proposed rules emphasize only the characteristics of the site and environs. They virtually ignore the other two aspects which determine suitability, namely, the characteristics of the facility itself, including the state of knowledge and past experience, and the safeguard features which are incorporated in the facility. It is probable true that, with sufficient knowledge of the potential hazards, any facility can be designed with appropriate safeguards to permit operation anywhere with acceptable risk.

As a rule, the prospective licensee selects a site for economic reasons and balances the cost of safeguard provisions against the added cost and inconvenience of a more isolated site. The AEC must similarly evaluate all factors to determine whether the overall hazard is acceptable. Because of the complex interplay of the many factors concerned, it is probably not practical to expect definitive standards. Some guide to the important factors considered by AEC and, if possible the probable relative weights to be applied would be welcomed by industry.

(3) *Jack K. Busby*, President of Pennsylvania Power and Light Co., wrote to the AEC:

We believe it most desirable that the Commission formulate and publish general site selection guides but, in our opinion, it is undesirable to designate minimum exclusion distances around power and test reactors, minimum distances of such reactors to the nearest town and city, and maximum offsite population densities. The problem is to establish reasonable assurance that there will be no hazards to the public. . . . We suggest that all minimum distances and maximum population densities be eliminated from the proposed regulation and that such factors be given consideration only in relation to the proposed type, design and safeguards of the particular reactor.

(4) *James F. Fairman*, of Consolidated Edison and Chairman of the Technical Appraisal Task Force on Nuclear Power of the Edison Electric Institute, said:

Indian Point, which is 24 miles north of New York City and on the east bank of the Hudson, was the most remote location we could find in our operating area. It is not only extremely difficult to acquire power plant sites within the area of New York City and Westchester County, but also expensive.

In the long term Con Ed will want to put nuclear power plants as close to its load centers as possible, which means, of course, right in the city limits. The setting of any arbitrary exclusion area limits would place a high cost premium on power plants in metropolitan areas and discourage the use of engineering ingenuity to find the most practical solutions to safety problems in built-up areas.

(5) *Patrick J. Selak*, Manager, Nuclear Engineering Development, Kaiser Engineers, proposed to the AEC:

Rather than establish a minimum distance from a "large city," perhaps a better criterion would be to establish a maximum number of people who might receive an overexposure in the event of a "maximum credible accident." Then the reactor builder could determine, subject to AEC approval, the optimum combination of exclusion zone, distance from populated areas, containment features, and inherent safety features in the reactor—which would provide adequate safety to the public at minimum costs.

The Maximum Credible Accident

It is not clear who originated the idea of "credible" or "maximum credible" accident, which was included in the proposed AEC rule. However, in June 1959, Dr. Clifford Beck of the AEC regulatory staff gave a paper entitled "Safety Factors to Be Considered in Reactor Siting" at a nuclear congress in Rome, in which he discussed credible accidents (and other things) as follows (Beck 1959):

It is an entirely different matter to evaluate the adequacy of a site-reactor combination for the accidental release of radioactivity which potentially could occur. Here a difficult dilemma is encountered. If the worst conceivable accidents are considered, no site except one removed from populated areas by hundreds of miles would offer sufficient protection. On the other hand, if safeguards are included in the facility design against all possible accidents having unacceptable consequences, then it could be argued that any site, however crowded, would be satisfactory . . . assuming of course that the safeguards would not fail and some dangerous potential accidents had not been overlooked. In practice a compromise position between these two extremes is taken. Sufficient reliance is placed on the protective features to remove most of the concern about the worst conceivable accidents, though there is seldom sufficient confidence in the facility safeguards to be sure that all hazards have been eliminated. Thus, a possible reactor site is reviewed against the possibility of credible accidents, and their consequences, which might occur despite the safeguards present.

It is inherently impossible to give an objective definition or specification for "credible accidents" and thus the attempt to identify these for a given reactor entails some sense of futility and frustration, and further, it is never entirely assured that all potential accidents have been examined.

It should be noted parenthetically, however, that this systematic search for credible accidents often contributes substantially to the safety of a facility. Potential accidents having substantial consequences and clearly credible possibility of occurrence may be discovered in this search. If such are found, safeguards against them of course are incorporated . . . and the evaluation then proceeds for the potential accidents remaining. In the plants finally approved for operation, there are no really credible potential accidents remaining against which safeguards have not been provided to such extent that the calculated consequences to the public would be unacceptable.

Despite these precautions, however, there always remains an uneasiness that these safety devices will not operate as expected or that something has been overlooked. After all, our cumulated reactor experience is quite small and, more significantly, our experience with any one type of reactor built to a standardized design, is a great deal smaller. Hardly any two reactor facilities are alike and the carry-over of detailed safety analysis from a component of one reactor to an approximately similar component in another is often surprisingly small. Finally, the consequences of a major accidental release of fission products are so great that the degree of confidence in the safeguards must far exceed that required in ordinary industrial processes.

These factors have led in the United States to the widespread use of external "vapor" containment vessels for power and test reactors built near populated areas. This practice originated in part from the characteristics of the earlier types of reactors which were built: pressurized water reactors with large excess reactivities and metallic fuel elements having aluminum or zirconium cladding. But the initiation of this practice and its extension to other types of reactors also reflects a special degree of concern for the protection of the public from the hazards of potentially dangerous devices which were not very well understood or extensively proven.

The external containment vessel, as a barrier of last resort against releases of radioactivity to public areas, offers a unique protection, completely independent of all other safety devices and engineering safeguards and its dependability is unaffected by errors in safety analyses and judgment of the reactor assembly. It stands as a visually obvious and intuitively attractive bulwark against the possible consequences of errors in reactor design, malfunction and misoperation which are admittedly present in any human undertaking.

It should be noted that the maximum credible accident approach was not universally accepted. For example, at the 1964 United Nations "Atoms for Peace" Conference in Geneva, F. R. Farmer of the United Kingdom gave a paper (Farmer 1964) in which he pointedly attacked the concept of the maximum credible accident, especially for use in any comparison of different reactor types. In particular, he emphasized the arbitrariness which is inevitably involved in the selection of an MCA. Farmer went on to emphasize the importance in the future of a comprehensive safety assessment and not merely a study of a consequences of a few selected major faults.

At the nineteenth meeting, September 10–12, 1959, the ACRS discussed the Pathfinder reactor and the Carolinas-Virginia Tube Reactor (CVTR) and put in writing a position it had taken on several previous cases, namely, that it lacked sufficient information with regard to certain design features to arrive at a conclusion concerning construction of these plants. When one considers the very limited information (by today's standards) which was presented for those reactors on which the ACRS decided it could act, the information on these reactors must have been very sketchy.

The First Review of California Sites

On March 5, 1960, the ACRS held a special meeting to consider a request by A. R. Luedecke, general manager of the AEC, for advice concerning the possibility of siting some relatively large LWRs (1,000 MWt) in California. Excerpts follow from the ACRS letter of March 6, 1960, to AEC Chairman McCone:

With respect to seismic considerations, we understand it is present utility industry practice in California to locate generating stations at least one mile from known surface faults; and to design and construct these stations using local codes supplemented by special analyses and increased seismic design factors for those critical plant components necessary to maintain the station on the line. In addition, in the case of a nuclear reactor facility, special analyses and increased seismic design factors are needed for those reactor plant systems whose failure could result in a release of radioactive material. With these precautions, the Committee believes the reactor facility would be adequately protected against seismic disturbance.

Referring to the frequency of inversion conditions, the situation of the Southern California coastal strip (south of San Francisco) is essentially unique in the United States. The semipermanent Pacific high pressure area induces a slow, large-scale, persistent subsiding motion in the atmosphere there. Air, warmed by this descent, contacts the coastal water surface which is cold as a result of upwelling. By this mechanism an inversion is formed; and the air layer extending up to a few thousand feet above the surface becomes a trap for air pollution.

Whereas persistent poor dispersion (stagnation) conditions of meteorology, lasting several days, may be expected on the average once per year anywhere east of the Rockies, the frequency of such episodes in the Southern California coastal strip is of the order of several per month. For example, during a two-year period, from July 1956 through June 1958, the Los Angeles weather was of the "smog warning" type 164 days.

In selecting a site for a high power reactor, consideration should be given to an adequate exclusion radius and the population density, not only in the immediate vicinity, five to ten miles, but also for greater distances. Obviously the lower the population density the better. The meteorology of the Southern California coastal strip is so unfavorable for dissipating pollutants that this area should be avoided if it is coupled with a high population density. In theory a reactor can be so designed, constructed, and operated that it will offset the unfavorable meteorology and high population density. Because of the present limited experience with the operation of power reactors and the large power level of the proposed reactors, the provision of an adequate degree of safety in practice may require an extreme of conservative design and containment.

Not surprisingly, the record of discussion within the ACRS itself indicates divided opinion as to how satisfactory the southern California coastal area was for reactors of appreciable power. Some members believed that very good containment, together with waste retention such that routine releases would occur only under ideal weather conditions, would eliminate restrictions due to unfavorable meteorology. However, this did not appear to be the consensus of the committee, which felt then that meteorology was a principal environmental consideration for a reactor accident in southern California. The ACRS of that day appeared to be fairly unconcerned about the difficulties of building a reactor only one mile from a fault.

ACRS Rejection of Two Sites

Soon after, at the twenty-fourth meeting, March 10–12, 1960, the ACRS reviewed and rejected the proposed 40 MWt reactor at Point Loma (San Diego), California. It appears from the ACRS minutes that Dr. Beck of the regulatory staff did not feel that the Point Loma site needed to be rejected, although he conceded that it was not a very good site. Beck appeared to feel that the unfavorable meteorology and the unfavorable hydrology (which related to the limited rate of ocean flow to remove routine radioactivity releases) could be dealt with by appropriate containment. However, the

ACRS wrote a report unfavorable to the proposed reactor, stating the following:

The Committee considers Point Loma to be a poor site because of unfavorable meteorology and high population density, aggravated by recreational and fisheries aspects, and lack of ocean dilution. The close proximity of the San Cabrillo Monument area with its numerous visitors and its proposed enlargement with the probability of an increased number of visitors add to the unfavorable features. The experimental nature of the proposed installation contributes to our lack of assurance.

The Committee believes it would be unwise at the present time from the safety point of view to locate this reactor at this site.

At the same meeting the ACRS also wrote a report unfavorable to the proposed 60 MWt Jamestown, New York, reactor. In this case the AEC regulatory staff took the same position as the committee. Excerpts from the ACRS report follow:

This 60 MW (thermal) pressurized light water moderated reactor is to be built and operated by the Commission on a site furnished by the City of Jamestown, New York, which will also provide the generating plant. The proposed site comprises thirty-five acres of city owned land located in the northwest corner of the city approximately 1.75 miles from the center.

The ACRS believes that such factors as the small size of the site; proximity to the City of Jamestown with its high population density; unfavorable meteorology; lack of control by the City of Jamestown over the area contiguous to the south and west boundaries of the site, which is located within the limits of the town of Celeron; and the long periods of low flow in the Chadakoin River with consequent adverse effects on liquid waste disposal all indicate that this site is not suitable for a power reactor of this size in the present stage of technology.

There was renewed discussion of the Jamestown reactor at the special ACRS meeting, June 7, 1960, and at the twenty-sixth meeting, June 22–24, 1960. The applicant advised that a larger site could be provided, and that 0.1% per day containment leakage rate would be the new design specification. In a letter to Chairman McCone dated June 30, 1960, the ACRS once again advised against the small PWR at the Jamestown, New York, site. In a rather strong policy type statement, the ACRS stated:

The Committee deplores the tendency on the part of some of those proposing reactor sites to place power reactors containing large quantities of stored energy in or near centers of population at this time to duplicate conditions for conventional power plants for the sake of demonstrating how near a population center such a reactor can be located. We believe that the Jamestown reactor is a case of this kind.

We wish to point out that the proximity to a population center would require more rigid specifications of all safety features including containment, leakage rate, power densities, ultimate power, shielding, etc.

The Committee can find no serious technical fault with the reactor, the containment, and the safety features proposed, insofar as the partial information supplied to date has presented the case. The Committee emphasizes, however, that power reactors are relatively new and untried, and that there exists a considerable degree of uncertainty in our knowledge of their long-term safe behavior. Accordingly, the Committee doubts that the new and relatively untried technical features for improved safety proposed by the applicant, since our last report, are a satisfactory substitute for the inherent safety implied by a greater distance from population centers.

We shall end here this partial recounting of light water reactor siting, as it took place, case by case, during the 1950s. Two decades later, finding a consistent thread is not a straightforward task. It is not clear that reactor-site combinations which were rejected would have posed greater societal risks than others which were accepted. Nor is it clear what level of societal risk was sought, except that it be less than from similar technologies. The advent of containment was clearly a decisive step in moving large reactors away from highly remote sites to populated areas. Just how much additional safety containment was providing was not quantified (or quantifiable, at that time). And not much was known about accident probabilities, both those for which the containment would function and those for which it would be violated.

The effects of natural phenomena and external events on safety received rather little attention. Meteorology was a big factor in California; hydrology seemed not to enter in any important way into the decision making.

As a point of interest, during many of the first two dozen ACRS meetings of the statutory ACRS, there were reports of incidents and accidents which were occurring at reactors. One example was the Windscale reactor accident in Great Britain, during which about 20,000 curies of radioiodine were released from the stack. During this period the ACRS was also occupied with the review of many test and research reactors and was trying to give advice to the AEC concerning a more acceptable future mode of operation for the relatively high power plutonium production reactors, which were built during World War II and which posed some major safety questions.

4

Development of the
Reactor Site Criteria,
10 CFR Part 100

Beginning in the spring of 1960, the reactor siting criteria, as they now appear in 100 CFR Part 100, began to take shape, with Dr. Beck and his aides on the regulatory staff taking the lead role in the actual drafting of a document and the development of a specific approach. There was a considerable pressure on both the regulatory staff and the ACRS, arising from the commissioners and from the nuclear industry, to provide some form of quantitative guidance by which the reactor designers could judge site suitability prior to regulatory review.

Preliminary Ideas

At the twenty-fifth ACRS meeting, May 5–7, 1960, Dr. Beck described his proposed philosophy (or assumptions). This was that the probability of a major accident was relatively small, that an upper limit of fission product release could be estimated, that reactors were expected to be in inhabited areas, and that the containment building would hold. The exclusion radius was defined as the distance at which not more than 25 rem whole body radiation would be received in two hours. The low-population zone surrounding the exclusion area had to contain less than 10,000 people, and evacuation should be possible prior to anyone receiving a dose of 50 rem (later reduced to 25 rem). The minimum acceptable distance to a city was to be set by a calculated dose at the city boundary that was less than 10 rem over the entire course of the maximum credible accident (MCA), with no evacuation. (The minutes of the ACRS subcommittee of August 23, 1960, note: "The real basis, however, for this criterion is an uncontained puff release of radioactivity resulting in an LD 50 [50% chance of death] dose at the city boundary." This means that the actual criterion in mind was that the distance to the nearest city would be large enough that if the core melted, the containment failed, and all the volatile fission products escaped with the wind blowing toward the city, the dose at the city boundary would be that which was estimated to kill half the people exposed to its full effect.) The recipe proposed by Beck was supposed to be equivalent in its effect, but it avoided discussion of such large doses from an uncontained release.

It was noted by Dr. Beck that the Indian Point 1 site would not meet several of these criteria; on the other hand, the rejected Point Loma site was satisfactory. Beck stated that the basis for acceptance of the Indian Point 1 site was its double containment.

On September 21, 1960, the ACRS received a letter from W. F. Finan, the AEC Assistant General Manager for Regulation and Safety, expressing the commissioners' desire to publish siting criteria for public comment and asking for ACRS comment on the draft criteria which had been prepared by the regulatory staff.

The draft criteria incorporated a detailed "sample calculation." It was assumed that 75%, 25%, and 1% of the noble gases, halogens, and nonvolatile fission products, respectively, were released to an intact containment building; however, the criteria permitted these percentages to be reduced if specially engineered safety features to remove radioactivity from the containment atmosphere were available. Instead of a fixed number of people in the evacuation area, a maximum population density of 100 people per square mile was now proposed, with additional limits on the population density in any angular sector around the reactor.

A table was included in the memo, showing how well the various reactors previously reviewed compared with the proposed criteria.* Of about 20 reactors, Indian Point, Pathfinder, Humboldt Bay, and Jamestown had city distances smaller than the minimum permitted by the criteria.

There was extensive discussion of these site criteria at the twenty-eighth ACRS meeting, September 22–24, 1960, and the ACRS prepared a long letter dated October 22, 1960, and signed by Chairman Leslie Silverman, to the Honorable John A. McCone, chairman of the AEC, in which it

---

*Reminiscing almost 20 years later, regulatory staff members, who had worked on this draft, recalled trying to find a set of parameters and assumptions which would fit essentially all the previously approved reactor site combinations, within some broader, generally acceptable framework.

reviewed its philosophy on reactor siting. The letter is reproduced below in its entirety.

You have asked that we supply you with criteria which could be used for judging the adequacy of proposed sites for reactors. The Advisory Committee on Reactor Safeguards has devoted considerable time to this problem. A large part of our delay in submitting site criteria stems from the fact that we believe it is premature to establish quantitative limits on the variables involved in site evaluations—especially if such limits will appear in Federal regulations, or otherwise be announced as Commission policy. We recognize that the correctness of the numbers which could be selected now cannot be proved by experimental or empirical data, and, therefore, these numbers would give a false sense of positiveness which could not be supported upon detailed scrutiny. Numbers chosen now will be expected to change as more information develops. For example, a quantitative calculation of dosage must include some estimate of the fraction of the total fission product inventory which may be air-borne. This fraction is currently under experimental examination and the estimate may be subject to change.

The Committee believes that the officially endorsed numbers could stifle progress toward a better selection of numbers. The ideas and interpretations from applicants themselves have played a major part in the formulation of the current bases for site evaluation. It would be a significant loss to stop the flow of new ideas from the applicants. The Committee also believes that it is possible that the appearance of quantitative numbers in a Federal regulation or policy statement will reduce the continual awareness of the applicant that he has assumed a responsibility to be alert and to act on unforeseen disadvantages of a site even after the site has been approved. The Committee, therefore, advises that a quantitative statement of site criteria not be included in Federal regulations.

These comments do not mean that the ACRS has no bases for judging the adequacy of sites. They merely emphasize that site selection is still largely a matter of judgment. Inasmuch as the ACRS has been making site and reactor evaluations, it may be helpful to review the framework on which these judgments are being made. It is a prerequisite, of course, that the reactor be carefully and competently designed, constructed, and operated. It should be inspected during all these stages in a manner to assure preservation of the intended protection of the public. Also, these factors are applicable only to those reactors on which experience has been developed. Reactors which are novel in design, unproven as prototypes, or which do not have adequate theoretical and experimental or pilot plant experience belong at isolated sites—the degree of isolation required depending on the amount of experience which exists.

Our site evaluations stem from several concepts. These are overlapping, but not conflicting:

1) Everyone off-site must have a reasonably good chance of not being seriously hurt if an unlikely but credible reactor accident should occur.

2) The exposure of a large segment of society in terms of integrated man–rems should not be such as to cause a significant shortening of the average individual

lifetime or a significant genetic damage or a significant increase in leukemia should a credible reactor accident occur.

3) There should be an advantage to society resulting from locating a plant at the proposed site rather than in a more isolated area.

4) Even if the most serious accident possible (not normally considered credible) should occur, the numbers of people killed should not be catastrophic.

Incidentally, the concept has been proposed by others that the damage to the people from reactor accidents can be accepted if it is no greater than that experienced in other industries. We reject this suggestion as premature, and follow rather the concept that the consequences of reactor accidents must be less than this. The reasons for this rejection are twofold: First, we do not have sufficient information on the probability of reactor accidents to make use of this concept in site evaluations. We do use, of course, the fact that the probability of a serious accident is very low. Second, we recognize that the atomic power business has not yet reached the status of supplying an economic need in a manner similar to that of more mature industries; and, therefore, arguments for taking conventional risks for the greater good of the public are somewhat weak. At the same time, we do not want to imply that the restrictions placed on site locations during the developmental period of atomic power will necessarily be carried over to the period of maturity of the atomic power industry.

The reduction of these concepts to a judgment as to the adequacy of a proposed site requires further logic and the introduction of some numerical estimates. We believe that the searching analysis which is necessary at this stage should be done independently by the owner of the reactor, using the characteristics which are peculiar to his site and to his specific reactor. This step, we believe, is essential in developing his continuing alertness to his responsibility to the community surrounding the site. However, in Committee deliberation, we balance his analysis against a generalized accident which serves as a reference point from which we can better understand the analysis submitted by the applicant.

Our generalized accident analysis assumes that a serious accident has occurred and predicts in rough terms the consequences of such an accident. It is obvious that the generalized accident is an arbitrary artifact subject to change and has value only so far as it aids judgment. As a matter of fact, for certain reactors and conditions judgment will indicate that the generalized accident is too severe. In the generalized accident, we must make numerical assumptions as to the amount, type and rate of radioactivity release (the source term), the dispersal of the radioactivity in the air and in the hydrosphere, and the effect of this radioactivity on people.

Source Term

An arbitrary accident is assumed to occur which results in the release of fission products into the outermost building or containment shell. About 100% of the total inventory of noble gases, 50% of the halogens, and 1% of the non-volatile products are assumed to be so released. It is then assumed that this mixture leaks out of the outermost barrier at a rate defined by the designed and confirmed leak rate. The reasoning back of this source term is admittedly loose. It stems primarily from a present inability to be convinced that coolant cannot be lost somehow from the reactor core, either by spontaneous fracture of some element in the primary system or a fracture caused by maloperation (instrumental or human) of the control rods. Admittedly, this assumed source term in large, but it thereby affords a factor of safety. In some cases it is justifiable to reduce this source term. It is also tacitly assumed that in this accident the outermost barrier will not be breached. The logic behind this assumption is that we require all of the components restraining the pressure of the primary system to be operating at temperatures above their nil-ductility temperature. We are, therefore, more confident, but not certain, that failure will occur by tearing rather than by brittle fracture and that the probability of ejection of missiles which penetrate the outermost barrier is low. The necessary supporting structures and shielding also protect against missile damage.

Dispersal of the Radioactivity

1) *Meteorology* We assume a dilution of air-borne activity using atmospheric diffusion parameters which reflect poor, rather than average, meteorological conditions. Choice of specific parameter values follows from a survey of meteorological conditions expected to apply at the site, primarily wind and stability distributions. To analyze the generalized accident, we use the standard diffusion calculation methodology outlined, for example, in AECU–3066 and WASH–740. The atmospheric diffusion phenomena is the subject of active research, and new results can be expected to firm up and improve the present methods, although we do not anticipate major revisions in this area.

2) *Hydrology* Considerations of hydrology are based on characteristics of surface and subsurface flow as they are related to the possible release of contaminated liquids to the off-site environment. Thus, the rate and volume of surface flow and the possible presence or absence of absorbing barriers of soil between the reactor complex and important underground aquifers should be taken into consideration. These factors must be favorable for restraining the flow of radioactive materials in case of accident. Design factors, including the capability of providing adequate hold-up in the event of adverse hydrology, are also significant.

Effect of Radioactivity on People

The upper limit to the exposure to a member of the public in the generalized accident should be no higher than the maximum once-in-a-lifetime emergency dose. Such a level has not been established by AEC. We are arbitrarily using a figure of about 25 r whole body or equivalent integrated dose for this level. This figure is mentioned in Handbook 59 of the National Bureau of Standards, pages 69–70. Since the iodine dose is often controlling, we are tentatively considering a thyroid dose limitation of 200–300 rads. The dosage so far mentioned refers to limits to people when the people are considered as independent individuals. We believe that it is essential that the Atomic Energy Commission attempt to confirm through its staff or its advisors in this field that this suggested value of 25 r whole body or equivalent is without significant biological effect on the individuals who might be subjected to this dose from the generalized accident.

When large numbers of individuals are exposed to radiation, another limit also exists because of genetic effects and because of the statistical nature of induced leukemia and the shortening of the life span. The limits of exposure to large groups of people are better expressed in terms of integrated man–rems. We are considering using a figure of 4×10^6 man–rems for this limit for the people who might be exposed to radiation doses falling between 1 and 25 rems. This figure of 4×10^6 man–rems is roughly equal to the dose received from natural background by a million people during their reproductive lifetime.

The implication of these numbers is this. About a reactor site, there should be an exclusion radius in which no one resides. Surrounding this, there should be a region of low population density, so low that individuals can be evacuated if the need arises in a time which will prevent their receiving more than a dose of 25 r. Beyond this evacuation area, there should be no cities (above 10,000 to 20,000 population) sufficiently close so that the individuals in these cities might receive more than the lower of the following: (1) $4 \times$ man rems in the generalized accident, and (2) 200 rems under the extremely improbable accident in which the outermost barrier fails completely to restrain all of the radioactivity of the generalized accident.

The Committee wishes to emphasize again that the numbers which have been used in discussion of the generalized accident should not be formalized into regulations or Commission policy. The Committee wishes to acknowledge the help it has received from the Hazards Evaluation Branch in this matter and suggests that these individuals be encouraged to present as technical papers, but not as regulations, a complete description of their working approach to making judgments on the adequacy of proposed reactor sites. Such a paper, of course, would have the status of the opinion of an informed technical individual, but would not imply Committee approval, nor would it have the rigidity of a Commission policy statement.

In this important letter, the ACRS proposed a basic approach to reactor siting and safety. However, it is not clear whether the ACRS looked in detail at the reactors already approved for construction, such as Indian Point 1, Dresden 1, or Shippingport PWR, to see whether the criterion of not exceeding 200 rem at a large population center under the worst accident conditions would be met. It is difficult also to tell from the minutes of previous ACRS meetings the extent to which the numerical guidelines presented in the letter of October 22 had actually been applied by the ACRS in its previous case reviews.

Following its letter of October 22, 1960, the ACRS received a memorandum from AEC commissioner Loren K. Olson, requesting that the committee summarize whatever general site criteria it considered appropriate, despite its reluctance to fix on specific numbers.

At its twenty-ninth meeting, December 8–10, 1960, the ACRS* prepared a reply dated December 13, 1960, which essentially repeated the recommendations and philosophy of the October 22 letter, including the ACRS reluctance to approve quantitative criteria at that time.

Proposed Site Criteria

Nevertheless, on February 11, 1961, the Atomic Energy Commission issued notice of its proposed reactor site criteria in the Federal Register for public comment. The complete notice, including both the statement of considerations and the site criteria appears as appendix A to this chapter.

The proposed criteria included the idea of an exclusion area and a low-population zone, with dose limits of 25 rem whole body and 300 rem to the thyroid, as suggested by both the regulatory staff and the ACRS. A population center distance of at least 1 1/3 times the distance from the reactor to the outer boundary of the low-population zone was proposed, together with a relatively vague statement that: "When very large cities are involved, a greater distance may be necessary because of total integrated population dose considerations." There was also an appendix to the criteria that spelled out a sample calculation, using what the regulatory staff considered to be reasonable assumptions.

The statement of considerations (which was eventually separated from the criteria) contained the following sentence: "Even if a more serious accident (not normally considered credible) should occur, the number of people killed should not be catastrophic." This statement reflected the philosophy proposed in the ACRS letter of October 22, 1960.

It is of some interest to note that at the hearings held by the Joint Congressional Committee on Atomic Energy (JCAE) on Radiation Safety and Regulation, June 12–15, 1961, Mr. Robert Loewenstein, Acting Director, AEC Division of Licensing and Regulation, specifically discussed the population center distance as follows:

If one could be absolutely certain that no accident greater than the "maximum credible accident" would occur, then the "exclusion area" and "low population" zone would provide reasonable protection to the public under all circumstances. There does exist, however, a theoretical possibility that substantially larger accidents could occur. It is believed prudent at present, when the practice of nuclear technology does not rest on a solid foundation of extended experience, to provide protection against the most serious consequences of such theoretically possible

*The ACRS members at this time were the following: L. Silverman, Chairman; W. P. Connor; R. L. Doan; W. K. Ergen; D. A. Rogers; R. C. Stratton; T. J. Thompson; C. R. Williams; and A. Wolman.

accidents. Consideration of a "population center distance" is therefore prescribed: this is a distance by which the reactor would be so removed from the nearest major concentration of people that lethal exposures would not occur in the population center even from an accident in which the containment is breached.

The AEC received a wide range of comments on the proposed site criteria. On July 31, 1961, a meeting was held between representatives of the Atomic Industrial Forum (AIF) and of the AEC. Mr. W. K. Davis made several comments on behalf of the AIF: "1) The Example given in Appendix A should be deleted; 2) The population center distance should be deleted since the 1 1/3 number is without technological basis; 3) If the AEC's policy is against the location of reactors in cities, it should be so stated as a matter of policy and not inferred by calculation."

In succeeding drafts of the reactor site criteria words like catastrophic were omitted, and the appendix was deleted, to be replaced by a separate AEC report, TID 14844 (DiNunno et al. 1962), which provided considerable detail on the methodology and parameters to be used in calculating accident doses to meet the requirements of the criteria.

The Adoption of 10 CFR Part 100

On April 12, 1962, the AEC published Part 100, reactor site criteria, to be effective one month later. The statements of considerations and the criteria are reproduced as appendix B to this chapter.

The new statement of considerations discussed the use of a minimum acceptable distance to the nearest population center as a way to limit the cumulative population dose (i.e., the sum of the individual dose received by each person) and to provide for protection against excessive radiation exposure to people in large centers, where effective protective measures might not be feasible. However, the rather specific criterion of the ACRS letters of October and December 1960 namely, that of no lethal doses at the population center for the most serious accident possible, had been dropped (and we shall see that it was not used in succeeding years).

It is of interest to note that the minutes of the thirty-ninth ACRS meeting, February 8–10, 1962, record dissatisfaction by several members concerning the 1962 version of 10 CFR Part 100. Member L. Silverman believed that the rewriting had eliminated some of the earlier significant ACRS ideas, which had been previously incorporated. Member W. K. Ergen was concerned over ambiguity in the criterion for distance from a large city. However, Part 100 was issued by the AEC as described above. And the ACRS lived with it.

The 1962 version of Part 100 specifically allows for sites having multiple reactor facilities. It asks that consideration be given to the possibility that an accident in one reactor might initiate an accident in the other(s). However, in practice, this has not normally been reviewed in the context of a class 9 accident, that is an accident whose consequences exceed Part 100, and which might impact adversely on the continuing ability of operators to remain on site and keep the other reactors safely shut down and adequately cooled.

In 1962 the containment building was looked upon as an independent bulwark, which should remain intact even if the core melted, thereby preventing any large release of radioactive fuel into the ground below the reactor foundation and limiting the radioactive material released to the atmosphere to a tolerable amount. It was recognized that failure of the containment building and melting of the core could both occur as a consequence of, say, gross rupture of the reactor pressure vessel, but such a rupture was deemed to be very unlikely. And containment failure was not expected to occur just because the core melted. This concept of the containment building as an independent bulwark had been originated for some relatively low power reactors. It was not until 1966 that it was recognized that this concept was probably invalid for the Indian Point 1 and Dresden 1 reactors, and would get progressively less meaningful as the power output of the upcoming reactors was increased.

What is, in a sense, remarkable is that until 1980 there has been very little change in this part of the reactor site criteria since their adoption as an AEC regulation in 1962. This could mean any of many things, including the following: (1) That Part 100, as originally written, was so well formulated that it has passed the test of time and can continue to be used for giving rather direct and appropriate guidance in the choice of sites; (2) That it was formulated in a sufficiently general fashion, or included enough permissible alternatives, that it permitted a wide range of interpretation, enough to cover all relevant site questions arising between 1962 and 1979; or (3) that it is not too meaningful in terms of our present degree of knowledge, but that it has been difficult to find a new set of reactor site criteria which can be considered a defensible improvement.

There are, undoubtedly, other possible explanations for the absence of important changes in Part 100. However, a combination of items (2) and (3) above probably comes close to reality.

The problems that the ACRS foresaw in October 1960, with the premature selection of numbers that would be specified as AEC regulations did not seem to arise. Rather, it appears that the flexibility of the site criteria permitted an encroachment by utilities on the stated AEC policy of keeping power reactors away from densely populated centers. This was done by substituting engineered safeguards for distance within the context of an independent, last-ditch protection afforded by the containment building. In some cases, like that of the large reactor proposed for Bolsa Island near Huntington Beach, California, in the mid-1960s, it was the development side of the AEC and some of the commissioners themselves who pushed hard for a power reactor very close to a large population center.

Chapter 4: Appendix A:
AEC Proposed Reactor Site Criteria, February 11, 1961

Statement of considerations: On May 23, 1959, the Atomic Energy Commission published in the Federal Register a notice of proposed rule making that set forth control criteria for the evaluation of proposed sites for power and testing reactors. Many comments were received from interested persons reflecting, generally, opposition to the publication of site criteria, as an AEC regulation, both because such a regulation would, to some extent, incorporate military limitations and because it appeared that in view of the lack of available experimental and empirical data specific criteria could not be established.

Judgment of suitability of a reactor site for a nuclear plant is a complex task. In addition to normal factors considered for any industrial activity, the possibility of release of radioactive effluents requires that special attention be paid to physical characteristics of site, which may cause an incident or be of significant importance in increasing or decreasing the hazard resulting from an incident. Moreover, the inherent characteristics and the specifically designed safeguard features of the reactor are of paramount importance in reducing the possibility and consequences of accidents which might result in the release of radioactive materials. All of these features of the reactor plus its purpose and method of operation must be considered in determining whether location of a proposed reactor at any specific site would create an undue hazard to the health and safety of the public.

Recognizing that it is not possible at the present time to define site criteria with sufficient definiteness to eliminate the exercise of agency judgment, the proposed guides set forth below are designed primarily to identify a number of factors considered by the Commission and the general criteria which are utilized as guides in evaluating proposed sites.

The basic objectives which it is believed can be achieved under the criteria set forth in the proposed guides are:

(a) Serious injury to individuals off-site should be avoided if an unlikely, but still credible, accident should occur.

(b) Even if a more serious accident (not normally considered credible) should occur, the number of people killed should not be catastrophic.

(c) The exposure of large numbers of people in terms of total population dose should be low. The Commission intends to give further study to this problem in an effort to develop more specific guides on this subject. Meanwhile, in order to give recognition to this concept the population center distances to very large cities may have to be greater than those suggested by these guides.

Notice is hereby given that adoption of the following guides is contemplated. All interested persons who desire to submit written comments and suggestions for consideration in connection with the proposed guides should send them to the Secretary, United States Atomic Energy Commission, Washington 25, D.C., Attention: Director, Division of Licensing and Regulations, within 120 days after publication of this notice in the Federal Register.

General Provisions

§ 100.1 Purpose

It is the purpose of this part to describe the criteria which guide the Commission in its evaluation of the suitability of proposed sites for power and testing reactors subject to Part 50 of this chapter. Because it is not possible to define such criteria with sufficient definiteness to eliminate the exercise of agency judgment in the evaluation of these sites, this part is included primarily to identify a number of factors considered by the Commission and the general criteria which are utilized as guides in approving or disapproving proposed sites.

§ 100.2 Scope

(a) This part applies to applications filed under Part 50 of this chapter for construction permits and operating licenses for power and testing reactors.

(b) The site criteria contained in this part apply primarily to reactors of a general type and design on which experience has been developed but can also be applied with additional conservatism to other reactors. For reactors which are novel in design, unproven as prototypes, and do not have adequate theoretical and experimental or pilot plant experience, these criteria will need to be applied more conservatively. This conservatism will result in more isolated sites—the degree of isolation required depending upon the lack of certainty as to the safe behavior of the reactor. It is essential, of course, that the reactor be carefully and competently designed, constructed, operated and inspected.

§ 100.3 Definitions

As used in this part:

(a) "Exclusion area" means the area surrounding the reactor, access to which is under the full control of the reactor licensee. This area may be traversed by a highway, railroad, or waterway, provided these are not so close to the facility as to interfere with normal operations, and provided appropriate and effective arrangements are made to control traffic on the highway, railroad, or waterway, in case of emergency, to protect the public health and safety. Residence within the exclusion area shall normally be prohibited. In any event, residents shall be subject to ready removal in case of necessity. Activities unrelated to operation of the reactor may be permitted in an exclusion area under appropriate limitations, provided that no significant hazards to the public health and safety will result.

(b) "Low population zone" means the area immediately surrounding the exclusion area which contains residents the total number and density of which are such that there is a reasonable probability that appropriate protective measures could be taken in the event of a serious accident. These guides do not specify a permissible population density or total population within this zone because the situation may vary from case to case. Whether a specific number of people can, for example, be evacuated from a specific area, or instructed to take shelter, on a timely basis will depend on many factors such as location, number and size of highways, scope and extent of advance planning, and actual distribution of residents within the area.

(c) "Population center distance" means the distance from the reactor to the nearest boundary of a densely populated center containing more than about 25,000 residents.

(d) "Power reactor" means a nuclear reactor of a type described in §50.21 (b) or 50.22 of this chapter designed to produce electrical or heat energy.

(e) "Testing reactor" means a "testing facility" as defined in §50.2 of this chapter.

Site Evaluation Factors

§ 100.10 Factors to be considered when evaluating sites

In determining the acceptability of a site for a power or testing reactor, the Commission will take the following factors into consideration:

(a) Population density and use characteristics of the site environs, including, among other things, the exclusion area, low population zone, and population center distance.

(b) Physical characteristics of the site, including, among other things, seismology, meteorology, geology and hydrology. For example:

(1)   The design for the facility should conform to accepted building codes or standards for areas having equivalent earthquake histories. No facility should be located closer than 1/4 to 1/2 mile from the surface location of a known active earthquake fault.

(2)   Meteorological conditions at the site and in the surrounding area should be considered.

(3)   Geological and hydrological characteristics of the proposed site may have a bearing on the consequences of an escape of radioactive material from the facility. Unless special precautions are taken, reactors should not be located at sites where radioactive liquid effluents might flow readily into nearby streams or rivers or might find ready access to underground water tables.

        Where some unfavorable physical characteristics of the site exist, the proposed site may nevertheless be found to be acceptable if the design of the facility includes appropriate and adequate compensating engineering safeguards.

0110C

(c) Characteristics of the proposed reactor, including proposed maximum power level, use of the facility, the extent to which the design of the facility incorporates well proven engineering standards, and the extent to which the reactor incorporates unique or unusual features having a significant bearing on the probability or consequences of accidental releases of radioactive material.

§ 100.11 Determination of exclusion area, low population zone, and population center distance.

(a) As an aid in evaluating a proposed site, an applicant should assume a fission product release from the core as illustrated in Appendix "A" of this part, the expected demonstrable leak rate from the containment and meteorological conditions pertinent to his site to derive an exclusion area, a low population zone and a population center distance. For the purpose of this analysis, the applicant should determine the following:

(1) An exclusion area of such size that an individual located at any point on its boundary for two hours immediately following onset of the postulated fission product release would not receive a total radiation dose to the whole body in excess of 25 rem or a total radiation dose in excess of 300 rem to the thyroid from iodine exposure.

(2) A low population zone of such size that an individual located at any point on its outer boundary who is exposed to the radioactive cloud resulting from the postulated fission product release (during the entire period of its passage) would not receive a total radiation dose to the whole body in excess of 25 rem or a total radiation dose in excess of 300 rem to the thyroid from iodine exposure.

(3) A population center distance of at least 1 1/3 times the distance from the reactor to the outer boundary of the low population zone. In applying this guide due consideration should be given to the population distribution within the population center. Where very large cities are involved, a greater distance may be necessary because of total integrated population dose considerations.

The whole body dose of 25 rem referred to above corresponds to the once in a lifetime accidental or emergency dose for radiation workers which, according to NCRP recommendations, may be disregarded in the determination, of their radiation exposure status. (See Addendum dated April 15, 1958 to NBS Handbook 59). The NCRP has not published a similar statement with respect to portions of the body, including doses to the thyroid from iodine exposure. For the purpose of establishing areas and distances under the conditions assumed in these guides, the whole body dose of 25 rem and the 300 rem dose to the thyroid from iodine are believed to be conservative values.

(b) (1) Appendix "A" of this part contains an example of a calculation for hypothetical reactors which can be used as an initial estimate of the exclusion area, the low population zone, and the population center distance.

(2) The calculations described in Appendix "A" of this part are a means of obtaining preliminary guidance. They may be used as a point of departure for consideration of particular site requirements which may result from evaluations of the particular characteristics of the reactor, its purpose, method of operation, and site involved. The numerical values stated for the variables listed in Appendix "A" of this part represent approximations that presently appear reasonable, but these numbers may need to be revised as further experience and technical information develops.

Chapter 4: Appendix B:
Part 100—Reactor Site Criteria, April 12, 1962

Pursuant to the Administrative Procedures Act and the Atomic Energy Act of 1954, as amended, the following guide is published as a document subject to modification, to be effective 30 days after publication in the Federal Register.

Statement of considerations. On February 11, 1961, the Atomic Energy Commission published in the Federal Register a notice of proposed rule making that set forth general criteria in the form of guides and factors to be considered in the evaluation of proposed sites for power and testing reactors. The Commission has received many comments from individuals and organizations, including several from foreign countries, reflecting the widespread sensitivity and importance of the subject of site selection for reactors. Formal communications have been received on the published guides, including a proposed comprehensive revision of the guides into an alternate form.

In these communications, there was almost unanimous support of the Commission's proposal to issue guidance in some form on site selections, and acceptance of the basic factors included in the proposed guides, particularly in the proposal to issue exposure dose values which could be used for reference in the evaluation of reactor sites with respect to potential reactor accidents of exceedingly low probability of occurrence.

On the other hand, many features of the proposed guides were singled out for criticism by a large proportion of the correspondents. This was particularly the case for the appendix section of the proposed guides, in which was included an example calculation of environmental distance characteristics for a hypothetical reactor. In this appendix, specific numerical values were employed in the calculations. The choice of these numerical values, in some cases involving simplifying assumptions of highly complex phenomena, represent types of considerations presently applied in site calculations and result in environmental distance parameters in general accord with present siting practice. Nevertheless, these particular numerical values and the use of a single example calculation were widely objected to, basically on the grounds that they presented an aspect of inflexibility to the guides which otherwise appeared to possess considerable flexibility and tended to emphasize unduly the concept of environmental isolation for reactors with minimum possibility being extended for eventual substitution thereof of engineered safeguards.

In consequence of these many comments, criticisms and recommendations, the proposed guides have been rewritten, with incorporation of a number of suggestions for clarification and simplification, and elimination of the numerical values and example calculation formerly constituting the appendix to the guides. In lieu of the appendix, some guidance has been incorporated in the text itself to indicate the considerations that led to establishing the exposure values set forth. However, in recognition of the advantage of example calculations in providing preliminary guidance to application of the principles set forth, the AEC will publish separately in the form of a technical information document a discussion of these calculations.

These guides and the technical information document are intended to reflect past practice and current policy of the Commission of keeping stationary power and test reactors away from densely populated centers. It should be equally understood, however, that applicants are free and indeed encouraged to demonstrate to the Commission the applicability and significance of considerations other than those set forth in the guides.

One basic objective of the criteria is to assure that the cumulative exposure dose to large numbers of people as a consequence of any nuclear accident should be low in comparison with what might be considered reasonable for total population dose. Further, since accidents of greater potential hazard than those commonly postulated as representing an upper limit are conceivable, although highly improbable, it was considered desirable to provide for protection against excessive exposure doses to people in large centers, where effective protective measures might not be feasible. Neither of these objectives were readily achievable by a single criterion. Hence, the population center distance was added as a site requirement when it was found for several projects evaluated that the specification of such a distance requirement would approximately fulfill the desired objectives and reflect a more accurate guide to current siting practices. In an effort to develop more specific guidance on the total man-dose concept, the Commission intends to give further study to the subject. Meanwhile in some cases where very large cities are involved, the population center distance may have to be greater than those suggested by these guides.

A number of comments received pointed out that AEC siting factors included considerations of population distributions and land use surrounding proposed sites but did not indicate how future population growth might affect sites initially approved. To the extent possible, AEC review of the land use surrounding a proposed site includes considerations of potential residential growth. The guides tend toward requiring sufficient isolation to preclude any immediate problem. In the meanwhile, operating experience that will be acquired from plants already licensed to operate should provide a more definitive basis for weighing the effectiveness of engineered safeguards versus plant isolation as a public safeguard.

These criteria are based upon a weighing of factors characteristic of conditions in the United States and may not represent the most appropriate procedure nor optimum emphasis on the various interdependent factors involved in selection of sites for reactors in other countries where national needs, resources, policies and other factors may be greatly different.

§ 100.1 Purpose.

(a) It is the purpose of this part to describe criteria which guide the Commission in its evaluation of the suitability of proposed sites for stationary power and testing reactors subject to Part 50 of this chapter.

(b) Insufficient inexperience has been accumulated to permit the writing of detailed standards that would provide a quantitative correlation of all factors significant to the question of acceptability of reactor sites. This part is intended as an interim guide to identify a number of factors considered by the Commission in the evaluation of reactor sites and the general criteria used at this time as guides in approving or disapproving proposed sites. Any applicant who believes that factors other than those set forth in the guide should be considered by the Commission will be expected to demonstrate the applicability and significance of such factors.

§ 100.2 Scope

(a) This part applies to applications filed under Part 50 and 115 of this chapter for stationary power and testing reactors.

(b) The site criteria contained in this part apply primarily to reactors of a general type and design on which experience has been developed, but can also be applied to other reactor types. In particular, for reactors that are novel in design and unproven as prototypes or pilot plants, it is expected that these basic criteria will be applied in a manner that takes into account the lack of experience. In the application of these criteria which are deliberately flexible, the safeguards provided—either site isolation or engineered features—should reflect the lack of certainty that only experience can provide.

§ 100.3 Definitions

As used in this part:

(a) "Exclusion area" means that area surrounding the reactor, in which the reactor licensee has the authority to determine all activities including exclusion or removal of personnel and property from the area. This area may be traversed by a highway, railroad, or waterway, provided these are not so close to the facility as to interfere with normal operations of the facility and provided appropriate and effective arrangements are made to control traffic on the highway, railroad, or waterway, in case of emergency, to protect the public health and safety. Residence within the exclusion area shall normally be prohibited. In any event, residents shall be subject to ready removal in case of necessity. Activities unrelated to operation of the reactor may be permitted in an exclusion area under appropriate limitations, provided that no significant hazards to the public health and safety will result.

(b) "Low population zone" means the area immediately surrounding the exclusion area which contains residents, the total number and density of which are such that there is a reasonable probability that appropriate protective measures could be taken in their behalf in the event of a serious accident. These guides do not specify a permissible population density or total population within this zone because the situation may vary from case to case. Whether a specific number of people can, for example, be evacuated from a specific area, or instructed to take shelter, on a timely basis will depend on many factors such as location, number and size of highways, scope and extent of advance planning, and actual distribution of residents within the area.

(c) "Population center distance" means the distance from the reactor to the nearest boundary of a densely populated center containing more than about 25,000 residents.

(d) "Power reactor" means a nuclear reactor of a type described in §50.21 (b) or 50.22 of this chapter designed to produce electric or heat energy.

(e) "Testing reactor" means a "testing facility" as defined in §50.2 of this chapter.

Site Evaluation Factors

§100.10 Factors to be considered when evaluating sites.

Factors considered in the evaluation of sites include those relating both to the proposed reactor design and the characteristics peculiar to the site. It is expected that reactors will reflect through their design, construction and operation an extremely low probability for accidents that could result in release of significant quantities of radioactive fission products. In addition, the site location and the engineered features included as safeguards against the hazardous consequences of an accident, should one occur, should insure a low risk of public exposure. In particular, the Commission will take the following factors into consideration in determining the acceptability of a site for a power or testing reactor:

(a) Characteristics of reactor design and proposed operating including:

(1)   Intended use of the reactor including the proposed maximum power level and the nature and inventory of contained radioactive materials;

(2)   The extent to which generally accepted engineering standards are applied to the design of the reactor;

(3)   The extent to which the reactor incorporates unique or unusual features having a significant bearing on the probability or consequences of accidental release of radioactive materials;

(4)   The safety features that are to be engineered into the facility and those barriers that must be breached as a result of an accident before a release of radioactive material to the environment can occur.

(b) Population density and use characteristics of the site environs, including the exclusion area, low population zone, and population center distance.

(c) Physical characteristics of the site, including seismology, meteorology, geology and hydrology.

(1)   The design for the facility should conform to accepted building codes or standards for areas having equivalent earthquake histories. No facility should be located closer than one-fourth mile from the surface location of a known active earthquake fault.

(2)   Meteorological conditions at the site and in the surrounding area should be considered.

(3)   Geological and hydrological characteristics of the proposed site may have a bearing on the consequences of an escape of radioactive material from the facility. Special precautions should be planned if a reactor is to be located at a site where a significant quantity of radioactive effluent might accidentally flow into nearby streams or rivers or might find ready access to underground water tables.

(d) Where unfavorable physical characteristics of the site exist, the proposed site may nevertheless be found to be acceptable if the design of the facility includes appropriate and adequate compensating engineering safeguards.

§100.11 Determination of exclusion area, low population zone, and population center distance.

(a) As an aid in evaluating a proposed site, an applicant should assume a fission produce release* from the core, the expected demonstrable leak rate from the

*The fission product release assumed for these calculations should be based upon a major accident, hypothesized for purposes of site analysis or postulated from considerations of possible accidental events, that would result in potential hazards not exceeded by those from any accident considered credible. Such accidents have generally been assumed to result in substantial meltdown of the core with subsequent release of appreciable quantities of fission products.

containment and the meteorological conditions pertinent to his site to derive an exclusion area, a low population zone and population center distance. For the purpose of this analysis, which shall set forth the basis for the numerical values used, the applicant should determine the following:

(1) An exclusion area of such size that an individual located at any point on its boundary for two hours immediately following onset of the postulated fission product release would not receive a total radiation dose to the whole body in excess of 25 rem* to the thyroid from iodine exposure.

(2) A low population zone of such size that an individual located at any point on its outer boundary who is exposed to the radioactive cloud resulting from the postulated fission product release (during the entire period of its passage) would not receive a total radiation dose to the whole body in excess of 25 rem or a total radiation dose in excess of 300 rem to the thyroid from iodine exposure.

(3) A population center distance of at least one and one-third times the distance from the reactor to the outer boundary of the low population zone. In applying this guide, due consideration should be given to the population distribution within the population center. Where very large cities are involved, a greater distance may be necessary because of total integrated population dose consideration.

(b) For sites with multiple reactor facilities consideration should be given to the following:

(1) If the reactors are independent to the extent that an accident in one reactor would not initiate an accident in another, the size of the exclusion area, low population zone and population center distance shall be fulfilled with respect to each reactor individually. The envelopes of the plan overlay of the areas so calculated shall then be taken as their respective boundaries.

(2) If the reactors are interconnected to the extent that an accident in one reactor could affect the safety of operation of any other, the size of the exclusion area, low population zone and population center distance shall be based upon the assumption that all interconnected reactors emit their postulated fission product releases simultaneously. This requirement may be reduced in relation to the degree of coupling between reactors, the probability of concomitant accidents and the

---

*The whole body dose of 25 rem referred to above corresponds numerically to the once in a lifetime accidental or emergency dose for radiation workers which, according to NCRP recommendations may be disregarded in the determination of their radiation exposure status (see NBS Handbook 69 dated June 5, 1959). However, neither its use nor that of the 300 rem value for thyroid exposure as set forth in these site criteria guides are intended to imply that these numbers constitute acceptable limits for emergency doses to the public under accident conditions. Rather, this 25 rem whole body value and the 300 rem thyroid value have been set forth in these guides as reference values, which can be used in the evaluation of reactor sites with respect to potential reactor accidents of exceedingly low probability of occurrence and low risk of public exposure to radiation.

probability that an individual would not be exposed to the radiation effects from simultaneous releases. The applicant would be expected to justify to the satisfaction of the AEC the basis for such a reduction in the source term.

(3)   The applicant is expected to show that the simultaneous operation of multiple reactors at a site will not result in total radioactive effluent releases beyond the allowable limits of applicable regulations.

Note: For further guidance in developing the exclusion area, the low population zone, and the population center distance, reference is made to Technical Information Document 14844, dated March 23, 1962, which contains a procedural method and a sample calculation that result in distances roughly reflecting current siting practices of the Commission. The calculations described in Technical Information Document 14844 may be used as a point of departure for consideration of particular site requirements which may result from evaluation of the characteristics of a particular reactor, its purpose and method of operation.

Copies of Technical Information Document 14844 may be obtained from the Commission's Public Document Room, 1717 H Street NW., Washington, D.C., or by writing the Director, Division of Licensing and Regulation, U.S. Atomic Energy Commission, Washington 25, D.C.

5

Siting of Large LWRs: 1961-1965

With adoption of the Part 100 reactor site criteria in 1962, guidance for the siting of power reactors existed in the form of a legal regulation, at least in principle. The 1950 rule of thumb had had no legal status and, of course, had quickly gone out of date with the use of containment buildings to "prevent" an uncontrolled release. The October 22, 1960, ACRS recommendation that a reactor be sufficiently far from a population center to avoid the radiation dose exceeding 200–300 rem at the city even for an uncontrolled radioactivity release, was, in a sense, an extrapolation of the rule of thumb philosophy, but it was here limited to large numbers of people, say 25,000 or more, for whom rapid evacuation might be difficult to accomplish. Of course, this aspect of siting philosophy was supposed to be covered in the 1962 reactor site criteria by the use of a limit on the minimum distance from the reactor to a city. However, the specific reasoning for this criterion was modified appreciably in the 1962 statement of considerations, and the original philosophic basis was lost. Reactors would be proposed which clearly violated this philosophic concept, and many would be approved.

The Part 100 site criteria, nevertheless, did impose the requirement of a large exclusion radius and a still larger outer radius for the low-population zone, because one had to assume in accident analysis that a large fraction of the radioactive iodine in the core was released to a high-pressure, contain-

57

ment atmosphere and remained there available to leak out until it disappeared by radioactive decay. The utilities had a considerable interest in using sites with a smaller exclusion radius, since this represented land they had to own or positively control. And for many sites of potential interest to utilities, a smaller permissible low-population zone distance was needed. Hence, there was pressure to develop an engineering design approach which would provide a basis for modifying or interpreting Part 100 in a way that relaxed its requirements on the distance between the reactor and people. These engineering changes took the form of containment buildings which could be designed to have very low leak rates, containment spray systems which would reduce containment pressure following a loss-of-coolant accident by condensing the steam and thereby reducing the driving force for leakage from the containment building, and containment atmosphere cleanup systems which could remove the iodine, say into a filter or into the spray water, so that it was unavailable to leak out. These systems, including the containment building, were called engineered safeguards.

After publication of the proposed Part 100 reactor site criteria in 1961 and the final criteria in 1962, a continuing series of decisions were made concerning proposals for constructing increasingly large LWRs at sites with larger surrounding population densities. These sites not only did not meet the 1950 rule of thumb, they also did not meet Part 100 without obtaining "credit" in calculating off-site doses; credit either for a reduction in the leak rate (by use of a double containment, for example), or by reduction of the postulated fission product source available to leak out of the containment building. There was also considerable pressure from the nuclear industry for the construction of large LWRs at sites far more populated than Indian Point, including New York City itself.

The ACRS and the regulatory staff both appear to have accepted fairly quickly the concept that engineered safeguards could be substituted for distance. Several high-power reactors were approved for construction in the next few years, using engineered safeguards to permit relaxation of the previous requirements for the size of the exclusion area and the surrounding region in which there had to be a low-population density. These included:

San Onofre 1 (Calif.)	1,347 MWt	(1963)
Connecticut Yankee (Conn.)	1,825 MWt	(1964)
Oyster Creek (N.J.)	1,930 MWt	(1964–1965)
Nine Mile Point (N.Y.)	1,850 MWt	(1965)
Dresden 2 (Ill.)	2,527 MWt	(1965)

Although each passing year saw the evolution of new safety requirements dealing with a wide range of technical issues, the principal focus for construction permit review during this period appears to have been on the efficacy of the engineered safeguards (containment buildings with filters and/or sprays) needed to meet the dose guidelines of Part 100. Accidents exceeding the MCA were not considered as part of the siting or construction permit review. Nor was much evaluation or emphasis placed on the potential effects on public health and safety of radioactive contamination of water supplies due to an uncontained reactor accident.

The Corral Canyon Site

We shall now examine this time-period in more detail.

At its thirty-sixth meeting, September 7–9, 1961, the ACRS reviewed the request of the City of Los Angeles for approval of eight proposed reactor sites. The sites were in three general areas of northwestern Los Angeles county, the San Francisquito Canyon area, the Green Valley area and the Fairmount area. The population distribution at the time of review was favorable, and the projected population growth appeared to be acceptable. For at least the San Francisquito Canyon site, it was stated that the meteorology could be unfavorable, and the ACRS advised that this area might require more engineered safeguards than reactors at the other sites. However, the ACRS concluded that it would be possible to locate reactors at any of these sites without undue risk to the health and safety of the public.

At the thirty-eighth ACRS meeting, December 7–9, 1961, the City of Los Angeles was back, this time for consideration of two new sites which were much more heavily populated. One of them, the Haynes Point site, was very near Long Beach. The other was the Corral Canyon site, west of Los Angeles near the ocean. The regulatory staff were of the opinion that the Haynes Point site was unacceptable because of the very large nearby population. The regulatory staff wished to discourage use of both the sites, but they did not believe it impossible to locate a safe reactor at the Corral Canyon location. The ACRS did not comment on the two new proposed sites at this meeting.

At the fortieth ACRS meeting, March 29–31, 1962, there was a meeting between the committee and AEC commissioner Loren K. Olson. The minutes note that commissioner Olson hoped that the project for reactors in the Los Angeles area could proceed. He asked for a positive approach by the safety review groups toward the recently proposed sites, and he asked if either an underground location or one in a hillside would be acceptable. At the same meeting, a member of the ACRS staff reviewed the congressional hearings related to the development of atomic energy. He reported that a representative of a Boston consulting engineering firm, Harold Vann, had impressed the Joint Committee on Atomic Energy. He advised that Vann had testified that the site criteria would deter the construction of reactors, and that Vann had proposed more development of iodine-removal equipment within containment buildings to help alleviate the situation posed by the Part 100 regulations.

Also at its fortieth meeting, the ACRS continued its review of the proposal by the Los Angeles Department of Water and Power for consideration of the Haynes Point and the Corral Canyon sites. Neither site met the new site criteria; the regulatory staff reported that there would have to be great dependence on engineered safeguards because of the lack of isolation.

C. Rogers McCullough, who had been an ACRS member and was now acting as a consultant, is quoted as corroborating fears expressed by Joseph DiNunno of the regulatory staff that the heat from the decay of fission products might breach the large amount of concrete in the General Electric reactor that was being proposed.

A Westinghouse group presented information on their proposed reactor for the Haynes Point site. The special features which were provided to make it acceptable included, in particular, a double containment system with twin liners surrounding an annular space to be held at negative pressure. Thus, we see here the proposal that the provision of very low leakage for the MCA is what is needed for urban (nearly metropolitan) siting, without specific mention of the question of containment integrity for more severe accidents.

General Electric proposed a 400 MWt direct-cycle boiling water reactor, using a pressure suppression containment building such as was being used at Humboldt Bay. Its safety objectives included low probabilities of an uncontained accident, but no details were given.

The ACRS also heard a presentation by Dr. George Housner, a consultant to the Department of Water and Power, in which he expressed confidence that a reactor could be satisfactorily designed for seismic conditions at either site.

The ACRS wrote a report at this meeting dated April 4, 1962, and made the following comments with regard to the two new sites:

In its most recent proposal, the City of Los Angeles presented two coastal sites which its representatives stated present appreciable economic advantages over the presently accepted sites. These two sites are a southern site now owned by the City, and a western site which could be obtained.

In regard to the two new sites proposed for reactors of the general concepts presented, the Committee has the following comments: Neither of the locations can meet the site criteria guidelines proposed in 10 CFR 100 for the power level requested. Both sites are within areas of high and increasing population. In this connection, it should be noted that power reactors of the size proposed have not yet been built and proved. Such reactors would contain larger fission product inventories than any licensed power reactor now operating or under construction.

If the sites proposed are to be considered acceptable, then reliance must be placed on *proved* engineering safeguards as a means of preventing exposure of significant numbers of people to possible radiation injury. The Committee believes that it is possible with present engineering technology to overcome the potential danger from serious consequences of a major earthquake.

The Committee has the following comment concerning the two reactor concepts proposed, and their respective containments: neither proposal provides proved assurance of satisfactory containment of an accident, such as a serious nuclear excursion, which releases radioactivity simultaneously with the release of pressure. The possibility of such an accident cannot be excluded on the basis of present knowledge.

Of the two coastal sites, the western site is in an area of lower population density and is further removed from large centers of population. Neither site is suitable for either of the proposed reactor facilities. The proposed plant designs might more readily be modified to a form suitable for the western site.

There is no indication how the ACRS dealt with the point made by Mr. McCullough concerning the possibility that heat from the decay of fission products could lead to a failure of the concrete containment building in the General Electric design. The question of why the General Electric design was the one discussed in this context is not clear from the minutes.

At the forty-third ACRS meeting, August 23–25, 1962, the Los Angeles Department of Water and Power was back to speak with the ACRS concerning the possibility of a newly revised design for the proposed boiling water reactor for the Corral Canyon site. The design for the 1,400 MWt reactor now included a suppression system and a confinement building around the suppression system, which also enclosed the refueling operations for the reactor. A fundamental problem facing the committee was whether it was acceptable to have part of the primary system leave the containment building, as was done in a direct-cycle boiling water reactor system. Some members felt that perhaps the turbine should be inside the containment building. Other members pointed out that moving machinery is a likely place of failure which could generate large missiles in the containment building. The applicant and his reactor designer proposed to put two isolation valves in the steam lines running to the turbine, so that if the turbine should fail, these valves could close and avoid a loss of coolant from the core. At the forty-third ACRS meeting, the Los Angeles Department of Water and Power (LADWP) also described a somewhat revised concept for the previously discussed PWR, employing a double containment vessel which completely enclosed the primary system and which included a feature involving the pumping back into the inner containment of any leakage into a porous "popcorn" concrete-filled space between the containment walls.* The PWR proposal also included temporary holdup of routine radioactive gas release, which would lead to reduced off-site effects.

The regulatory staff said that they had concluded that the proposed reactors could be built and operated safely at the Corral Canyon site. It is not clear from any of the meeting minutes whether consideration was given at this time to the general recommendation of the October 22, 1960, ACRS letter about site criteria concerning the limitation that, even in the event of the worst possible accident, there should not be a catastrophe. The committee did not take action on the proposed design at the forty-third meeting.

It is clear from the minutes of the forty-fourth ACRS meeting, October 4-6 and 12, 1962, and from other meetings in that time-period, that, during ACRS review of the Corral Canyon site (or possibly the Haynes Point site), strong differences of opinion developed within the committee. (There is even a discussion in the minutes of the existence of proposed majority and minority letters.) However, detailed differences, as they might have appeared in majority or minority letters, are not available in the minutes.

Following its forty-fourth meeting the ACRS issued a report dated October 12, 1962, on the reactor proposals for the Corral Canyon site. The ACRS report expressed the opinion that the PWR containment system was adequate, but had some reservations concerning the proposed boiling water reactor, particularly its dependence upon the rapid closure of isolation valves in the event of an accident involving the rupture of the steam line outside the containment building. However, in a major decision on siting, the committee concluded in favor of either reactor at the Corral Canyon site, if it were provided with adequate containment. Thus, the ACRS (and the regulatory staff) approved rather large reactors (1,600 MWt) for a site not very distant from a large population center (say 10 miles). The letter says that: "The Committee in its reviews has focused its attention on the adequacy of engineered safeguards for the containment of any signifant potential releases that might affect the health and safety of the public." What appears to have been emphasized in the review was the containment system (evaluated only in terms of the MCA). Other engineered safety features which would keep the core from melting, and mea-

---

*At the 1962 IAEA Symposium on Reactor Safety and Hazards Evaluation Techniques, W. E. Johnson of Westinghouse gave a paper in which he described this proposed containment design, which had been developed by the Stone and Webster Company, as an absolute containment or no-leakage concept, with absolutely no leakage even for the worst hypothetical accident including core meltdown. He suggested the concept would permit siting of nuclear power plants within large populated areas (Johnson 1962).

sures, such as improved integrity of a primary system, to prevent serious accidents from occurring, received rather little attention. In a sense, this seems to have established a trend for the focus of regulatory review from 1962 until the middle of 1966, when a major change in the requirements for accident prevention and mitigation occurred. We will come to this point in Chapter 8.

At a special meeting, November 9–10, 1962, the Los Angeles Department of Water and Power was back once again to discuss additional safety features in the design for the proposed General Electric boiling water reactor (BWR) for possible use at the Corral Canyon site. The ACRS wrote a report concluding that a boiling water reactor of the type proposed, with adequate engineered safety features, could be located at the Corral Canyon site. This left the Los Angeles Department of Water and Power free to choose either a PWR or BWR.

There is no record in the minutes of the many meetings on the Corral Canyon site that either the ACRS or the regulatory staff suggested consideration of alternate, less-populated sites. The question addressed was: Is the proposed site acceptable? For Haynes Point, an answer of no was indicated. For Corral Canyon, the answer was yes.

The Los Angeles Department of Water and Power next applied for a construction permit for a 1,300 MWt PWR at Malibu, California (formerly called the Corral Canyon site). The ACRS reported favorably upon the construction permit proposal in a letter from the Chairman, H. Kouts, to AEC Chairman Glenn T. Seaborg, dated July 15, 1964, which is reproduced in the appendix to this chapter.

The ACRS had endorsed the Malibu site in its letter of October 12, 1962, on Corral Canyon. The review performed in 1964, however, was done in greater detail because it was a formal construction permit application.

The reactor at the Malibu site was never built. As is discussed in Chapter 17, questions were raised concerning the adequacy of its seismic design by an intervenor group at the ASLB hearing, and the intervenor's position was upheld first by the ASLB and then by the AEC commissioners themselves.

San Onofre Unit No. 1

At its forty-ninth, meeting, September 5–6, 1963, the ACRS reported favorably on the application by Southern California Edison for construction of a 1347 MWt PWR at a site in northern Camp Pendleton, close to San Clemente, California.

The regulatory staff noted that this site could not tolerate 100% meltdown of the fuel and full release of the fission products to the containment building, as it would have to were it to meet Part 100 regulations. Credit had to be given for an emergency core-cooling system, so that only 6% of the core was assumed to melt with reduced release of fission products to the containment building. This represented a major departure from the postulated maximum credible accident release in Part 100.

The ACRS accepted the approach discussed above without making a detailed review of the actual effectiveness of the core-cooling system. Excerpts from the ACRS report of September 12, 1963, on San Onofre Unit No. 1 follows:

The applicants propose to contain the reactor in a spherical steel structure designed for a maximum leakage rate of 0.1% per day at pressure and with critical penetrations designed to permit frequent leak testing. Additional engineered safeguards are required for this site. Such safeguards proposed include a multiple borated-water injection system to prevent extensive core meltdown in the unlikely event of a major break in the primary water system, a containment spray system, and an internal air cleanup system.

A meteorological factor favorable to the proposed reactor locations is the fact that air movement from the site toward San Clemente occurs, at most, only a few percent of the time.

The ACRS has emphasized that the engineered safeguards must be designed and reviewed with great care for both adequacy and reliability. Special attention should be directed to the safety injection system which must perform as proposed to validate the applicants' assumption of low release of radioactivity to the containment under accident conditions. A halogen removal system may be required. Design details of the holdup system for reactor off-gases resulting from routine operation will also require careful attention. The ACRS has recommended study of the consequences of rain-out following an accident; the results of this study should be taken into account in the final design of the engineered safeguards.

In view of the favorable prevailing wind direction, conservative seismic design approach, and with engineered safeguards of the type proposed, it is the Committee's opinion that a pressurized water reactor of the type and power level proposed can be designed, constructed and operated at the site without undue hazard to the health and safety of the public.

Connecticut Yankee Reactor

At its fifty-third meeting, February 13–15, 1964, the ACRS completed review of the proposal for construction of the 1,473 MWt Connecticut Yankee PWR. Excerpts from the ACRS report of February 19, 1964, follow below:

In its previous report to the Commission on February 6, 1963, the ACRS pointed out that the Haddam Site did not meet the present site distance criteria, and hence reliance must be placed upon engineered safeguards to reduce off-site exposures in the unlikely event of a serious accident. Because of otherwise favorable site location, low population density and meteorological characteristics, a reduction factor of about 6 in addition to that provided by containment is needed to bring the potential dose from a maximum hypothetical accident to guideline limits.

The proposed design has the reinforced concrete containment detailed below. The design includes the following additional engineered safeguards: an internal recirculation containment spray system; a continuously operated air recirculation system with cooling, involving four independent systems which can remove halogens and other fission products. The plant is also to be provided with a safety injection system having three independent pumps and a large supply of borated water.

A reinforced concrete containment vessel with a steel inner liner is proposed. Containment leakage is specified to be not more than 0.1% per day and penetration leakage rates will be monitored. The proposed containment is designed for the use of stainless steel clad fuel elements in the reactor. If, for instance, Zircaloy cladding is used, it may be necessary to increase the design pressure or volume of the containment.

The reactor is to be a pressurized water system of proven operating characteristics with cluster type control rods. The use of four separate steam generator loops decreases the significance of a major primary coolant line rupture. Details of the reactor physics behavior will be resolved during the design phase. The Committee believes the possibility and effects of control rod ejection deserve further evaluation and documentation.

The Committee considers that the proposed engineered safeguards provide the necessary redundancy and reliance to assure reduction of releases to below guideline values in the unlikely event of a reactor accident. The filter-absorber systems, while not finally selected as to performance characteristics, should be protected against steam and water releases, and may require capability for various forms of halogens. These factors should be reliably established before the facility operates.

It is the opinion of the ACRS that the proposed engineered safeguards, including the containment as proposed, will provide the necessary protection in the unlikely event of an accident. On this basis, the ACRS believes that there is reasonable assurance that the general type of reactor proposed for the Connecticut Yankee Atomic Power Company, including engineered safeguards, can be constructed at the Haddam Site with reasonable assurance that it can be operated without undue hazard to the health and safety of the public.

The Connecticut Yankee letter continues the emphasis on containment and on engineered safety features to reduce the fission product concentration in the containment building following the postulated release from the MCA.

A safety injection system is mentioned; however, little evaluation was made of the design basis or efficacy of this emergency core-cooling system during the review. The allusion to a potentially larger containment design pressure, if Zircaloy cladding was used instead of stainless steel, arose from the concept that a large fraction of the Zircaloy would undergo metal-water reaction in a core meltdown, releasing heat and hydrogen (which could burn, adding more heat), leading to a higher peak accident pressure. As studies showed some years later, this represented an incomplete evaluation of metal-water reaction problems, since steel could also react chemically with steam to form hydrogen.

This letter was the first to point out the requirement for study of an accident involving control rod ejection. This requirement led to design changes in large LWRs, either to limit the maximum effect on the neutron chain reaction of each control rod (and hence to keep the resulting power rise tolerable in such an accident) or to add an additional mechanical restraint to control rod ejection, thereby making the probability acceptably low (the approach taken in BWRs).

Oyster Creek Reactor

At its fifty-seventh meeting, August 24–26, 1964, the ACRS completed a report favorable to the construction of a 1,600 MWt BWR at Oyster Creek, New Jersey. The brief ACRS letter noted that, "Many details of the proposed design have not yet been completed. The applicant is continuing to study the limitation of maximum reactivity of individual control rods and the design of the reactor protection system." The letter also stated, "Provision should be made to prevent any hydrogen-oxygen reaction that would disrupt the containment."

The letter makes no mention of the emergency core-cooling system (ECCS) proposed for this reactor. However, the minutes of the meeting indicate that there was discussion of the fact that the reactor included a duplicate core-spray arrangement to limit melting of the core in the event of a loss-of-coolant accident. Edson Case of the regulatory staff had indicated that a hydrogen explosion in the containment building following a possible zirconium-water reaction was the chief remaining problem, and that an inert atmosphere, that is nitrogen, could prevent this.

At that time the experimental data and analytical methodology did not exist to fully analyze the efficacy of the core-spray system, and the ACRS was reluctant to give general "credit" for emergency core-cooling systems, although it apparently had done so for the San Onofre site.

What is interesting is that there had by now been a succession of reviews of rather large LWRs, but none of these were operating, none were fully designed, and none had received detailed regulatory analysis and evaluation. The actual amount of information available for the decision-making process in 1964 was quite limited, compared to what would be available in the 1970s.

Engineered Safeguards

Early in 1964, the Atomic Energy Commission had asked the ACRS to put into a report the manner in which it was permitting engineered safety features to be substituted for distance in meeting Part 100 regulations. At the fifty-ninth meeting, November 12–14, 1964, the ACRS prepared a letter entitled "Report on Engineered Safeguards" (issued November 18, 1964). The things that received approval were: containment and certain confinement systems, the pressure suppression method, containment building sprays to reduce containment pressure following a loss-of-coolant accident (LOCA), heat-exchange methods of limiting containment pressure, and containment air-cleaning systems following a LOCA.

The letter was ambivalent concerning core-spray and safety injection systems, saying that they cannot be relied upon as the sole engineered safeguards since "they might not function for several reasons such as severed lines and low water supplies. Nevertheless, prevention of core melting after an unlikely loss of primary coolant would greatly reduce the exposure of the public. Thus, the inclusion of a reactor core fission product heat removal system as an engineered safeguard is usually essential."

The body of the report noted the need for adequate emergency power sources of an ECCS. What seems to be lacking from the report is any identification of engineered safety systems for residual (fission product decay) heat removal from a core in which the power has been shutdown, although attention was given to methods of removing heat from the containment building.

Thus, the period from 1960 to the end of 1964 represented a time during which quantitative siting criteria evolved and then were relaxed as the acceptance of engineered safety features, in place of distance, became part of the regulatory process. It was also a period in which there was a strong beginning of looking at other things in addition to the maximum credible accident. However, there was not a comprehensive, systematic look at all (or most) accidents which might, in fact, represent a threat to containment and have consequences far exceeding Part 100. And at least with regard to siting, the approach developed in the period up to 1964 in large part represents the approach used by the regulatory staff up to 1979.

Chapter 5: Appendix A:
ACRS Report to the AEC on City of Los Angeles—
Malibu Nuclear Plant—Unit No. 1

At its fifty-sixth meeting at Brookhaven National Laboratory on July 9–11, 1964, the Advisory Committee on Reactor Safeguards reviewed the proposal of the City of Los Angeles to construct and operate a 1473 MW(t) pressurized water reactor, Malibu Nuclear Plant—Unit No. 1, at Corral Canyon, twenty-nine miles west of Los Angeles. The Committee had the benefit of discussions with representatives of the Department of Water & Power of the City of Los Angeles, Westinghouse Electric Corporation, Stone & Webster Engineering Corporation, the AEC Staff, their consultants, and of a Subcommittee meeting on June 18, 1964. The Committee also had the benefit of the documents listed below.

The proximity of large population centers and the probable growth of population in the vicinity of the proposed reactor site require dependence on engineered safeguards to limit the consequences in the unlikely event of a major credible accident. For this reason, safeguard provisions more extensive than those normally employed in nuclear power reactor plants must be provided in lieu of the distance factor to protect the public.

The applicant has proposed as engineered safeguards a novel containment structure intended to prevent any leakage to the environment, and additional features consisting of:

1. A reinforced concrete containment structure.
2. A containment volume spray system, and
3. An emergency borated-water injection system.

The total containment feature of the building is to be achieved by providing two complete steel liners separated by a layer of porous concrete. The space between the liners will be maintained at a sub-atmospheric pressure by continuously pumping back air to the containment volume. An air recirculating and cooling system is required to remove any heat that is generated within the containment volume. Power and water to assure operation of these systems under all conditions must be provided.

Detailed design of the reactor core has not been established yet, but the general features will be similar to those of other nuclear plants proposed for construction by the same nuclear contractor, and expected to be tested in operation prior to completion of the Malibu plant. Nuclear reactivity coefficients are expected to be negative in this reactor. The probability and effects of control rod ejection require further evaluation. The applicant has suggested several possible means of limiting the consequences of such an accident, and the Committee believes that this question can be resolved satisfactorily during the design stage.

Although stainless steel cladding is planned for the first core, it is anticipated that zirconium alloys may be used in future cores. Complete information on the effect of a possible zirconium–water reaction on the course of accidents is not available. Hence, further review will be needed prior to use of zirconium alloy clad cores.

The Committee was informed that the geology of the site was suitable for the proposed construction. It was reported that no active geological faults are present at the site. Grading of the canyon slopes is proposed to ensure that potential landslide motion does not present a hazard to the plant. It is proposed that critical structures be designed for a suitable response spectrum associated with an earthquake which has a maximum acceleration of 0.3 g. occurring when the containment is under the pressure associated with an accident. The resulting stresses will not exceed 80% of the minimum yield value. Components within the building will be designed to withstand 0.3 g. acceleration acting simultaneously in horizontal and vertical planes.

The ability of the plant to withstand the effects of a tsunami following a major earthquake has been discussed with the applicant. There has not been agreement among consultants about the height of water to be expected should a tsunami occur in this area. The Committee is not prepared to resolve the conflicting opinions, and suggests that intensive efforts be made to establish rational and consistent parameters for this phenomenon. The applicant has stated that the containment structure will not be impaired by inundation to a height of fifty feet above mean sea level. The integrity of emergency in-house power supplies should also be assured by location at a suitable height and by using water-proof techniques for the vital power system. The emergency power system should be sized to allow simultaneous operation of the containment building spray system and the recirculation and cooling system. Ability to remove shutdown core heat under conditions of total loss of normal electrical supply should be assured. If these provisions are made, the Committee believes that the plant will be adequately protected.

The applicant has proposed to deny entrance to the containment while the reactor is operating. This mode of operation does not permit frequent surveillance of equipment and prompt detection of incipient defects. Operating experience at other power plants has demonstrated the value of accessability for inspection. The Committee suggests that the applicant reconsider this question and explore design modifications which will allow entrance without violating the containment integrity.

As the Committee has commented in its earlier letters, the hold-up of routine gaseous and liquid releases may be necessary during unfavorable conditions. In this connection, it will be necessary to conduct additional preoperational meteorological and oceanographic survey programs.

The Advisory Committee on Reactor Safeguards believes that the items mentioned above can be suitably dealt with during construction, and that the proposed Malibu Nuclear Plant can be constructed with reasonable assurance that it can be operated at the site without undue risk to the health and safety of the public.

6

The Ravenswood Reactor
and Metropolitan Siting

In the late 1950s, the nuclear industry jointly with the reactor development side of the AEC had proposed to build several small (60 MWt)
reactors in or next to small cities; they were rebuffed in large part but not
completely by the ACRS and the regulatory staff. (The Piqua organic
cooled and moderated reactor was built.)

The Part 100 reactor site criteria specifically notes the "current policy of
the Commission of keeping stationary power and test reactors away from
densely populated centers." However, it goes on to say immediately thereafter, "It should be equally understood, however, that applicants are free
and indeed encouraged to demonstrate to the Commission the applicability
and significance of considerations other than those set forth in the guides."

Thus, Part 100 reflected an ambivalence on the part of the AEC. Strict
adherence to the criteria would have made it difficult, if not impossible, to
consider metropolitan siting favorably. However, utilities were encouraged
to propose reactors with improved engineered safety features for consideration at urban or city sites. The term "stationary" power reactors was used
so that bringing the nuclear-powered ship N.S. *Savannah* (or its possible
successors) into densely populated ports would not be ruled out.

The nuclear industry challenged or encroached on Part 100 restrictions
on siting in two ways. One way was to move in stepwise fashion toward
increasingly more-populated sites. This approach was initiated successfully

with the San Onofre, Connecticut Yankee, and Oyster Creek reactors. Similarly, the Los Angeles Department of Water and Power first received approval for some relatively remote sites, then obtained regulatory staff approval of the much more populated Corral Canyon (Malibu) site. The other approach was a direct attack on metropolitan siting restrictions, namely, the application for a construction permit for the Ravenswood reactor essentially in the heart of New York City.

We shall see that the AEC regulatory staff rejected this application before the ACRS had provided a committee recommendation to the AEC, and that there appeared to be mixed sentiment among the ACRS members concerning the Ravenswood reactor, although the site clearly violated a basic precept of the October, 1960, ACRS letter on siting. This ACRS ambivalence on metropolitan siting again appeared in 1965 when the ACRS was asked for its opinion concerning a proposed moratorium on metropolitan siting. The ACRS opposed a formal moratorium and instead recommended safety requirements for a metropolitan reactor which could not be met by existing designs. An important consideration that probably entered into this thinking was that, if metropolitan siting were not categorically denied, designers might be able to develop fairly drastic new approaches that would make the risk acceptable. In fact, many years later a conceptual study was performed in Sweden of a possible underground reactor, the so-called Vartan reactor, which was intended for consideration at an urban site and which was to include safety features that went far beyond those on existing plants.

Application for the Ravenswood Reactors

On December 10, 1962, the Consolidated Edison Company submitted an application for a construction permit for two reactors to be located at the Ravenswood site in New York City. These two Westinghouse pressurized water reactors each were to have a thermal power output of 2,030 MW; and the estimated completion date was between October 1, 1969, and October 1, 1970, which would have represented a design and construction period similar to that for other reactors of that day. The summary of the construction permit application stated that the two containment vessels for each reactor would prevent the release of *any* radioactive material to the surrounding area in the event of an accident. There was to be double containment for each of the reactor systems and for the common storage facility for spent fuel. These containment structures would consist of two steel shells completely surrounded by 5½ feet and 2½ feet of concrete, respectively, with pervious concrete occupying the annulus between the inner and outer steel membranes. This was like the containment described in the paper by Johnson of Westinghouse at an IAEA Symposium in 1962 (Johnson 1962). It was stated that with double containment there would be no leakage of radioactivity even under the worst conceivable accident conditions. A research and development program was outlined, and it was stated that this program was to be completed prior to January 1, 1967, which indicated that no difficult or long-term research problems were expected. The containment included a spray system to reduce the pressure within the containment vessel in the event of an accident. There was also a pump-back system by which the pressure between the two steel membranes was to be maintained at slightly below atmospheric pressure by pumping air from this region into the interior of the reactor containment. A safety injection system was included, one which would supply borated water to the reactor core following a loss-of-coolant accident (LOCA). The design basis for the emergency core-cooling system (ECCS) was stated to be a rupture of the largest connecting pipe to the main pipes in the primary system, rather than of the main pipes themselves. This was the ECCS design basis used in 1963. In 1966 the design basis for the ECCS was made much more stringent.

Consolidated Edison defined the "worst conceivable" accident as one caused by the instantaneous release of the entire contents of the primary cooling system into the containment building, followed shortly thereafter by melting of 100% of the core. From this event the hazards to the environment were stated by Consolidated Edison to be less than the hazards accepted for routine releases in 10 CFR Part 20, which are very small.

The regulatory staff stated that relatively less information was available concerning the design of this reactor than for other reactors recently reviewed. The site, which was in the Borough of Queens, was 8.7 acres in size, bounded on the west by the East River, on the north by Thirty-sixth Avenue, and on the east by Vernon Boulevard. The minimum distance from the reactor containment building to the fence around the site boundary was approximately 90 feet. The population within a circle of a radius of ½ mile was estimated to be 19,000 at night and 28,000 at day; within a circle of 5 miles radius, it was estimated to be 3 million people at night and 5½ million during the day. Although proposals had been made previously for the siting of relatively small reactors within small cities, for example, the Jamestown reactor, this was, by far, the most difficult reactor site proposed to the Atomic Energy Commission.

The regulatory staff began to analyze the reactor and site with the information that it had. In view of the fact that the site clearly would not meet the normal conditions of the site criteria with regard to exclusion distance and low-population zone, or even the distance to a population center, the staff decided to see whether the site could be made acceptable on a so-called engineered safeguards basis. This was prior to looking at accident analysis, etc. On August 9, 1963, the regulatory staff sent out a set of questions, only 13, in which it requested additional information from the Consolidated Edison Company. The bulk of these questions related to the design of the containment, the ability to measure leakage rates in the containment, the way in which penetrations through the containment could be monitored, and things like filter systems which could remove radioactivity from the containment. The Consolidated Edison Company responded to these questions in a letter dated November 14, 1963. The utility proposed to use redundant systems for the safeguards features employed in the Ravenswood plant; and the use of a single-failure criterion* was planned in order to provide adequate reliability. Also, the utility mentioned that in the layout of the plant careful consideration would be given to insuring that an initiating accident would not impair operability of safety components. For example, proper separation of piping and cabling would be used, and it was stated that the components themselves would be designed to operate under the temperatures and other environmental conditions which would exist in the plant containment building following a loss-of-coolant accident. It was not planned to protect against gross failure of the reactor pressure vessel, pressurizer, or steam generator. While there was frequent reference in the application to the containment design, and to the engineered safety features which would keep the dose to the public below the very low levels permitted for routine releases in 10 CFR Part 20, given the fission products corresponding to full-scale core melt, nowhere was mentioned the possibility that full-scale core melt might lead to failure of the containment.

On September 25, 1963, the regulatory staff sent a report to the ACRS concerning the Ravenswood reactor in which the staff outlined the features of the plant at that time. The report described the ways in which the site departed greatly from what was permitted by the AEC site criteria, and discussed an approach in which they (the staff) would try to see whether it was possible that a containment scheme could be devised which would permit the site to yield acceptable doses of radioactivity after an MCA. The regulatory staff announced their intention first to complete an evaluation of proposed engineered safeguards systems, and then pursue further analysis of accident consequences.

---

*This criterion states that, if safety equipment is required to terminate an abnormal change in power or flow or to cope with an accident such as a LOCA, the safety equipment will perform adequately even if one failure occurs in some component of the safety systems, such as a pump and valve, in addition to the failure which initiated the event.

The results of this phase of the evaluation were to be presented for review by the ACRS in time for the December 1963 meeting of the committee.

An ACRS subcommittee meeting was held with the Consolidated Edison Company on September 11, 1963, concerning the Ravenswood reactors. It is not possible from the meeting minutes to tell what was in the mind of the members. ACRS member F. A. Gifford did make a comment to the applicant that "since the site is lousy, questions would center on containment and engineered safeguards." At the meeting, the Consolidated Edison Company stated that they had specified that radiation from the proposed plants under any and all conditions must not exceed 10 CFR Part 20. This statement was not changed when ACRS member K. R. Osborn indicated that there was not protection for reactor vessel rupture, or when ACRS member D. A. Rogers asked about the adequacy of missile shielding.

There were stated to be four independent outside sources of electrical power for the plant. When ACRS member Gifford asked whether failure of the steam generator was included in setting the containment design pressure in connection with a LOCA, Dr. R. L. Wiesemann of Westinghouse stated it was cheaper to support the pipes mechanically so that primary system rupture would not lead to a secondary system failure.

The bulk of the discussion during the meeting related to the possibility of providing containment with the very low leakage rates that Consolidated Edison was seeking.

Another ACRS subcommittee meeting was held on October 21, 1963. The preliminary discussion by the subcommittee members prior to their meeting with the regulatory staff indicated that the primary emphasis was on what degree of assurance could be credited to the proposed design of a plant with a zero leak rate. It was noted that many paths existed by which a double containment scheme might be bypassed, and that even a small release might be intolerable at this site. At least from the minutes, there does not seem to have been much discussion about accidents involving a large release of radioactivity and a loss of containment integrity.*

At this meeting, Consolidated Edison noted that the design basis was that maloperation of no single component would cause damage to the reactor core, and that gross failure of no single component would impair the capability of containing fission products within the plant. However, the reactor pressure vessel and certain other components could not really be included in that statement.

ACRS member W. K. Ergen summarized his concern over location of the reactor at this site by stating that it was unrealistic to design and operate

---

*In 1977, Dr. H. Kouts, an ACRS member during the Ravenswood review period, recalled that the principal problem with the Ravenswood site was its vulnerability to "small" accidents.

a plant with assurance that the required very low leak rates could be met in the event of an accident. He had calculated that the required leak rate would have to be of the order $10^{-3}\%$ per day of the contained volume, in order not to produce more than 300 rem to the thyroid to 4,000 people around the plant. He said it would be easy to have any number of "Achilles' heels," like a minor steam generator tube leak.

The minutes of the fiftieth ACRS meeting, Octobr 10–11, 1963, record further discussion of the Ravenswood reactor. ACRS member Osborn appeared to consider that this plant had had a more comprehensive study at this stage than is usual; to him the general ideas of design, construction, and safety were good. Member T. J. Thompson felt that the ACRS could approve such a plant, provided it had all the safeguards then proposed. On the other hand, member Rogers would have required the applicant to guarantee that no accident would affect the public, before ACRS approval were granted. He had doubts about the effectiveness of the secondary containment.

In the discussion between the ACRS and the regulatory staff, Mr. R. Lowenstein, the AEC Director of Licensing, indicated that computer studies showed that even if all the engineered safeguards operated, leakage must still be limited to the order of 10^{-4} cubic feet per minute. The regulatory staff saw this as impossible with the design proposed. Mr. Lowenstein indicated that, if it were to make a decision then, the regulatory staff would reject the application because the proposal entailed too much of an advancement in reactor technology for this location.

Application for the Ravenswood Reactors Is Withdrawn

At the fifty-first meeting, November 7–8, 1963, the ACRS had some further discussion of the Ravenswood site in an executive session. The bulk of the opinion seemed to be that the regulatory staff was going to reject the site. The consensus of the committee seemed to be that more time should be given for review of the matter, but this was not unanimous.

Between the fifty-first and fifty-second ACRS meetings, Consolidated Edison withdrew its application for consideration of the Ravenswood reactors.* It is not possible to reconstruct the overall position of Consolidated Edison with regard to the safety of the reactors. The minutes indicate

*The ACRS minutes have no record of any negative (oral) opinion concerning the Ravenswood site being forwarded from the AEC to Consolidated Edison. However, it is the author's understanding from recent discussion with a senior member of the regulatory staff that Consolidated Edison was told by the AEC that their application would receive an unfavorable response, if it were not withdrawn.

that they considered that, for any situation that was "credible," they would be protected. They did not indicate any concern that building those reactors at that time in the heart of New York posed an undue risk. It is also not clear what the basis for their judgment was. The official reason given for withdrawal of the application of the Ravenswood reactor by Consolidated Edison was the availability of cheaper power from Labrador, 1100 miles away.

Although it was withdrawn, the Ravenswood reactor application had forced the regulatory groups to consider very specifically the question of metropolitan siting. And in a sense, the application was one form of pressure on the regulatory groups to see in what way, if any, metropolitan siting of reactors could be approved. As we shall see, the industry continued to propose sites involving a large surrounding population, not quite like that of Ravenswood but still considerably beyond what had been accepted before.

The minutes of the fifty-second ACRS meeting, January 9–10, 1964, show that the committee once again discussed reactors in populated areas even though the application for the Ravenswood reactor had been withdrawn only a few days before. Chairman Kouts asked if the ACRS wished to prepare a letter, for example, a letter drafted by member Thompson, regarding the location of reactors in populated areas. Briefly, this letter considered such sites acceptable provided there were (adequate) engineering safeguards and a reactor of the same type and power level had operated safely elsewhere. However, some members felt the committee should wait, giving advice if and when a particular case were before the committee.

This admittedly incomplete record of the ACRS review of the Ravenswood site indicates the fairly wide range of opinion on the committee. There was clearly no unanimous consensus that the Ravenswood site should be rejected out-of-hand. The seeming emphasis on "small accidents" is somewhat curious, since this site represented such a strong contradiction to the October 22, 1960, ACRS siting recommendation that "Even if the most serious accident possible (not normally considered credible) should occur, the numbers of people killed should not be catastrophic."

During the period of the review of the Ravenswood reactor, the ACRS felt that large-scale core meltdown could not be ruled out as a possible, although improbable, accident. However, as of the November 1963 ACRS meeting, they had not, as a matter of policy, recommended against the Ravenswood reactor. It is not clear from the minutes how the possibility of releases, very much larger than Part 100, in the heart of a large population center was affecting the review. It is clear from the discussion that reactor vessel failure and other modes of containment failure were considered. It does not appear that in 1963 the ACRS envisaged that core meltdown would probably lead to containment failure for a reactor of the size of Ravenswood.

The Edgar Station Site Is Proposed

Another interesting sequence of events concerning metropolitan siting began at the special ACRS meeting held on February 6, 1965, when Harold Price, the Director of Regulation, reported to the ACRS that the Boston Edison Company had made preliminary studies of six reactor sites and the preferred one was at the Edgar site near Quincy, Massachusetts, in a rather highly populated area. The utility desired commission advice on its siting policy regarding large reactors in or near big cities.

The regulatory staff had formulated a draft position which would recommend a moratorium on city locations for reactors. The consensus of the ACRS, however, seemed to be that this was a very important matter, and that no hasty conclusions should be drawn. Siting of each reactor was seen as an individual case by the committee.

Discussions ensued with Price in which he related the results of recent meetings regarding the public acceptance of large reactors. He said that the proposed reactor near Boston and the possible reactivation of the Ravenswood case, both high-power reactors, were putting pressure on the regulatory staff.

The minutes indicate that a complex group of factors entered into the decision-making process, including the timing and nature of any announcement. Part of the problem appeared to be that the proposed extension of the Price-Anderson Act was soon to be before the Joint Committee on Atomic Energy, and it was expected that the coal companies would object to this favor to the nuclear industry. Price advised the ACRS that AEC approval of reactors in large cities could be argued as being in conflict with AEC support for the Price-Anderson arrangement. When the Ravenswood project had been withdrawn, the regulatory staff was in the midst of a study on siting of large reactors. Although no such large reactors had operated, the design engineers claimed that safe designs could be built. Dr. C. Beck of the regulatory staff believed that the engineering ideas on engineered safeguards were quite good and would be proven in years, but that their reliability at the time was questionable.

ACRS members asked the regulatory staff for an opinion as to what degree of redundancy in engineered safeguards would make a metropolitan site acceptable, but received no answer. Dr. R. L. Doan of the regulatory staff (a former ACRS member) considered isolation of reactors from a city as allowing much more maneuverability in operations, latitude in power levels, and in requirements of engineered safeguards. Price believed that assuring 100% operability of engineered safeguards was impossible. But ACRS member L. Silverman considered certain items, e.g., the containment, the filtering and the air cleaning, to be quite reliable.

The ACRS held a subcommittee meeting on siting of large power reactors in metropolitan centers on March 20, 1965. According to the minutes of this meeting, the draft regulatory staff paper concluded that "the public interest can best be served by continuing to exclude large cities as permissible locations for nuclear power plants". Various ACRS members made comments on the paper. ACRS member Osborn noted that the staff position appeared to reflect a concern with the smaller, more probable accidents (e.g. fuel-handling or fuel-shipment accidents), in addition to the traditional maximum credible accident. ACRS member H. Etherington said that there was no design which would eliminate the question of "operator error" with possible violation of containment. Member Rogers stated that the location of a reactor (city versus country) made little difference if a very large accident (e.g., breached containment with fission product release) were to occur. He concluded, therefore, that the smaller, more probable accident was the situation of concern. Member Gifford observed that a "formal" moratorium of metropolitan siting would discourage any new developments or improvements in reactor safety over the designs that were then available.

During the ensuing discussions, Price said he was trying to avoid a drawn-out, detailed design review which would finally end in a policy decision to turn down the application. ACRS member Etherington told Price that a strict interpretation of the Part 100 siting guide would preclude the siting of reactors in cities without any further action on the part of the AEC. Price agreed but indicated such an interpretation would need an endorsement from the commission itself.

Subcomittee members expressed a concern that a formal moratorium would stifle any further development in the field of engineered safeguards. Member Rogers expressed the opinion that routine operation of reactor plants alone would not provide information about performance of engineered safeguards under accident conditions. Dr. Beck pointed out that even routine operation, testing, and maintenance of engineered safeguards were not then well developed, but that operation for several years would help to identify and correct system deficiencies.

Dr. Beck confirmed that the regulatory staff was concerned about smaller accidents than the MCA, since larger accidents would have very serious consequences no matter where the location.

Several subcommittee members suggested that applicants should be advised of appropriate criteria on the basis of a case-by-case review. Mr.

Price maintained, however, that this would leave applicants in a state of confusion and uncertainty.

Dr. Beck suggested that a logical set of criteria should be developed before the door was opened to metropolitan sites, since industry was eager to move into cities with facilities similar to those of the Malibu and Oyster Creek designs. ACRS member Rogers agreed that utilities were not likely to propose any additional safeguards which were not required by the AEC.

In response to a question from Dr. Beck, member Osborn stated that the ACRS review of the Ravenswood reactor had not been completed, but that there was no indication that it would have been turned down by the ACRS on the basis of a policy decision. It seemed generally agreed that the Malibu and Oyster Creek designs were not yet acceptable for use in densely populated areas, and that more stringent criteria should be developed for city reactors.

The matter of metropolitan siting was discussed at considerable length at the ACRS special meeting, March 26–27, 1965. Most of the ideas mentioned above entered into this discussion. Member Etherington recalled that no large reactors in cities were contemplated at the time of development of the reactor site criteria (Part 100). Etherington observed that for more frequent situations where both the probability of an accident and its consequences could be obtained, the reactor designer could evaluate alternative approaches to limit the consequences or avoid the event. However, if the accident were of extremely low probability and its consequences were very large, no such analytical approach was open (to designer or regulatory staff). In summary, Etherington felt the problem lay with the accident that had exceedingly low probability and very high consequence.

According to Price, the commissioners were not inclined at that time to take any stand against metropolitan reactors. However, the regulatory staff needed a public posture, and had to respond formally or informally to recent inquiries from utilities and reactor vendors.

ACRS Opinion on Metropolitan Siting

Following the special meeting, March 26–27, 1965, the ACRS released for the information of the AEC commissioners a draft report on metropolitan siting, which is reproduced below.

The Advisory Committee on Reactor Safeguards has been informed by the Director of Regulation that representatives of the nuclear power industry have, in recent weeks, visited him to explore the possibility of locating large power reactors in metropolitan areas. This letter is in response to your inquiry as to the views of the ACRS on this subject.

This subject was discussed with members of the AEC Regulatory Staff during the 62nd ACRS meeting on March 11–13, and the Special ACRS Meeting on March 26 and 27, 1965. In addition, a subcommittee met with the Regulatory Staff on March 20th.

The ACRS offers the following comments on the question of locating large power reactors in metropolitan areas:

1. The engineering of reactor safety has been a process of evolution. Much has been accomplished; more remains to be done. The larger power reactors now under construction or described in current license applications represent a large step in this process of evolution. However, considerable further improvements in safety are required before large power reactors may be located on sites close to population centers. None of the large power reactor facilities now under construction or described in current license applications are considered suitable for location in metropolitan areas.

2. A flexible position with respect to locating reactors close to cities should be maintained. License applicants should be encouraged to use imagination and to employ improved provisions for safety. A suitable channel for the early consideration of new facility concepts should continue to be available.

3. Designers should be encouraged to develop engineered safeguards of extremely high reliability and with provision to assure that such reliability can be demonstrated at all times.

4. The quality of operation, maintenance and administrative control, upon which dependability of engineered safeguards relies, must be further improved.

5. Guidance for designers and operators should be developed for location of reactors in cities.

In connection with the last item the ACRS is considering the following tentative points:

(1)   The design goal for reactors being considered for metropolitan use should be the elimination of any possibility of a severe reactor accident.

(2)   It would seem prudent to operate in metropolitan areas only reactors of a proven type, which do not represent a large extrapolation in power, involving radical changes in reactor design from reactors already in service. In other words, reactors in metropolitan areas should closely duplicate reactors with demonstrable and favorable operating experience.

(3)   It must amount to a practical certainty that under no circumstances will significant amounts of fission products reach the public. Provisions taken should include containment of the refueling operation, spent fuel storage area, and radioactive waste.

(4)   In order to assure a reliable containment, it is necessary to establish in some way an upper limit to the energy release in any possible accident. This energy release should include nuclear excursion energy, stored thermal energy and chemical reaction energy.

(5)   The containment should be adequately protected from missiles both from within and without, including those arising from the disintegration of equipment.

(6)   Reliance should not be placed on valves to effect isolation of normally operating ventilation systems.

(7)   A design goal for instrumentation and control systems including all electronic and mechanical devices should be that all safety systems are fail-safe including consideration of effects of fire, steam, and other possible environments.

(8)   Improved reliability of emergency power supplies appears required. Or, in lieu of that, the facility should require no emergency electric power.

(9)   The possibility of simultaneous independent failures should not be neglected in evaluating engineered safeguards.

(10)   Primary reliance for safety should not be placed on procedural control methods.

This draft letter was never issued; however, it was discussed in detail with members of the regulatory staff, and with the AEC commissioners at the sixty-third ACRS meeting, May 13–15, 1965. At the May meeting, the commission indicated a desire to avoid any action which might preclude large reactors near large cities. The issuance of guidelines for the siting of such reactors seemed to be favored by the commissioners.

The ACRS draft stated that only reactors of proven design were suitable for city locations. AEC Chairman Glenn T. Seaborg commented that he hoped the ACRS meant an applicant could anticipate favorable operating experience of a reactor at a remote site prior to operation of a metropolitan reactor; this would allow construction to proceed and save time, but with the applicant accepting a risk.

Some of the thoughts included in this draft letter on metropolitan siting were presented publicly in the testimony of the ACRS to the Joint Committee on Atomic Energy in the hearings on extension of Price-Anderson indemnity legislation, June 22, 1965. This ACRS testimony was presented by Chairman W. Manly and past Chairman H. Kouts. They reiterated item 1 of the draft letter and went on to say:

To put the matter in a different way, the devices and safeguards that prevent all accidents, large or small, must be made even more reliable than they are now and the consequences limiting safeguards must be made even more fool-proof. The questions to be settled are complex ones, and resolution would depend on the nature and details of each proposal. It also appears that novel reactor systems, or reactors of considerably higher power level than previous ones, should not be operated in population centers.

Review of the Edgar Station Site

Depending on how one reads the draft ACRS letter of March, 1965, one might or might not read into it a moratorium or semimoratorium on metropolitan siting. Nevertheless, the review of the proposed Boston Edison BWR for the relatively populated Edgar station site continued for several months. One finds in the information submitted by the applicant, statements such as the following:

The integrated dose 600 ft. from the reactor or any point beyond shall not exceed 2.5 rem to the whole body and 30 rem to the thyroid in the event of the design basis, coolant-loss accident, coincident with 100% core melt.

One cannot ascertain unequivocally from the written material if the applicant actually thought that the containment building would remain intact in the event of a core melt and if, in fact, he considered the calculated doses valid for that event; or if core melt was being treated as a generalized accident in which fission products were postulated to be released to a containment building which was assumed to remain intact. However, in more than one place the statement was made that the containment building would be designed on the basis that significant core melting and metal water reactions would occur; hence, a logical deduction to the reader is that the applicant was proposing a containment system that could handle core melt.

The ACRS minutes of a meeting between the Boston Edison Group and the AEC Division of Reactor Licensing, September 20, 1965, give further information on the proposed reactor. It was stated that the plant was basically an Oyster Creek type reactor with double pressure-suppression containment and several other improvements. These improvements included use of internal jet pumps for reactor circulation (which make it a Dresden 2 type reactor). The design provided a secondary shell around the core which could be reflooded following a LOCA. Both core-spray and core-flooding systems, as well as duplicate spray systems, for containment vessel cooling following a LOCA were to be provided. Special features were also proposed to make highly improbable a reactivity insertion accident, which could cause large, sudden bursts in core power. Flow restrictors were to be provided in main steam lines to limit steam flow to 200% in the event of a steam-line break.

Some of the special features were not basically new to this plant, since questions concerning reactivity accidents had been raised earlier in connection with review of Oyster Creek and some other plants. However, this appears to be the first mention of use of both a core-spray and the core-flooding system.

Dr. Doan expressed concern over basic questions that must be faced when the location of a reactor in heavily populated centers is considered (e.g., Do we really know that a collapsed core can be cooled?) There was no response given to that question, and it is not clear whether Dr. Doan had something more specific in mind.

ACRS member Silverman agreed that several improvements had been incorporated in the Boston Edison design, but noted they were based on a specified series of events which were defined as the maximum credible accident. He mentioned a fire in the control room as an example of an accident which must be considered. The representatives of General Electric stated that they considered multiple, independent failures highly improbable. They said that the plant could accommodate two independent failures and suggested that the need for additional containment was questionable on the basis of probability.

The ACRS minutes of a further meeting between the Division of Reactor Licensing and the Boston Edison Company, January 26, 1966, provide further insight into the thinking of the time. Dr. Doan identified the principal problem as being how to evaluate the reliability of the proposed reactor and its safeguards when there had been no operating experience on reactors of the same power and design. He pointed out that at the time Part 100 was written, a different situation prevailed regarding the siting of reactors, namely, reactors were being located at some distance from cities. He did not know how to interpret population center distance for a metropolitan site.

Dr. Doan felt that a new (safety) approach had not been taken in the design of the proposed reactor facility for the Edgar station site, and that only relatively minor safety improvements had been made over past designs. He said that the matter of whether the proposed reactor facility was sufficiently safe could not be based on experiments to be performed at some time in the future. In his opinion, until more information was available regarding experience with engineered safeguards, there was no chance that the Edgar station site might be approved by the AEC for the proposed reactor. He did not see how there could be a breakthrough of knowledge regarding safety of large reactors until some of the proposed large reactors had obtained operating experience.

Dr. Doan stated that no one wanted to say that metropolitan sites were unsuitable for reactors. However, acceptability of metropolitan sites would come sooner if the industry recognized that present designs were not acceptable for metropolitan sites, and put some effort into the matter. In regard to the proposed reactor design for the Edgar Station site, Dr. Doan pointed out that the containment building could not withstand rupture of the pressure vessel. (An ACRS letter regarding pressure vessel failure and the possible need to design for its failure, especially for more-populated sites, had just been issued two months earlier.)

Dr. Doan concluded by stating he could not base a decision regarding issuance of a construction permit on conceptual design and that "hard" information would be required.

At some point after the January 26th meeting, the Boston Edison Company decided no longer to propose construction of the nuclear plant at the Edgar station site: instead the Pilgrim reactor near Plymouth, Massachusetts, was proposed.

The N.S. *Savannah*

One more example of how the metropolitan siting issue was dealt with during this same time-period arises in connection with the 70 MWt nuclear-powered ship N.S. *Savannah*. The AEC and the ship operators wished to bring the N.S. *Savannah* into port in or near the heart of large cities, including New York City. The ACRS had not been generally happy with the design and operation of the N.S. *Savannah*, and they had to face the matter of how or whether such coming into port would be acceptable. It was a controversial matter among the committee members, some of whom were reluctant to rely on the engineered safety features supplied on the ship.

Only after very many meetings on the N.S. *Savannah* and a near-dissent was an ACRS recommendation developed which required that the operating power be limited the day before entering port so that a two hour "time-to-melt" would exist if a rupture in the primary system led to a loss of coolant from the core and no cooling were available. Tugs were to be on call so that in the two hours before fuel melting and a large release of fission products to the containment vessel, the ship could be towed to a more remote place in the harbor.

In the next two chapters we shall look at two safety issues, pressure vessel failure and the China Syndrome matter, both of which represented mechanisms by which an accident would probably lead both to core melt and containment vessel failure. Consideration of these issues led to important changes in reactor design and siting. The principal siting question of the next decade changed from consideration of the possible use of truly metropolitan sites to the acceptability or not of sites similar to or somewhat worse than the most populated site already approved and used for a power reactor, namely Indian Point.

7

The Question of
Reactor Pressure Vessel Integrity:
1965-1966

The importance of reactor pressure vessel integrity to light water reactor safety was not a new concern. The minutes of the thirty-third ACRS meeting in 1961 show member W. Connor raising a question concerning the need for improved inspection for reactor vessels, and in the minutes of the thirty-sixth ACRS meeting member K. Osborn noted that pressure vessel rupture could lead to failure of the containment building. In a report to the AEC dated September 11, 1961, the ACRS recommended the development of adequate codes and standards for the pressure vessel and other parts of the primary system of reactors. In a report to the AEC dated May 20, 1961, the ACRS discussed matters related to the possible embrittlement of reactor vessels due to irradiation by neutrons over the lifetime of the reactor vessel. However, for the pressurized and boiling water reactors reviewed prior to 1965 failure of the pressure vessel of the reactor was treated as "incredible". There was no protection provided against gross vessel failure, and were it to occur, it would have been expected to cause both core meltdown and a violation of containment integrity, leading to a major, uncontrolled release of radioactivity.

Actually, concern about pressure vessels had been growing among some ACRS members during the year 1965. In 1964 there had been a failure at a temperature near the nil ductility temperature (the temperature below which gross brittle failure can occur) of a very large heat exchanger under test by the Foster Wheeler Corporation. On April 23–24, 1965, the ACRS held a subcommittee meeting on pressure vessel integrity. A range of questions arising from the very high requirement for pressure vessel integrity were left unanswered at that subcommittee meeting, including adequacy of fabrication and inspection techniques, ability to ascertain the brittle-ductile (or nil ductility) temperature transition region, and the behavior of thick-walled sections. These concerns were added to by some reports published in 1964 and 1965 by British research workers concerning the possible rapid failure of steel reactor vessels at temperatures above the nominal brittle-ductile temperature transition range; that is, in the operating temperature range where sudden failure was not supposed to occur because the material was so tough (Irvine, Quirk, and Bevitt 1964; Nichols et al. 1965).

The existence of a serious safety concern among some or many ACRS members did not automatically lead to a committee recommendation on the subject. The pressure vessel issue was particularly difficult in that there did not exist a clearly feasible way to design engineered safety features that would prevent core melt and maintain containment integrity if catastrophic pressure vessel failure occurred. And, while the probability of reactor pressure vessel failure was thought to be very low, no reliable failure statistics were available.

The Pressure Vessel Issue Is Posed for the Dresden 2 Reactor

One ACRS member (D. Okrent) decided to force the issue in the fall of 1965. Since it was not clear what action, if any, the ACRS was going to take on pressure vessels as a generic issue applicable to all LWRs, he chose the ongoing Dresden 2 construction permit review as the vehicle for addressing the pressure vessel question.

The Commonwealth Edison Company had applied on April 15, 1965, for a license to construct and operate Dresden 2, a 2,255 MWt BWR, at the site of Dresden Nuclear Unit 1. The largest thermal power level previously approved for a construction permit was that of the 1,600 MWt Oyster Creek reactor; hence, the Dresden 2 reactor represented a large jump. While the surrounding area was relatively rural, the city of Joliet, Illinois, was about 14 miles from the site, and the city of Chicago, Illinois, was about 40 miles from the site. Many members of the ACRS saw the Dresden 2 reactor as a probable prototype for other reactors which would be proposed for metropolitan areas; for this and other reasons, the Dresden 2 reactor received extra emphasis during ACRS review, and the potential resolution of certain generic matters became tied to the case.

The continuing pressure from industry (and, in a sense from the reactor development side of the AEC) for metropolitan siting was evident in a variety of ways. For example, in 1965, the AEC established a Steering Committee on Reactor Safety Research, consisting of members from both the development and regulatory sides of the commission; one function of this new committee was to define the safety research and development needed to allow reactors into metropolitan areas.

The minutes of the ACRS subcommittee meeting held on September 1, 1965, to discuss the Dresden 2 reactor show there was considerable discussion of pressure vessels with the members of the regulatory staff and with the applicant. D. R. Muller of the staff is listed as saying that pressure vessel failure was incredible. When asked by an ACRS member what would happen if the pressure vessel did fail, A. P. Bray of General Electric replied "it would depend on the energy rate." He stated, "The containment could withstand a larger break than the maximum credible accident [which was rupture of a large pipe] but not a complete break of the pressure vessel." During the executive session of this meeting one member of the ACRS stated that he felt that reactor pressure vessel failure was credible. Another member agreed, but suggested that the matter should be handled in a generic way by developing appropriate design criteria for all LWRs, rather than by dealing specifically with this particular applicant.

The Possibility of a Dissent Looms

During the November 10–12, 1975, portion of the sixty-eighth meeting, ACRS member Okrent took the position that, while it was acceptable for the Dresden reactor to be constructed at the site selected, in view of the current state of the art of pressure vessel fabrication and inspection, improvements were needed in the assurance of pressure vessel integrity, and consideration should be given to the possibility of major pressure vessel failure. He proposed that it would be desirable, even prudent, to restrict pressurized and boiling water reactors not designed to cope with this extremely unlikely accident to relatively remote sites such as that proposed for the Dresden 2 reactor; also, that future large reactors of these types should incorporate appropriate protective design features if intended for sites closer to population centers. During the November 10–12 period the ACRS discussed this matter extensively but did not arrive at a decision as to whether it wished to prepare a letter of approval regarding the Dresden 2 reactor, with additional remarks by a member, or whether it wished to deal with the issue in some other way, for example, by writing a general (non-case-specific) letter concerning pressure vessel safety for future reactors. It was decided to continue the sixty-eighth meeting and to hold another discussion of pressure vessels as part of the extended meeting.

To help the ACRS in dealing with its problem, two members, T. J. Thompson and N. J. Palladino, each prepared rather lengthy letters to all the other committee members in which they summarized the state of knowledge as they saw it, and tried to pose possible points of view and possible approaches. These letters provide a good example of how difficult it is to deal with a problem such as possible failure of the pressure vessel of a reactor; they also provide considerable insight into the ways in which the ACRS tried to develop varying points of view in approaching such a problem (Okrent 1979).

Following the meeting on pressure vessels of November 23–24, 1975, the ACRS decided to issue a letter (without any dissent) favorable to the construction of the Dresden 2 reactor and at the same meeting to write a general letter to the AEC concerning reactor pressure vessels. The ACRS members at the sixty-eighth meeting were W. Manly, Chairman; H. Etherington; F. Gifford; S. Hanauer; J. McKee; H. Newson; D. Okrent; N. J. Palladino; L. Silverman; T. J. Thompson; and C. Zabel. The pressure vessel report was sent to the Honorable Glenn T. Seaborg, chairman of the AEC, on November 24, 1965, and is reproduced here.

The design of pressurized and boiling water nuclear power plants has undergone many improvements with regard to safety, improvements which markedly reduce the risk of significant radiation exposure to the public in the unlikely event of certain accidents or system failures in such reactors. There is a facet of current pressurized and boiling water reactor design practice which should be recognized, however. Containment design is generally predicated on the basis that a sudden, large-scale rupture of the reactor pressure vessel or its closure is incredible. Reactor designers have supported this view by detailing the extreme care to be taken in design, fabrication, and inspection of a vessel, and by specifying pressurization only at temperatures above the nil ductility transition temperature. They further cite the excellent record for large pressure vessels which comply with the ASME Boiler and Pressure Vessel Code.

The Committee believes, with the industry, that the probability of a sudden major pressure vessel failure leading to breaching the containment is very low. Nevertheless, it seems desirable and possible to make some provisions in future designs against this very unlikely accident.

1. To reduce further the already small probability of pressure vessel failure, the Committee suggests that the industry and the AEC give still further attention to methods and details of stress analysis, to the development and implementation of improved methods of inspection during fabrication and vessel service life, and to the improvement of means for evaluating the factors that may affect the nil ductility transition temperature and the propagation of flaws during vessel life.

2. The ACRS also recommends that means be developed to ameliorate the consequences of a major pressure vessel rupture. Some possible approaches include:

(a)   Design to cope with pressure buildup in the containment and to assure that no internally generated missile can breach the containment.

(b)   Provide adequate core cooling or flooding which will function reliably in spite of vessel movement and rupture.

(c)   If breaching the containment cannot be precluded, provide other means of preventing uncontrolled release of large quantities of radioactivity to the atmosphere.

In view of the very small probability of pressure vessel rupture, the Committee reconfirms its belief that no undue hazard to the health and safety of the public exists, but suggests that the orderly growth of the industry, with concomitant increase in number, size, power level, and proximity of nuclear power reactors to large population centers will in the future make desirable, even prudent, incorporating in many reactors the design approaches whose development is recommended above.

Strong Industry Reaction

The reaction of the industry to this letter is well illustrated in the following article from the January 1966 issue of the trade journal *Nucleonics*, Volume 24, No. 1.

ACRS Qualms on Possible Vessel Failure Startle Industry

The community of power-reactor designers, suppliers, and operators was taken aback last month by a terse six-paragraph letter from the Advisory Committee on Reactor Safeguards to AEC, bearing recommendations quite unexpected at this time, and whose effect on the nuclear industry may take weeks to evaluate. The recommendations, in a nutshell, are 1. that sudden, catastrophic failure of a pressure vessel—since the start of the power reactor program classified as an incredible accident, one that need not be taken into account in reactor safety analyses, be reclassified as a possible accident; and 2. that future nuclear power station plans design against the possible consequences of such an accident.

Industry reaction was sharp and dismayed. It ranged from resignation to protestations that the desired levels of quality control, stress analysis etc., were already being met, to comments of it's "almost impossible to design against complete separation of the vessel," as ACRS asks, nor is it necessary, and that "this kind of thing, done this way, borders on the irresponsible."

One of the aspects of the matter that troubled industry was that ACRS apparently acted without prior consultation with the technical safety experts on AEC's regulatory staff, and indeed gave AEC only the most cursory informal advance notice. If AEC felt at all uncomfortable about the ACRS letter, it did not dispel such an impression when it took the unprecedented step of attaching a covering statement to the ACRS report. In this, AEC called attention to the positive ACRS comments on safety of water reactors; pointed out that ACRS was recommending "that additional work be done," and recalled that AEC had already launched an "augmented and reoriented" safety program which would include work on the ACRS suggestion.

The blow was all the sharper because only two weeks earlier AEC had issued a set of design criteria for power reactors as guides to applicants for construction permits (NU Wk [*Nucleonics Week*], 25 Nov. '65, 1; test in AEC press release H-252). An immediate favorable response from industry met issuance of the criteria, hailed as a good step, a step in the right direction (NU Wk, 2 Dec. '65, 1). Two weeks later, publication of the ACRS report brought wry comments from industry that every step forward on the licensing front seems to be accompanied by two steps backward.

Commented one industry official: "If you assume ductile metals can fail catastrophically from brittle failure, you must also assume the properties and behavior of ductile metals are not what we have assumed over the years, and that you cannot design anything: why do you assume a bridge will stay up?... The same thing applies to the possibility of a guillotine break in a pipeline; there has never been one in history; ductile materials don't fail in that way. Should we assume concrete will no longer support a building, or that glass will no longer insulate?"

Said another: "All the pressure vessels for any water reactor operate in the ductile range and therefore couldn't fail in a brittle manner... I think it's wise that we look at everything, that ACRS recognize its responsibility for safety and not permit anyone to build anything that isn't safe. The question is—you can analyze things forever and never get anything built... I wonder if it isn't coming to a ridiculous point. You could analyze if the reactor operator should ever step out in the street: something might happen to him and he might not be able to come to work."

Observed another: "You can't argue against beefing up quality control, stress analysis, etc.—we go along with that. But that's quite different from saying you should design for a failure."

At Burns & Roe an official said, "We've studied this and concluded that it's not possible: we could see no conceivable possibility of a full vessel rupture. Now you get to the numbers game—you can't say it's zero, but it's so small that it's not a postulated accident as far as we're concerned."

And at General Electric there was open talk about attempting to get the report's recommendations reversed by AEC, and about the existing system [never used yet by a nuclear industry member] of review by the courts, in case AEC did not act responsively.

Vessel Fabricator's Reaction

At Babcock & Wilcox, one of the only two U.S. firms that fabricate the huge reactor pressure vessels, an official said, "I believe the possibility of massive failure of reactor vessels has been reduced to zero—and the more we build, the better they will be. In time we will reduce the possibiity of failure to nil to the sixth power . . . As for ACRS' three points: we are already doing this, increasingly and constantly."

Combustion Engineering, in a more formal statement, said: "The materials, the design criteria and methods, the inspection techniques and testing of nuclear components result in quality that vastly exceeds that which was previously obtainable for pressure vessels. The design specifications have and do take into consideration the known effects on the material exposed to radiation environments . . . We are certain that evaluation of present and past practice will show that the pressure-containing components of the primary system are quite adequate for the operating conditions for which the unit was designed."

Industry spokesmen did not fail to point out that the recent report of the Mitchell Panel on streamlining AEC reactor-licensing procedures had urged elimination of public disagreements between AEC licensing bodies by the making of "every effort to reconcile differences" in joint meetings. But it appeared that ACRS not only had not done so, but had in fact given AEC only the barest advance notice of its letter report. One industry official felt that in matters of this nature ACRS should have gone even further in the opposite direction; not only consulted with AEC in advance, but put out a draft version of its report for industry comment—as AEC did with the design criteria—prior to issuance in final form.

Three Principal Issues

There were three principal issues contained in the ACRS report to which industry reacted. One was ACRS' concern over the possibility of vessel cracking and failure due to changes in the nil ductility transition temperature. A second was ACRS' feeling that each future reactor should design against the possibility of gross vessel failure. Thirdly, it is understood, one of the Committee's concerns is the possibility of vessel head bolts shearing off and becoming missiles that might breach the containment.

The first, no new issue, is the effect of radiation on the nil ductility transition temperature, that is, the temperature at which a change to brittle behavior occurs in a ductile metal. This temperature normally is in the range of 10–40 F in steels such as those used in reactor vessels, or safely below operating temperatures. But under irradiation the temperature at which the phenomenon occurs may increase until it can approach—or so ACRS fears—the 300–400 F range of vessel operation.

Said one industry official with long familiarity with vessel fabrication; "Based on all data we have yet seen, as the nil ductility transition temperature shifts, the metal yield strength also increases, so that I'm not sure that we don't end up with a vessel that's safer." Another recalled that considerable investigation had been done on this at Argonne a few years ago, with no procedural changes resulting.

Declared another: "If after 10 years of operation you had a crack, your vessel might conceivably break under hydrostatic test; while this would be embarrassing as hell, it would not be a catastrophe since it would be inside containment, and the safety of the public simply would not be involved."

Improved materials of today were stressed. "Materials [for reactor vessels] are undergoing much better inspection than for a power boiler. We know really that the materials that went into power boiler steam drums 40 years ago were nothing like the quality we now have, yet we have had no power-boiler drum failures in all that time." And: "We are building vessels of better steel than ever before, doing better stress analysis than ever before, using better inspection methods than ever before and using them more extensively than ever before . . . It is awfully late in the game to be coming up with this kind of judgment." And again: "It's almost preposterous to postulate a gross vessel failure, as opposed to perhaps a nozzle crack."

Secondly, on designing against failure, there were mutually self-contradictory views opposing ACRS from opposite extremes. On one hand, it was said, "We now have to design against instantaneous severance of a major recirculation loop in a BWR or of a major primary coolant loop in a PWR, and complete loss of all fluid [in either case]. It's difficult for me to realize that anything arising out of a small crack or flaw in a pressure vessel would exceed that requirement. We are now designing to handle the complete loss of fluid and fission products from the reactor."

On the other hand were those who said, "This is an intolerable requirement—we can't live with it"; or "I doubt personally that it is credible to completely design, at least economically, for this massive failure of a pressure vessel. You just keep piling things one on another and you get to the point where you can't do it. If the pressure vessel can fail, the containment can fail; if you contain that containment, that can fail, and so forth . . . Back of this has been a steady increase and pyramiding of the number and severity of accident modes we are supposd to take into consideration. It would seem that these things are changing as the makeup of ACRS changes: I have the impression that membership of ACRS is moving more and more toward an academic, a college physics professor, type of person."

Academic Trend?

The feeling that ACRS' requirements are trending toward the academic was fairly widespread. Said another experienced industry man: "This [requirement] clearly says, 'Six months, bud, on top of any completion date you got.' Like Rickover, they say 'prove it.' So you do, and they say, 'prove that'—and you can go on doing arithmetic forever . . . They should have to prove the justification of their question. We always have to prove—they can just think, opine. Maybe they should be asked 'where?' 'when?' 'why' "

Finally, as to the possibility of missiles breaching containment, one industry man countered: "We have a very good story on this, because we do actually stress each bolt with a bolt tensioner, so we know the actual stress on each bolt as we close it, and the bolt is easy to examine."

As to designing containment against internally generated missiles, one architect-engineer commented, "The question is how big a missile and how much force? Containments are now designed to take rupture of a primary pipe and the hydraulic head if a pipe breaks. Now they ask, suppose a bolt flies off. Suppose the head flies off? I don't know what the intent is. Suppose you say the bottom head comes off? How far can you push these things? If you continue, the design problems will become virtually insurmountable.... We don't know how to design a structure against the kinetic energy of a 100-ton vessel trying to blast its way through.... Is this just thinking up other things to protect against? If so, where is it going to stop? Why bring it up now? Do they know something we don't know? I question it."

Quality Control or Failure Protection?

The pressure vessel report clearly introduced considerable consternation in the industry and the AEC, and at the next ACRS meeting, the sixty-ninth, January 6-8, 1966, the committee gave much attention to the question of how it would propose to implement the letter. The ACRS had various questions to consider. Was it to be applied to successive reactors in terms of improved quality? Was it to be applied after a suitable time-interval to all reactors in terms of protection against certain vessel failure modes? Was it to be applied in terms of certain or all vessel failure modes only for reactors of very high power at relatively highly populated sites? At the sixty-ninth meeting, the ACRS discussed the matter in executive session, with the regulatory staff, with members of the Division of Reactor Development, and with the commissioners.

The commissioners were told by the ACRS that one intent of the pressure vessel letter was to encourage improvement in the quality of the workmanship and in the extent of inspection, and to promote better surveillance throughout the reactor's life. The ACRS also suggested to the commissioners that design steps to limit the consequences of pressure vessel failure might be needed for favorable consideration of the possibility of a metropolitan site.

The ACRS advised the regulatory staff that the committee planned to implement the quality control and inspection aspects progressively rather than suddenly, and that reactors at more-populated sites would receive increased ACRS attention with regard to pressure vessel safety. The committee felt that the Boston Edison reactor, which had been proposed for the densely populated Edgar station site near Boston, might soon provide a situation where full precautions against vessel failure would be required. H. Price, the Director of Regulation, responded that if any applicant for a commercial power reactor construction permit were forced to protect against pressure vessel failure, all other commercial reactors would have to comply regardless of location. This position by Price was philosophically different from that of the ACRS, since the committee clearly believed that additional protective measures were appropriate for "poor" sites. This position by Mr. Price also acted as a brake on any tendency of the ACRS to recommend early implementation of measures to protect against vessel failure, since the committee was not inclined to recommend such a requirement for lightly populated sites.

Review of the Brookwood, New York, and Millstone Point, Connecticut, reactors was facing the ACRS in the immediate future, and it was decided to pursue with these two reactors various aspects of the pressure vessel question. The discussion would include the consequences of various failure modes and the probabilities of these types of failure, as well as the things that could be done to improve the quality of reactor pressure vessels.

It was with some anguish that the utilities discussed, on the record and in writing, questions concerning pressure vessel failure. It was a topic not dealt with in this manner previously; it was a topic relating to accidents for which the reactor was not protected.

Industry Concern with Public Discussion of Serious Accidents

The minutes of the seventy-first ACRS meeting, March 10–12, 1966, show that Mr. Roger Coe, a representative of the Yankee organization appearing on behalf of the Millstone Point reactor, joined the committee in executive session and read a statement regarding the recent trend of written questions from regulatory groups to applicants regarding very serious postulated accidents, which involved the potential for core meltdown and failure of the containment building. Mr. Coe stated that the correspondence becomes a public document and, for example, in one case, the Brookwood reactor (later renamed the Ginna reactor), this correspondence was quickly collected and aired by the trade press. (See the excerpt from *Nucleonics Week*, February 17, 1966, that is reprinted below.)

ACRS is pressing its concern over pressure-vessel failure in reviewing Brookwood, the first reactor project to come up for a construction permit since the Advisory Committee on Reactor Safeguards in December first postulated as a serious possibility a large-scale vessel rupture (NU Wk, 9 Dec '65, 1). The 450-MWe Brookwood station is to be built for Rochester Gas & Electric by Westinghouse. In a letter to RG&E earlier this month, AEC's reactor-licensing chief, R. L. Doan, submitted two series of questions, one on behalf of the licensing staff, the other on behalf of ACRS. The ACRS questions bearing on the pressure-vessel-rupture issue include the following:

• If one postulates the rapid propagation of a crack circumferentially, with the contained energy of the system, what would happen to the upper section of the vessel including shearing the primary pipes, etc.?

• Can you visualize any problem from the propagation of a crack from top to bottom of the vessel but not though the head?

• Do your calculations confirm that the steam-generator tube sheet will withstand shock loading by abrupt loss of primary coolant, or will the head go instead?

• Are you considering procedures for detecting the propagation of cracks within the pressure vessel wall, i.e., acoustic emission?

• Define the pressure vessel flaw size and type that is accepted in the specifications. What flaws larger in size or of special significance might not be detected, particularly in zones of irregular geometry?

• What flaw size is accepted in the studs of the pressure vessel? What frequencies of study, inspection or replacement is planned? How many studs can fail without threatening the integrity of the closure?

• Please describe requirements concerning the support structure for the pressure vessel, including the degree of levelness over reactor life, which are needed to insure no problems due to local overstressing of the pressure vessel.

• Describe how small leaks in the pressure vessel would be detected, and the action to be taken should such occur. How is adequate response assured in the event of a previous existence of small leaks in other parts of the system?

In addition to these, ACRS asked five questions on nil ductility, including validity of neutron flux dose predictions for the pressure vessel, weld regions and heat-affected zones. The AEC staff questions submitted by Doan were largely answered orally by RG&E and Westinghouse at a meeting Jan. 26–28, but the staff wanted a "written response . . . to confirm the oral information."

Mr. Coe referred to private discussions in the past with applicants where the presentation of information was on a very candid basis. He predicted that this public method of communication would lead to less frankness and perhaps to intervention by the opponents of nuclear power in order to delay private nuclear development. Mr. Coe was quoted as desiring an arrangement by which such questions could be raised without the public being informed. He said that verbal requests and informal replies were a possible way to do this.

These comments by Mr. Coe were triggered by the written questions, transmitted by the regulatory staff on behalf of the ACRS,* to the applicants for the Brookwood and the Millstone Point reactors, concerning possible modes of pressure vessel failure and possible means to deal with pressure vessel failure, as well as the probability of differing types of failure. Questions of this type had not been asked in writing frequently in the past, although discussion of such questions certainly did occur from time to time during the meetings between the ACRS and the regulatory staff, or the ACRS and the various applicants.

It is clear from the minutes of the seventy-first meeting, March 10–12, 1966, as well as the minutes of the previous two meetings of the ACRS, that

---

*The regulatory staff was the legally constituted body which granted construction permits. Applicants filed their requests to the staff; it was the staff who normally formulated and sent written questions to an applicant. The ACRS received copies of all material filed in connection with an application and normally conducted most of its questioning orally at committee meetings.

the committee had decided in the cases of the Brookwood and Millstone Point reactors that additional design measures to protect the public against postulated pressure vessel failure would not be recommended. Instead, improved quality control during fabrication of the vessel and improved surveillance methods during operation were going to be pursued.

Pressure Vessels and the Indian Point 2 Reactor

The Indian Point Unit 2 reactor posed a somewhat more complicated problem. This was the highest power PWR to be reviewed to date. There was already a smaller reactor at the site (which was the most-populated site approved for a reactor having a power of several hundred megawatts). To some, Indian Point represented a nearly metropolitan site because New York City was less than 30 miles away and there was a considerable population density in the region between the reactor and New York City. To others, the Indian Point 2 site represented a better location than the Edgar station site recently proposed by Boston Edison; it was very much better than the Ravenswood site in New York City. The application for the Indian Point reactor was submitted in December, 1965, and it was hoped by the applicant and the AEC to get completion of regulatory staff and ACRS action by about June, 1966; this would have been rather fast pace, but one which had been met for the Dresden 2 reactor, for example.

The reactor had a minimum exclusion distance of only 0.3 miles; the nearest boundary of Peekskill, the closest population center, was 0.87 miles. In view of the short distances involved in this case, it was evident to the AEC regulatory staff that the specifics of Part 100 were not too meaningful. They elected to evaluate off-site doses for an exclusion distance and low-population zone of 0.32 miles and 0.67 miles, respectively, to see if the expected leakage rate and the proposed measures to remove radioactivity from the containment building (sprays and/or filters) would enable a meeting of the dose guidelines. They were following the traditional approach of Part 100, but with a very small, low-population zone. If it passed this test, it appeared that they would approve the reactor.

To the members of the ACRS, Indian Point looked like a site much worse than the others they had been considering for large reactors. A major question facing the ACRS members was: Should measures to cope with pressure vessel failure be applied in some way for Indian Point 2?

At the seventy-second ACRS meeting in April 1966, there was considerable discussion with the representatives of the applicant for the Indian Point 2 reactor concerning pressure vessel quality and failure modes. It was stated by Westinghouse that the reactor design could probably withstand the forces of a longitudinal failure of the pressure vessel, but it was not clear as to whether a circumferential break or loss of the vessel head could be also handled.

In executive session, ACRS member Palladino was inclined to require the Indian Point group to show that any pressure vessel failure could be withstood. Member Etherington saw as the biggest question, "what reactor design the Committee is willing to accept for a metropolitan site; the measures for the much higher hazards of this site are not clear." Member S. H. Bush sensed that the engineered safeguards were either insufficient for this site or overdesigned for some other sites.

At its seventy-third meeting in May 1966 the ACRS was divided with respect to recommending that the Indian Point 2 containment building be protected against failure due to gross pressure vessel failure. Consolidated Edison was advised that the committee was still considering the matter.

At its seventy-fourth meeting, June 8–11, 1966, the ACRS agreed by a preliminary vote to recommend protection against the mechanical forces of a longitudinal pressure vessel split for the Indian Point 2 reactor, although some members wanted more stringent requirements, namely, to protect the containment building against the missile formed if the vessel head broke loose. This remained the ACRS position for the Indian Point 2 reactor and, in fact, represented the continuing future position of the committee for the Indian Point 3 reactor, for the Zion 1 and 2 reactors which were proposed for a site having a surrounding population density similar to that of the Indian Point site, and also for the Midland reactor which had a relatively large nearby population.

The proposal was that the integrity of the reactor vessel cavity was to be maintained in the unlikely event of a longitudinal vessel split, but there was no accompanying requirement that the core be kept from melting in connection with this unlikely event. In part, the ACRS recommendation that protection be provided against the forces involved with a longitudinal vessel split appears to be related partly to the higher probability of this type of failure, compared to failure of all the vessel head studs or to a circumferential vessel failure. Partly, it appeared to be more practical to design against this particular set of forces. And perhaps, partly, it was a way of initiating what might later be more comprehensive protection against vessel failure for still more highly populated sites.

However, by the time the ACRS arrived at its decision in June 1966, concerning a recommendation about pressure vessel failure for the Indian Point 2 reactor, this issue had been overtaken and surpassed by a still greater issue, namely, that of the "China Syndrome."

Concurrent with its review of the Indian Point 2 reactor during the spring of 1966, the ACRS had held further subcommittee meetings on matters related to metropolitan siting. On May 4, 1966, the subcommittee devoted considerable time to a discussion of the probabilities and consequences of very serious accidents as a function of surrounding population density. Although only rough estimates were available, it appeared that for the extreme accident involving essentially no containment, with the wind blowing in the worst direction and with unfavorable meteorological conditions, there might be 5,000 early fatilities at sites like Brookwood or Millstone Point, and 13,000 early fatalities at the Indian Point site. The estimated fatalities were a factor of five higher for the proposed Boston Edison site at Edgar station. Very rough estimates of the probability of an accident involving core meltdown concurrent with containment failure due to missiles or to overpressure because of a failure in the containment cooling system were made by some members; these estimates lay in the range 10^{-5} to 10^{-6} per reactor-year for each of several possible accident paths.

There existed a school of thought that an uncontained reactor accident at a site 10–15 miles outside a city could produce consequences as bad as if the same accident occurred within the city, because the meteorology could be such as to encompass much of the city within a highly radioactive plume from the reactor site 10 miles away. No comparison of the overall risk for two such reactors was available at that time, but this argument about the qualitative similarity between rural and city sites surfaced frequently.

The subcommittee agreed that both early fatalities and the integrated man-rem dose had to be considered in evaluating the consequences of severe reactor accidents. The subcommittee trend was toward disapproval of the proposed Boston Edison site, and toward a graded set of safety requirements, which increased as the population density increased.

Brookhaven National Laboratory Reevaluation
of the WASH–740 Report

At its seventy-third meeting, May 5–7, 1966, the ACRS, at its own request, was provided with a briefing about the reevaluation of the WASH–740 report, which was originally published in 1957 (AEC 1957), that had recently been performed by Brookhaven National Laboratory for the AEC. The ACRS minutes of the session follow below:

Mr. K. Downes and Miss A. Court, Brookhaven National Laboratory, joined the Committee to comment on the draft report entitled "Exposure Potential and Criteria for Estimating the Cost of Major Reactor Accidents." Mr. H. Price, Dr. P. Morris, and Dr. C. Beck of the Regulatory Staff also joined this session. The study was originally motivated by an interest in economics for insurance purposes. Extensive computer codes were arranged to assess the hazards from the many isotopes (55 fission products were assumed) which might be released from a very large reactor accident. Both volatile and non-volatile fission products were considered. Many parameters, e.g., meteorology, population, biological effects, were factored into the code.

The radiation exposure to the lower large intestine, perhaps 1000 rem, from ingested fission products was found to be controlling as far as fatalities were concerned. The external dose, e.g., from a cloud, might be only 100 rems and hence less important. Iodine was not considered controlling for such a large accident, because thyroid loss can be compensated for by medicines or surgery.

The reactor accident of the study was somewhat different than what the Committee normally assumes; the containment was assumed to have a large (few square meters) leak. Of course, if the reactor containment holds, there is no hazard to the public. Even with a sizeable fission product release, enough non-volatile materials, such as strontium and the rare earths, would be left to keep the fuel molten; with no core cooling, melting through the bottom of the reactor appears likely if the power level is as high as 100–1000 MW(e). If fuel is beyond a few months in age, the fission product content doesn't change very much, and fuel a few years old was assumed. Fission products from such large power reactors was shown by the model to spread widely in an accident. A Gaussian probability distribution of the fission products in the plume was used and a uniform population was assumed to vary exponentially radially. With a two mile per hour wind, fatalities might extend to thirty miles.

Dr. Beck considered the BNL accident analysis techniques as needing much more editing before publication is possible. He doesn't want the conclusions written down. A paper on the meteorology aspects of the codes is now ready for a journal; these have already been used for a Savannah River plant analysis. The Staff time of the BNL group is limited, and the AEC has not asked for a concluding report. The BNL group does not wish the chore of preparing a final report; the effort appears to be a large one and is not considered as fruitful as other BNL projects, e.g., building reactors. Only two BNL individuals have had much contact with this accident study.

Dr. Hanauer and Dr. Okrent expressed a desire for conclusions, but Dr. Beck would promise only to have a methods report prepared. Dr. Beck said that this accident study information is being transmitted to the Joint Congressional Committee on Atomic Energy (JCAE); the Regulatory Staff will further explore the results of the study.

The ACRS exerted direct pressure on Dr. Beck and he agreed to provide the ACRS with some representative results. This was done at a subcommittee meeting on June 3, 1966, prior to the full committee meeting in June. Excerpts from the minutes of this subcommittee meeting follow, beginning first with an executive session and then a discussion with Dr. Beck.

Mr. Etherington noted that very bad reactor accidents (e.g., uncontained accidents of the WASH–740 type) can result in higher doses than 10CFR Part 100 limits. He suggested that the ACRS should establish an acceptable limit for the consequences of this type accident. Ten thousand (10,000) fatalities was discussed as a possible limit.

Dr. Ergen (a former ACRS member, now an ACRS consultant) suggested that the reactor industry has come of age and the hazards should be considered in light of the hazards represented by other competing industries. For example, if one becomes too restrictive in safeguards requirements for reactors it will force construction of other types of power producers such as dams. Since dams do fail occasionally with resulting casualties, one has accomplished little in overall safety.

Subcommittee members questioned that reactors had come of age, however, and proposed a study to determine what the public reaction might be from a major reactor accident.

Dr. Bush noted that the timing of this accident would have a major effect on the public reaction. For example, an accident in the first year would have a much more serious effect than the same accident if it occurred in the 1,000th year. There was also discussion of the effects that frequency vs. consequences might have. It seemed generally agreed that infrequent but serious accidents would have a more adverse effect than frequent but less serious accidents.

There was considerable discussion of the consequences (e.g., number of fatalities, man-rem) which the ACRS should consider acceptable vs. the probability of such an accident. It was also suggested that additional information is needed concerning the effect of the site location on the consequences of an accident of this type. For example, does a metropolitan site really make a difference for an accident of this magnitude? It has been proposed on several occasions that the site, especially when city distances of 20–30 miles, are involved, makes little difference in the consequences of a severe accident; although the probabilities may be changed somewhat by the wind rose, etc. Dr. Okrent suggested that some thought should be given to the development of a nation-wide evacuation plan for reactors similar to that, and perhaps associated with the plan for civil defense.

Dr. Okrent described the British system wherein the engineers assign appropriate probabilities to the various accidents considered and the hazards evaluators then decide if these probabilities and the related consequences are acceptable from a national risk standpoint.

Dr. Ergen suggested, however, that there is a lower limit which can be assigned to a catastrophic accident which would preclude reactors from cities unless one were willing to accept some serious consequences or the possibility of evacuation. This lower limit on probability would be determined by the lack of 100% assurance that engineered safeguards would function as designed when needed.

The Subcommittee was unable to reach a decision on the number of fatalities which can be accepted from a serious reactor accident and how this might be adjusted for the frequency of accidents vs. the consequences. It was agreed that this is a very basic question which must be resolved by the Full Committee.

Dr. Ergen was to develop additional information concerning the ability to evacuate heavily populated areas. Sabotage, plant deterioration and/or sloppy operation and exposure of components to conditions beyond design limits were suggested as possible causes of major accidents in addition to those previously listed.

Presentation by Representatives of AEC Staff

G. Wensch and C. Beck described some of the recent BNL work related to re-examination of the consequences of a serious reactor accident.

BNL studies were based on a 3,200 MWt reactor of a type similar to a typical water power reactor. A Loss-of-Coolant Accident was assumed to occur and no credit was given for engineered safeguards or phenomena unless there is information available to prove they will function. The plant blows down and all water is boiled off in about 6 hours. No metal-water reaction was considered. The core then melts and fission products are released from the fuel. The core will melt through the pressure vessel and become molten in about 10 hours. Calculations indicate that the core from a 3,200 MWt reactor would melt through the pressure vessel, the concrete of the containment floor and the containment vessel into the earth until enough material was involved to dissipate its heat. For a 1,000 MW(th) reactor the core may actually solidify before it melts through the containment if enough foreign material is incorporated to increase its heat transfer characteristics.

In any event, fission products will escape from the molten mass until all of the significant isotopes are released (10–12 hours).

The rare earths and alkaline earths are released slowly but the volatile oxides (ruthenium, molybdenum, and technetium), cesium and the other more volatile isotopes are evolved more rapidly. Plating-out of fission products may occur inside the biological shield but since the fission product heat would probably revaporize it, no reduction factor was taken for plate-out. Once the fission gases pass out into the larger cooler air mass of the containment they are transformed to aerosols by cooling and agglomeration so that plate-out would not occur in this area either.

Doses to the intestine, whole body and lung were then calculated based on the assumption that all of the fission products released inside the containment were breathed by an individual with no credit for atmospheric diffusion, decay in transit, fall-out, etc.

Dr. Okrent noted that some work had also been done by BNL to take into account the effects of diffusion, etc. on a recipient some distance from the reactor. Dr. Wensch agreed that some preliminary work had been done by BNL in this area and agreed to check to determine what information is available for use by the Committee. Dr. Beck suggested that if the ACRS desires that a report be prepared on this topic, the Committee should request it formally from the Commission since BNL is not committed to do any additional work on this subject. He noted that the conclusions from the BNL study are contained in letters from Chairman G. T. Seaborg to Representative Chet Holifield and from Commissioner J. Palfrey to Mr. David E. Pesonen.

The item of special interest in the minutes of the June 3, 1966, subcommittee is that Dr. Wensch and Dr. Beck reported that core melt in a 3,200

MWt reactor would not only lead to melt-through of the reactor vessel but that calculations indicated the core would melt through the concrete of the containment floor into the earth (China Syndrome) until enough material was involved to dissipate its heat.*

This appears to be the first unequivocal statement by the regulatory staff to the effect that containment failure was a likely result of core melt in large LWRs. However, this conclusion was not drawn in the letter from Chairman Seaborg to Representative Holifield.

*The term "China Syndrome" was quickly coined to characterize a core melting its way into the earth (on its way to China from the United States).

8

The China Syndrome—Part 1

The advent of the idea of the China Syndrome in 1966 led to major changes in the AEC safety requirements for large light water reactors. Prior to this time, the containment building had been viewed as an independent bulwark capable of limiting the external release of radioactivity to modest amounts, even if the core melted, so long as there were no large internal missiles such as might arise from gross failure of the pressure vessel, or other forces far beyond the design basis of the containment vessel. If the core melted through the thick steel pressure vessel of the reactor, the very thick concrete foundation, covered with a layer of water, was pictured as holding intact; the containment building was assumed to be kept acceptably cool by the engineered safeguards that were provided.

Actually, if the molten fuel went through the concrete foundation, it was estimated to solidify in the ground after penetrating less than 100 feet, and this in itself did not pose any immediate threat to the public health and safety. The noble gas fission products (xenon and krypton) would be expected to escape through the hole in the foundation to the atmosphere, but these gases did not represent the major threat and were not likely to cause lethal doses of radioactivity off site. The radioiodine and less volatile fission products (all more dangerous than xenon and krypton) would be expected to be retained to a very great extent in the ground.

The concern that arises in the event of core melt is that the containment building might fail *above* ground, prior to the core melting through the concrete foundation. Such failure might occur if the pressure and/or temperature in the containment building became too high, or possibly (and this is still speculative) if a very rapid transfer of large amounts of thermal energy from the molten fuel to the water in the reactor vessel or on the containment floor could generate pressure waves large enough to rupture the containment building (the so-called steam explosion, which occurs infrequently in foundries and paper mills).

Were the containment vessel to lose its integrity—say a hole equivalent to an open door—the off-site effects would depend on the weather, the population distribution, the time and ability to evacuate, etc. In principle, the number of lethal radiation doses (about 400 rem per person) could range from none up to ten of thousands. The summation of the smaller individual radiation dose to large numbers of people could lead statistically to latent effects ranging from tens to tens of thousands of delayed cancers. Assuming a large accident had occurred, the probability that any specific exposed individual would die from cancer many years later as a result of this radiation exposure might be between one in a hundred and one in ten thousand. This is not a large increment over the normal chance of dying from cancer due to all causes (say, one in five). However, many individuals might have been exposed to this risk.

It is clear from the records that at its seventy-fourth meeting, June 8–11, 1966, the ACRS arrived at the conclusion that full-scale core melting must be strongly correlated with a loss of containment integrity, and that, as a result, the then proposed Dresden 3 reactor was not acceptable, nor was the Indian Point 2 reactor. It is not clear just when, prior to June 1966, it became evident to various individuals and groups that containment failure would be the probable consequence of the full meltdown of the core of a large light water reactor which had previously built up a large inventory of fission products and their associated radioactive decay heat. In 1963, when large reactors such as the Ravenswood reactor, the San Onofre reactor, and the Connecticut Yankee reactor were all being proposed, there was no mention in the available review material for these reactors of the possible connection between full-scale core meltdown and a loss of containment integrity. This was despite the fact that the generalized accident (or maximum credible accident), which served as the basis for evaluating the acceptability of containment design and of engineered safety features that were intended to limit the release of fission products, assumed full-scale core melt and the release of the bulk of the volatile and gaseous fission products to the containment building.

In 1963, the loss-of-flow test (LOFT) experimental program was initiated by the AEC. In this safety research program it was proposed to build the 50 MWt LOFT reactor at the National Reactor Testing Station in Idaho, run it at power for a period of time sufficient to build up a sizeable fission product inventory, and then deliberately incur a loss of coolant accident which would lead to full-scale melting of the core. This was to provide a large-scale experimental basis for describing the course of migration of fission products first from the fuel to the containment building and then leaking from the containment building out into the environment. It was expected, correctly, that the molten fuel would be held within the LOFT containment building.

At its November 1963 meeting, the ACRS wrote a letter concerning the AEC safety research program in which it said,

> The Committee believes it is of primary importance to determine to what extent engineered safeguards can be relied on in relaxing reactor site restrictions. In the light of present knowledge, it seems unlikely that general principles will render incredible the possibility that high power nuclear reactors can have large power excursions, or that they can have substantial core meltdown. Therefore, it must be expected that the safety analysis for locating and designing nuclear reactors will continue to assume such accidents to be possible, even if only remotely so.

The letter went on to rule out accidents leading to explosions resembling nuclear weapons. The ACRS supported a research program on fission product release transport, but gave only lukewarm support to the large-scale test in the LOFT program since it was not expected to contribute significantly to basic understanding of the phenomena involved.

In any event, this letter shows that in November 1963 the ACRS did not correlate large-scale core melt directly with a loss in containment integrity for the LWRs then under review.

Metal-Water Reaction

With the advent of the use of Zircaloy* cladding, instead of stainless steel cladding, concern arose for possible Zircaloy-steam reactions in light water reactors as a consequence of the postulated core meltdown. Heat and hydrogen were products of this reaction, and the hydrogen could burn, adding to the heat load on the containment vessel. At that time it seemed plausible that differences in containment design might be required when Zircaloy was used. (The stainless steel-steam reaction was discounted at that time but later was found to progress rapidly near the melting point of steel with the generation of hydrogen but less heat.)

---

*Zircaloy is the trade name for a zirconium base alloy which absorbs fewer neutrons than steel and thus improves the reactor performance.

The AEC regulatory staff held a symposium (AEC 1965a) on possible zirconium-water reactions in water reactors, in Germantown, Maryland, on April 29, 1965. It was chaired by Dr. Clifford Beck. In his introductory remarks at the symposium, Dr. Beck pointed out that the regulatory staff was faced with the possibility of having to review a large number of reactors using Zircaloy cladding. This raised the possibility of the occurrence of a zirconium-water reaction with, as its consequence, a substantial release of energy and hydrogen, which could burn or explode.* The regulatory staff had to decide if safeguard systems must be designed for the above situation.

A speaker for the Westinghouse Atomic Power Division, R. Weisemann, concluded that, if engineered safeguard systems worked properly, there would be essentially no zirconium-steam chemical reaction. He went on to discuss the effects of large-scale zirconium-steam reactions, the resultant hydrogen generation, and the potential effects on pressure in the containment building. He concluded that containment integrity could be maintained in the face of substantial zirconium-water reactions, although he did not allude to other difficulties that would be associated with any situation wherein the temperature of the core was so high that a large-scale reaction occurred. L. Epstein of General Electric gave a talk in which he discussed the possible course of events if one assumed a loss-of-coolant accident with no emergency coolant available, and concluded that the range of metal-water reaction would be limited to about 15% to 20% of the total zirconium in the core. This analysis assumed that the zirconium would be quenched when it fell into cool water below the core; it did not follow the accident along through core melting and the subsequent effects, including more metal-water reaction and containment failure. Time has shown Epstein to have been overly optimistic.

Dr. Beck asked the representatives of Westinghouse and General Electric if they had considered the situation wherein a loss-of-coolant accident occurred, the coolant left the core, no metal–water reaction occurred, and then the core-spray system functioned late, supplying water to a core which was now very hot from decay heat. Beck wondered if this wasn't a more dangerous situation than if water had not been added by the core-spray system at all. S. Levy of General Electric responded that the answer was to design a good core-cooling system; the best way to prevent the metal-water reaction was to keep the fuel rods cool. This meant that one had to design an injection system that did the job in all instances. The Westinghouse representative, Wiesemann, agreed about design to prevent the metal--water reaction; however, he stated that the containment design had to be

*The industry often argued that if hydrogen were formed, it would be hot and ignite spontaneously on escaping from the primary system into the containment building through a postulated pipe rupture.

such that even if safeguards failed, the metal–water reaction could be handled by the containment system.

For several of the reactors that were reviewed in the period of time before, around, or shortly after the April 29, 1965, meeting on metal–water reactions, the regulatory staff used a figure of 25% of the Zircaloy cladding as being involved in metal–water reactions. They also assumed that unless the containment atmosphere was inerted (that is reduced in oxygen concentration so that hydrogen would not burn), the building had to be designed to accept the energy generated from the combustion of the hydrogen (as it was formed, not in an explosion).

The AEC formed a study group on metal-water reactions in nuclear reactors. In early 1966, the study group reported that "if one assumes the safeguards systems do not function in such a manner as to prevent or limit the possible metal-water reactions, such reactions could encompass most of the metal available in the core and also produce adequate quantities of hydrogen to result in an explosive hydrogen-air mixture in the containment or confinement system." This last conclusion suggests a possible containment failure mode arising from large-scale overheating of the core; however, it is not clear how the regulatory staff evaluated this in their licensing review of PWRs at that time. There appears to have been no consideration of a corresponding change in the approach to containment design. At least for the PWRs reviewed for construction permits in the few months following February 1, 1966, namely, the Brookwood and Indian Point 2 reactors, the regulatory staff did not assume 100% metal-water reaction and the accumulation and possible explosion of hydrogen gas in the containment. Nor did the ACRS.

The Revision of the WASH-740 Report

A review of the work performed by Brookhaven National Laboratory on a possible revision of the WASH-740 report (AEC 1957) also yields an insight into the development of thinking at this time about the relationship between full-scale melting of the core of a large LWR and a correlated containment failure. This work was initiated by the Atomic Energy Commission at Brookhaven National Laboratory (BNL) in 1964. It was expected that the very large consequences resulting from the postulations used in the WASH-740 report could now be reduced in the light of recent knowledge, and that this would be of value in consideration of extension of the Price-Anderson law on liability limits and insurance for nuclear reactors. The BNL study was performed under the guidance of a steering committee chaired by Dr. Beck of the AEC, and which included among its members. W. D. Claus, R. L. Doan, A. P. Kenneke, W. J. McCool, J.

McLaughlin, U. M. Staebler and F. Western, all of the AEC, and F. Gifford of the Weather Bureau, and D. Okrent of Argonne National Laboratory. We shall not herein examine the details of how decisions were made on what to report concerning the work performed by Brookhaven National Laboratory. Nor will we look at the controversy which has arisen concerning whether or not the Atomic Energy Commission should have released more information in 1965, when it chose to publish only a very brief summary of the results of this work, except to note that because the newer reactors were larger, the maximum consequences were larger. Rather, we shall look at selected pages in some of the minutes and reports prepared in connection with this work for insight into the technical knowledge about the China Syndrome which was available, or at least was portrayed in writing, at that time.

Beck's Position in 1965

On January 22, 1965, Dr. Beck wrote to the members of the Atomic Industrial Forum Safety Committee (a nuclear industry group) and sent them a working draft of Chapter 1 and 2 of the proposed report resulting from the reexamination of the WASH–740 report (AEC 1957). We first quote from page 8 of draft Chapter 2, "Loss of Coolant with Containment."

The Emergency Core Cooling System cannot be made foolproof. It must be turned on and must have an adequate water supply in order to operate effectively. Thus, if one of the major coolant pipes fails and the emergency core cooling system also does not perform adequately, then the fuel element temperature would rise, the elements would melt and the fission products would be released from the fuel matrix. An aerosol of fission products could be swept out of the vessel and into the containment shell by convection currents.

If the containment is effective, that is, if the leakage rate is less than or equal to the design leak rate, then the fission product aerosol will deposit within the containment shell at the rate of 50% per day and will leak out of the system at a very low rate. The design criteria for maximum leak rate from the containment system is that the leakage of fission products resulting from an accident of this nature will not subject anyone beyond the reactor site boundary to more than 25 R. Personnel exposure levels in this region produce essentially no damage. Thus, if the containment system is effective, a loss of coolant accident in which the emergency core cooling system also fails, would result in essentially no damage to the public.

This statement seems to mean that the authors expected it was possible that one could melt the core of a large reactor and maintain the containment system effective. There is no suggestion that the authors expected a strong correlation between melting of a large core and a loss of containment integrity. Again, on page 13 of the same draft, Chapter 2, prepared by Brookhaven and forwarded to the Atomic Industrial Forum by Dr. Beck, we have:

If there was an *unimpeded* path for convection of the air in and out of the vessel, then most of the fission product aerosol would be dragged out of the vessel. Calculations have shown that the particle size of the aerosol under such conditions would probably not exceed 1 micron. At the end of 4 hours the molten fuel will have melted its way through the bottom of the ,pressure vessel and quickly will have gotten to a concrete floor. Fission product afterheat contained within the molten fuel would spall the concrete until such time that a large enough area for conduction of heat to the ground has formed and the fuel solidifies, thus terminating any further fission product release. The fission product aerosol which was dragged out of the pressure vessel by convection currents would enter the containment shell and would begin to deposit in the shell at the rate of 50% per day. Since the majority of surfaces available for deposition are painted surfaces, no one fission product group would be preferentially deposited within the shell.

Here we see mention of attack of the concrete floor of the containment building by the molten fuel; however, there is no direct statement that a loss of containment integrity is expected. On page 14 of the same draft report we find the following statement:

Release from Containment

At this time we would normally expect the containment shell to be intact; and the containment spray and/or filter system to be effective and to trap most, if not all, of the fission product aerosol.

The containment shell itself is the last of the present day safeguards. If the containment system does work, that is, its leakage rate is as designated in the hazard summary report, then, as has been shown, very small damage to the public will ensue. There remains, however, the small but finite probability that the containment will be breached by an open door or other mechanism. Thus, we must assume that the containment is not complete and depend only on the natural deposition mechanisms for depletion of fission products from the air. An opening the size of a door will have an exhaust time due to wind action which is short compared to the fission product deposition time. Under these conditions most of the fission products would be released to the atmosphere.

Again there is no direct suggestion that the melting of the core would be expected to lead to a breach of containment integrity.

Another example can be derived from a letter dated May 14, 1965, from Dr. Beck to the members of the steering committee, which forwarded some possible drafts of a report on the new Brookhaven work. Included for consideration was a memorandum by K. Downes and A. Court of BNL entitled "Theoretical Consequences of Hypothetical Accidents in Large Nuclear Power Plants", dated May 5, 1965. On page 3 of this memorandum it says,

> However, to achieve the purpose of the present study, which involves identifying the point at which damages and public injury would occur, it is necessary to suppose that all the means of assuring safety have failed to function. For instance, the emergency cooling system is simply supposed not to operate as it should. In addition it must be assumed that some large penetration in the reactor containment building is open at the time of the accident or that the containment building is damaged by a missile from the accident, so that containment is violated. At this point, the hypothetical accident would become a hazard, and its consequences would be severe. In order to identify the point in the spectrum of hypothetical reactor accidents where public injury and financial damage would begin, it has been necessary to assume that a very improbable event is followed by a failure of a complete set of safeguards that are engineered to prevent hazards to the public or to reduce these hazards.

Another bit of evidence on the thinking at the time comes from a packet sent by Dr. Beck to the members of the steering committee on April 21, 1965. The packet included a possible draft letter to Representative Holifield, Chairman of the Joint Committee on Atomic Energy, which was to review the results of Brookhaven's reevaluation of the WASH–740 report (AEC 1957). On page 3 of the draft letter it states,

> The preliminary results of Brookhaven's reevaluation can be summarzied by noting that in the first two cases postulated, i.e., where the emergency cooling system and/or the containment system function as designed and tested, a loss of coolant accident, irrespective of the degree of fuel melting, will not result in substantial injury to the public or damage to offsite property. It is only in the highly improbable instance where these and all other engineered safeguards failed simultaneously that a loss of coolant accident could result in a public hazard.

It seems fair to assume that in the spring of 1965, if there were groups or individuals who made a direct connection between large-scale core melting and a loss of containment integrity, it was not a widely held piece of knowledge; it seems not to have been a part of the thinking process undergone in the re-review of the WASH–740 report (AEC 1957); and it apparently was not a part of the information presented to the Atomic Energy commissioners or to the Joint Committee on Atomic Energy in 1965.

The minutes of a meeting held November 2, 1965, between the regulatory staff and Consolidated Edison concerning the Indian Point 2 reactor, reaffirm that containment was considered an independent safeguard at that time. The exerpt which follows quotes Dr. R. Doan, the Director of Reactor Licensing, and Mr. R. A. Wiesemann of Westinghouse.

Dr. Doan indicated that the Staff and the ACRS would look at the reactor system itself to see that meltdown is prevented for all including the biggest rupture. Then, the containment is put around it in order that there be protection even if the others fail, and metal-water reactions occur. Dr. Doan indicated that it was no more credible that safeguards in the containment work than those in the primary system, but he believed that the containment had to be designed to contain something.

Mr. Wiesemann stated that this had been done; that the containment was satisfactory if the safety injection system does not work at all. A low head pump will be available with high reliability, but this is not being depended on for containment integrity.

As we saw at the end of the last chapter, the minutes of the June 3, 1966, meeting of the ACRS subcommittee on Reactor Design and Operating Criteria included the description given to the ACRS by Dr. Beck and Dr. Wensch about the BNL redo of the WASH-740 report (AEC 1957). This appears to be the first ACRS document to unequivocally relate core melt to containment failure.

Sometime prior to June 1966, both Brookhaven and Dr. Beck had reached the conclusion that for the large water reactors, full-scale core melt would be associated with the loss of containment integrity; that is, at least the core would melt through the bottom of the containment building. Nevertheless, there had been no change in the regulatory staff approach to the acceptance of reactors based on this knowledge, nor was it mentioned as part of the safety evaluation issued by the regulatory staff for any of the reactors reviewed during that time-period, for example, Dresden 2, Brookwood, Millstone Point, Indian Point 2, and Dresden 3. However, Mr. Price and Dr. Beck did take a rather negative attitude toward the possibility of reactors being constructed at metropolitan sites during this time-period. (Indian Point was not categorized as such a site by the staff.)

The China Syndrome Enters the Regulatory Process

The question of core melt and its adverse affect on containment integrity finally came to a head as part of the licensing process at the seventy-fourth ACRS meeting, June 8-11, 1966. The minutes show that there was considerable discussion among the ACRS members concerning the possible ultimate fate of this large amount of molten fuel. There were varied opinions ranging from the possiblity that there was insufficient information available to portend a serious safety problem to the point of view that fuel melting was an unanalyzed safety problem (which required resolution before proceeding). ACRS Chairman Okrent took the position that the matter had to be faced.

The matter was discussed in detail at the June 1966 ACRS meeting with Commonwealth Edison, who were applying for a construction permit for the Dresden 3 reactor. At the request of Commonwealth Edison, the ACRS met with them again on Saturday (which was not usual); at this session General Electric presented a preliminary analysis of what might happen in the core melt situation for a boiling water reactor. The minutes indicate that Commonwealth Edison and General Electric were suggesting that possibly the fuel might be retained in the concrete base of the containment vessel, if there were sufficient water above the molten fuel to remove a large amount of the heat from the fuel by radiation. At the conclusion of this meeting, the ACRS decided that more information was needed on the subject of core melt as part of its construction permit reviews of the Dresden 3 and Indian Point 2 reactors.

When we review later in this chapter what transpired during the long, hot summer of 1966, beginning in June and ending in October, we shall see that during the first two months there was intensive examination by the ACRS, and by Westinghouse and General Electric, of the possibility of providing engineered safeguards which would maintain containment integrity in the presence of large-scale core melt for the large reactors being considered. On its own, the ACRS arrived at the conclusion that it was very difficult at that time to provide a feasible solution for coping with core melt for the Indian Point 2 PWR, which had a large dry containment vessel, and that it was relatively impossible, or nearly so, to do this for the Dresden 3 reactor design with its smaller, pressure-suppression-type containment.* The applicant and reactor vendor for the Dresden 3 reactor, General Electric, presented information in support of the thesis that maintaining containment integrity in the face of core meltdown was not feasible for their design. They did not propose to try to design to cope with core melt; they also believed that their existing emergency core-cooling system was adequate to prevent core melt in the face of a loss-of-coolant accident. Consolidated Edison and their reactor vendor, Westinghouse, proposed that they could supply, for the Indian Point 2 reactor, a structure below the reactor vessel which should be able to hold the molten core and keep containment integrity intact. The regulatory staff took the position that each reactor

---

*In the General Electric pressure-suppression containment, the steam escaping from the primary system (if there were a large pipe rupture) is directed into a large pool of cool water which condenses it, reducing the total pressure rise in the containment building.

design, as proposed prior to the June ACRS meeting, was acceptable, although they acknowledged that some further study was warranted on both the question of emergency cooling systems and on problems associated with core melt.

The pressures on the ACRS were very great, and in the midst of this extremely complex discussion, evaluation, and review, the AEC gave public notice that the beginning of hearings on the Indian Point 2 reactor by the Atomic Safety and Licensing Board would begin, despite the fact that the ACRS had previously requested that the AEC refrain from doing this before the committee had completed its review.

A wide range of opinions existed among the members of the ACRS, but they eventually arrived at a consensus and decided that they could write letters favorable to the construction of the Indian Point 2 reactor and the Dresden 3 reactor on the basis of greatly improved emergency core-cooling systems and much greater emphasis on primary system integrity to reduce the probability of a LOCA. The ACRS decision on these two reactors included an agreement that the ACRS would write a general (non-case-specific) letter to the AEC in which the ACRS was to make strong recommendations concerning the rapid development and future implementation of further engineered safety features to cope with problems associated with core melt; this was the basis on which several members agreed to the issuance of letters favorable to the construction of the Indian Point 2 and Dresden 3 reactors. Such a letter was prepared by the committee and submitted for comment to the regulatory staff who submitted it to the commissioners themselves. However, at the September 1966, meeting of the ACRS, the commissioners urged that the committee, rather than send such a letter, await the report of a task force that the commission would establish to study and quickly report on problems associated with core melt. We shall see at the end of this chapter that the majority committee opinion was to go along with the proposal by the commissioners and that a letter specifically recommending the implementation of safeguards to cope with core melt was never sent. However, the ACRS did write a safety research letter issued October 12, 1966, recommending that the safety research program of the Atomic Energy Commission give emphasis to problems associated with phenomena related to large-scale core melting as well as to improvements in our knowledge of emergency core cooling.

In a later chapter we will discuss the report of the Ergen task force (or task force on emergency core cooling), ACRS reaction to the report, and the action, or lack thereof, by the AEC and the industry on the general problem of core meltdown in the ensuing months and years.

It must be recognized that during this same period, there were many other reactors being reviewed by the regulatory staff and by the ACRS, and that for the Indian Point 2 and Dresden 3 reactors there were many other technical questions under review in addition to core melt. For example, questions were raised concerning the adequacy of fire protection at the Indian Point 2 reactor, as well as the reliability of various systems such as the emergency power supplies.

Looking back a decade later, one may well ask, "Why was the review procedure pushed at so rapid a pace with so serious a question involved?" And "Why was resolution accepted based on partial information and on general criteria?" It is clear that there was considerable pressure from the industry not to impose further delays on the start of construction of these plants, and that in the particular time-period 1965–1966, the commission and the AEC regulatory staff were very sensitive to the question of delays arising from the regulatory process. Curiously, this was the period during which the time between application for a construction permit and issuance of a construction permit was perhaps the shortest it has ever been.

Also, looking back with the hindsight of more than ten years, it is clear that the loss-of-coolant accident (LOCA) was uppermost in the minds of the ACRS and the regulatory staff as the most probable source of core meltdown. They were not ignoring other accident sources, and as time passed, because of the clear relation between core melt and the loss of containment integrity, all possible sources of core melt began to be searched out, evaluated, and modified, as possible, to reduce the probability that any particular source would be an important contributor. Nevertheless, the emphasis during the next decade was on the loss-of-coolant accident, the adequacy of the emergency core-cooling system, and on means greatly to reduce the probability of a LOCA.

Events of June and July 1966

Now we return to a relatively detailed history of events following the June 8–11, 1966, ACRS meeting at which the ACRS decided to hold up completion of its review of the Dresden 3 and Indian Point 2 reactors in order to examine the issues arising from the China Syndrome. On June 14, 1966, the regulatory staff contacted the ACRS office to ascertain if the committee desired that the applicants submit written information regarding the course of a core meltdown accident and the reliability of emergency cooling systems for the Dresden 3 and Indian Point 2 reactors. The regulatory staff then advised the applicant for each reactor that neither the ACRS nor the regulatory staff was requesting written information in regard to the above items. Commonwealth Edison indicated that no written information would be submitted regarding the Dresden 3 reactor. Consolidated Edison stated that they might provide a limited amount of information for the Indian Point 2 reactor in their third supplement to the Preliminary Safety Analysis Report (PSAR) which was in the process of preparation for submission.

A meeting of the Indian Point 2 subcommittee was held on June 23, 1966. The major discussion during the meeting related to consequences of pressure vessel failure and to the question of core melt. Mr. Roger Boyd of the regulatory staff stated that the staff believed the proposed Indian Point 2 plant was acceptable. They considered Indian Point 2 to be a suburban rather than urban reactor. The staff indicated they did not believe there was a significant difference between the Indian Point 2 and Dresden 3 reactors regarding the consequences of a core meltdown accident. Mr. Boyd said he recognized the inconsistency of the staff's present position that, on the one hand, the core would melt, leading to fission product release and metal-water reaction considerations, but that, on the other, core melt would not cause a problem from a loss of containment integrity.

Westinghouse made a presentation of the heat-transfer calculations they had made concerning the ability of the reactor vessel itself to hold molten fuel; they had concluded (erroneously) that rather large fractions of core (as much as 60%) could be held in the vessel itself. They went on to present the heat-transfer analysis of a stainless steel core-catcher device which was lined with refractory material and cooled with water. They had described the device in the recent supplement to the PSAR and proposed to place it below the vessel, in case the core melted through the vessel. Westinghouse indicated they believed that, if the core melted through the pressure vessel, there would be rapid chilling and formation of solidified uranium oxide when the molten fuel hit the water located below.

The Indian Point 2 subcommittee emphasized the need to provide adequate information to the full ACRS regarding the core melt-through accident. The Westinghouse presentation on their core-catcher device did not include considerations of the possible generation of large amounts of hydrogen and its effect on containment integrity, or the possibility of a steam explosion, or several other phenomena relevant to the reliability or the effectiveness of such a system.

The Westinghouse presentation on the ability of the containment vessel to withstand several modes of gross pressure vessel failure, including longitudinal splitting or circumferential rupture of the vessel below the flange, was relatively optimistic concerning its ability to withstand such failures. Westinghouse also classified any such failure modes as being of extremely low probability.

A subcommittee meeting on the Dresden 3 reactor was held July 7, 1966. Prior to this meeting, on June 25, 1966, ACRS member Etherington provided a memorandum to other ACRS members giving the results of a very quick and crude analysis he had performed on core melt. Using a simplified spherical model, he estimated that the molten fuel would penetrate the bottom of the building and form a liquid pool that would continue to grow at a decreasing rate by fusion of the concrete and earth. The simple model gave liquid pool diameters of 32 feet, 54 feet, and 80 feet, after one day, one week, and one month, respectively. Containment integrity above ground would be jeopardized by the possibility of building collapse or by rupture of the containment liner. Also, high pressure might open fissures in the soil through which material could be forced to the surface. After the molten mass froze, slow leaching of the fission products by ground water and their subsequent appearance in wells, rivers, etc., would remain to be considered.

A set of brief memoranda prepared for ACRS members H. O. Monson and D. Okrent by Messrs. J. Hesson, R. Ivins, and A. Tevebaugh of Argonne National Laboratory was another source of information which helped the ACRS to form opinions concerning the likelihood of being able to deal in a feasible fashion with core meltdown, assuming the current reactor design approach. It appeared to be very difficult to maintain containment integrity with any degree of assurance, given a core meltdown.

At the July 7, 1966, subcommittee meeting on the Dresden 3 reactor, General Electric presented a detailed analysis of the course of full-scale core melt; containment failure was the consequence. Various potential design modifications intended to enable the containment to maintain its integrity were discussed. These included a fire brick lining, cooling coils in the concrete, a thick, cooled, steel plate, increased containment volume, etc. General Electric felt that none of these could be adequately engineered; they believed the answer lay in prevention of melting rather than holding the molten material.

In executive session, several ACRS members were pessimistic about the possibility of containing a core melt. Some members apparently felt that the ACRS should proceed with the Dresden 3 reactor on the basis that melt-through was not credible. Some felt that improving core cooling was a proper solution. Still others thought that more study might produce workable ideas for containing the molten core. General Electric was asked to give further consideration to ways of containing a melted core even if design changes were involved beyond those they had suggested thus far.

On July 7, 1966, the regulatory staff issued their third report to the ACRS on the Indian Point 2 reactor. At the beginning of this report the staff noted that Indian Point 2 has special siting considerations and that the applicant had proposed a containment and engineered safeguard system which the staff believed to be superior to those provided at facilities in less-populated areas.

The staff report included a topic called "Items Requiring Further Study" which included "Melting of core through reactor vessel" and "Consequences and causes of reactor vessel rupture" as two of nine matters. It was not clear how the regulatory staff planned to deal with these two questions. Nor was the basis clear for the staff conclusion that a construction permit could be issued for the Indian Point 2 reactor as then proposed.

At the seventy-fifth meeting, July 13–15, 1966, the ACRS met with both the Dresden 3 and the Indian Point 2 applicants. Consolidated Edison repeated its proposal to supply a water-cooled, refractory-lined core catcher below the vessel, but Westinghouse noted the lack of data in support of their design and stated they were not proposing an experimental program to validate it. They would put primary reliance on core cooling.

General Electric described the brief preliminary studies they had made on possible design changes to handle a molten core and concluded that all the concepts they had examined would fall short. General Electric believed that the basis of design, therefore, must be the prevention of core melting by emergency cooling in case of a LOCA. They proposed to install either two core-spray systems or a core-spray and a high-flow-rate, core-flooding system.

With regard to the Indian Point 2 reactor, some ACRS members felt that the proposal for a core catcher was a step in the direction of safety. However, the ACRS noted that Westinghouse's view that a molten core could be contained was not consistent with the General Electric position that melt-through was inevitable. The ACRS did not accept the Westinghouse core catcher as providing any guarantee of retention.

Subcommittee Chairman Etherington reported that with regard to the Dresden 3 reactor, as proposed, core melting would probably result in pressure vessel failure, overpressurization of the containment vessel, and deterioration of the concrete underneath. He stated that many high power level reactors presented the same problem and that the only clear solution at that time was to provide adequate emergency cooling to prevent melting, but that such cooling was not then required as a design criterion. ACRS member S. Hanauer pointed out the uncertainty in the succession of operations which are needed to assure the effectiveness of the emergency core-cooling system.

Some members favored accepting the Dresden 3 reactor as proposed and deferring any further consideration of the China Syndrome issue to future cases.

The committee agreed not to ask for provisions similar to the Indian Point 2 core catcher (by a soft vote—four in favor of such a request, five against, three abstentions). Although he was not opposed to the Consolidated Edison proposal, one member suggested that Westinghouse had proposed a core catcher only to get committee approval.

ACRS Chairman Okrent listed several alternatives to the committee: approval of the same reactor as had been accepted previously for the Dresden 2 reactor; approval with more core sprays; approval but with reservations about the design adequacy of the core sprays; approval with measures to handle a molten core; and rejection of the proposed reactor.

ACRS member Palladino felt that the recent ACRS approval of the Dresden 2 reactor implicitly included acceptance of Dresden 3. However, member Hanauer stated that the committee must be receptive to new information which might allow acceptance of risks with past plants but suggest that changes might be needed for future reactors.

Member Etherington suggested the use of a probabilistic approach in which a graph was plotted with the logarithm of the consequences of a reactor accident as the ordinate and the logarithm of the probability of an accident as the abscissa. Perhaps the loss of a thousand lives with a probability of one in a million per reactor-year might be acceptable, while a consequence beyond this might not be. The difficulty of defining an acceptable relation was noted.

A separate ACRS letter to the AEC regarding core melting was proposed by member Hanauer, rather than including such recommendations in the report on the Dresden 3 reactor. However, a delay in release of any general ACRS letter on the core melting problem was predicted by Dr. Ergen (an ACRS consultant).

Dr. Beck told the ACRS that while the regulatory staff considered a meltdown accident to present a difficult situation, the low probability of the event led the staff to accept this hazard. Dr. Doan said that any conclusions on core melting which were applied to the Dresden 3 reactor would also apply to Dresden 2; although the core melt issue was important, he saw no reason why this possible accident should delay these two reactors.

It is rather clear from the minutes of the seventy-fifth meeting in July, 1966, that the ACRS was fairly well convinced that means of coping with a fully molten core within the framework of the General Electric suppression pool design, and probably within the framework of the Indian Point 2 design, were not readily available and would be difficult to demonstrate on a short time-scale. What is less clear is what direction the committee thought the review of these two construction permit applications should take. There was a considerable divergence of opinion among the ACRS members, and the committee arrived at no conclusions at the seventy-fifth meeting.

Development of a Tentative ACRS Position

In the days which followed, some members wrote memoranda or draft ACRS letters that either expressed the point of view of the individual member or might provide the basis for a consensus decision. On July 16, 1966, member H. Newson circulated the following memorandum entitled "Report on Dresden 3" to his fellow committee members:

> The applicant and the General Electric Company have informed the Committee that no provision had been made to maintain containment in the unlikely event of a core melt-down following loss-of-coolant.
> Under these circumstances, the applicant must design for one of the following criteria to be applied at the time of issuance of an operating license.
> 1. The licensed operating power of the reactor must be reduced to the point where a molten core may be cooled efficiently enough to prevent melt-through of the bottom of the containment vessel or
> 2. An emergency cooling system, far more reliable than any which now exists, must be designed and installed with sufficient precautions to convince the AEC Staff and this Committee that after a loss-of-coolant accident at design power, there is as much assurance that core melt-down can be prevented as the reliability of conventional containment for relatively low-power reactors.
> At the present time, neither an efficient cooling system for a melted core nor a highly reliable emergency cooling system have been invented so that it is impossible for the Committee to advise that Dresden III reactor may be built and operated at the proposed site and power level without undue risk to the health and safety of the public. If a construction permit is issued before the development of these novel engineered safeguards, the safe operating power can only be determined after their construction and the demonstration of their capabilities.

Newson's memorandum takes a fairly stringent position. However, it does not necessarily propose rejection of the Dresden 3 construction permit.

Since it did not appear practical in July 1966 to recommend that a reactor soon to begin construction must include provisions which had a good chance of maintaining containment integrity following core melt, one possible position for the committee to take was the following. For Indian Point 2 and Dresden 3 reactors (and for other reactors to be proposed for construction at acceptable sites in the next couple of years), to recommend measures to greatly reduce the probability of core melt arising from a LOCA followed by a failure of the ECCS to function or perform properly; and at the same time, in a general letter, to recommend the rapid development and early implementation on reactors to be constructed (after, say, two years) of other measures which could cope with or ameliorate the consequences of core melt. Drafts of possible ACRS letters which followed this philosophy were prepared and distributed to the ACRS members for consideration and comment during July.

On July 28, ACRS member Hanauer sent a memorandum to the other members, which included the following points:

> The primary system rupture is one of the design-basis accidents for hazards analysis for Dresden 3, Indian Point 2, and lots of other reactors. In discussing its consequences, therefore, or the adequacy of safeguards to cope with this accident, it seems to me that adversion to the improbability of the accident is out of order. We should indeed require steps to be taken to reduce the probability, but that is a separate subject.
>
> In view of the potentially serious consequences of primary-system rupture, it is my present view that each reactor should have two defenses against this design-basis accident. One of these defenses might be a well-engineered, redundant system to put water onto or into the core. However, because of the many uncertainties regarding the functional adequacy of pouring water (emergency core coolant pipe rupture as a result of the accident, fuel melting in spite of success in getting water to core, loss of cladding strength leading to blockage of coolant channels or drooping fuel to the bottom of the vessel, steam explosions, steam blanketing, etc.), a second defense of a different species should be provided....
>
> What I really think we have to have is a good core cooling complex with an analysis which shows that if it works the vessel won't fail, plus some other complex with an analysis that shows a reasonable probability of averting danger to the health and safety of the public if the core cooling does not provide the expected protection.

Events in August, 1966

The ACRS scheduled a special meeting, August 4–5, 1966, in order to pursue its review of the Indian Point 2 and Dresden 3 reactors. At the beginning of this meeting, the regulatory staff still proposed to accept these two reactors as they were proposed prior to eruption of the China Syndrome issue. Despite the complex new issues which had been raised, the regulatory staff, furthermore, gave public notice that the public hearing of the ASLB on the construction permit application for the Indian Point 2 reactor would be held on August 31, 1966. Since such a hearing could not proceed without an ACRS letter on record, which represented completion of committee review, this action was overt pressure by the AEC on the ACRS to finish its review in August.

Consolidated Edison repeated the position that the core catcher was intended as a reserve safeguard only, and that emergency core cooling was the primary safety mechanism. For rupture of the largest pipe in the primary system, they estimated that 25% to 30% of the fuel cladding might melt leading to a 12% to 15% metal-water reaction; they implied that this would not threaten containment. Other studies have since introduced doubt as to the validity of this position.

General Electric and Commonwealth Edison rejected the idea of a core catcher. They said such a scheme would not be acceptable to Commonwealth Edison and would not be marketed b, General Electric.

Several ACRS members again expressed the opinion that the demonstration of a workable arrangement for the retention of a molten core was impossible. They felt that core cooling offered the best chance for a solution. Others thought that since a core catcher had been proposed for the Indian Point 2 reactor, the committee had no choice but to require such a device on Dresden 3. ACRS consultant Ergen suggested that a venting and filtering system would make the question of containment integrity less important.

ACRS member Thompson did not attend the meeting but was reported to have said that if the containment were lost for the Indian Point 2 reactor, the consequences to the public would be the same as for a Dresden 3 reactor accident. ACRS Chairman Okrent disagreed, saying this conclusion could be extended to state that the Dresden 3 reactor presented the same hazards as the previously proposed Ravenswood reactor, a conclusion which he doubted. However, consultant Ergen offered the opinion that, although for intermediate size reactors sites offered differences in hazards, for very large reactor accidents there was probably little difference in consequences among sites.

ACRS member Zabel suggested the possibility that above some power level, say 750 or 1,000 MWe, core-retention measures might be required. Such a recommendation would have permitted the Dresden 3 and possibly Indian Point 2 reactors to be approved, since they would fall below the threshold power level.

Hence, there were still many differences of opinion among the members. Nevertheless a tentative committee consensus was emerging to approve the Indian Point 2 and Dresden 3 reactors on the basis of improved emergency core-cooling systems (ECCS) and to write a general letter to the AEC with recommendations for the future implementation of mitigation measures for core melt.

Consolidated Edison was told at the conclusion of the August 4–5, special meeting that the ACRS thought it would be able to write a letter favorable to construction of the Indian Point 2 reactor and that the letter would include reservations about the adequacy of the ECCS.

For the Dresden 3 reactor, the committee was not satisfied with only two core sprays for the ECCS; however, if core flooding were provided as an independent emergency cooling system, Dresden 3 would be acceptable.

A Political Aside

As an interesting historical aside, we see that the minutes of the ACRS special meeting, August 1966, also refer to some important procedural matters, as follows:

A proposed letter to Mr. Farmer, of the British reactor safety group, was reviewed by the Committee. The attempt by the Committee to make private arrangements with the British safety groups has disturbed Dr. Beck; he sees a possibility of the ACRS opinions expressed being contrary to the AEC policy. He prefers that any such arrangements be made through the Regulatory Staff. This view recalled early conflicts between the ACRS and the Staff which was at that time under the General Manager, e.g., the ACRS having consultants was not favored by the Staff, and the Staff was against Executive Sessions for the Committee.

At a recent Commission meeting, Dr. Okrent was asked about Committee measures to handle the increasing work load; apparently, this reflects comments from applicants, e.g., the Dresden group, on delays. Dr. Okrent replied that he was attempting to keep the ACRS at full strength and hopes for an increase in Regulatory Staff responsibility. He made comments to the Commission on the core melting accident.

These minutes mention that Dr. Beck was disturbed by an attempt by the ACRS to make private arrangements to meet with British safety groups, and they also note a meeting between ACRS Chairman Okrent and the AEC commissioners. What became publicly known much later, from a brief history of the ACRS prepared by the AEC Chief Historian, E. G. Hewlett (Hewlett, 1974), was that during this very difficult period in which a course of action was being sought by the ACRS about loss-of-coolant accidents and the China Syndrome, Dr. Beck was interceding with the AEC commissioners concerning what he considered inappropriate activities by the ACRS. Hewlett wrote as follows:

With the rapid expansion of projects utilizing nuclear energy, the role of the ACRS became increasingly amorphous and expansive. This trend prompted a letter from Dr. Clifford Beck, Deputy Director of Regulation, dated July 19, 1966, to the files entitled "Current Trends in ACRS Activities." Beck* argued that the present trend was similar to the situation in the late 1950s when the ACRS was heading toward a role independent of the AEC, with its own expanding staff, a proliferation of

*The existence of Beck's letter was not disclosed to Okrent nor were Beck's concerns.

consultants, and direct lines of communication with applicants and others outside the agency.

In particular, Beck attributed to these tendencies a lack of common basis of technical understanding between staff and the ACRS because of differing consultant sources, a needlessly increasing ACRS staff, and an increase in ACRS involvement in such matters as design details, operation, inspection, and compliance activities. Among other recommendations, Beck requested an *ad hoc* task force, similar to that established in 1954, made up of AEC and ACRS personnel, to examine the present and future relationships of the two organizations.

Subsequently the Commission met with Dr. David Okrent (Chairman of the ACRS) on August 3, 1966 to discuss these "current trends." No solutions were proposed to the problems Beck had enumerated in his letter of July, 1966, but the statutory guidelines on ACRS activities were under continual review. In November of 1967 an amendment to 182b of the Atomic Energy Act was circulated by the Director of Regulation, proposing a modification of the requirements for mandatory reviews and reports by the ACRS. The Chairman of the AEC, Glenn T. Seaborg, sent a final legislative package to the Bureau of the Budget on December 20, 1968. The expressed purpose of the new legislation was to make the statute flexible enough to permit the ACRS to omit its review if the Commission and the Committee agreed. This amendment would assist the ACRS in its deliberations as the standardization of designs increased.

Dr. Zabel's* appointment as Chairman of the ACRS in 1968 seemed to herald a new era of cooperation between the AEC and the ACRS. Mutually satisfactory liaison precedures were established between the two bodies. In January 1969 it was decided by the commission that the Director of Regulation was to be responsible for resolving serious deficiencies in the Quality Assurance programs of applicants, not the ACRS.† This decision was part and parcel of the move to free the ACRS of responsibility for routine matters.

The ACRS Agrees on a Package

At its seventy-sixth meeting, August 11–13, 1966, the ACRS completed action on both the Indian Point 2 and Dresden 3 construction permit reviews.‡ As an integral part of this action, the committee also completed preparation of a general letter concerning problems of primary system integrity, the loss-of-coolant accident, and the possible consequences of failure to cool the core.

*The ACRS chairman is elected by the ACRS members themselves, and not appointed by the commissioners.

†As Chapter 15 will discuss, in 1968 the ACRS called attention to some major deficiencies in quality assurance for the Oyster Creek plant.

‡The ACRS members present at the July meeting were D. Okrent, chairman, S. Bush, H. Etherington, F. Gifford, S. Hanauer, H. Manglesdorf, J. McKee, H. O. Monson, H. Newson, A. O'Kelly, and N. J. Palladino. C. Zabel participated in the August meetings; H. Newson did not attend. Members H. Kouts and T. J. Thompson missed all the July and August meetings.

At the August meeting, General Electric had proposed for the Dresden 3 reactor, and for Dresden 2 as well, two core-spray systems and a flooding system, any of which would meet a criterion that no clad should melt in a loss-of-coolant accident, and each of which was operable with emergency on-site power. The ACRS agreed that a letter could be written on the Dresden 3 reactor, although the final design of the emergency cooling system would require additional review.

The ACRS reports on the Dresden 3 and the Indian Point 2 reactors were sent from ACRS Chairman David Okrent to AEC Chairman Glenn T. Seaborg on August 16, 1966, and are appendixes A and B to this chapter. The draft general ACRS letter to Seaborg on problems arising from primary system rupture is appendix C to this chapter.

In its reports on the Indian Point 2 and Dresden 3 reactors the ACRS emphasized the need for additional measures to reduce the probability of a loss-of-coolant accident and the need for improved emergency core-cooling systems. With regard to the ECCS, the ACRS took the unusual step of recommending that, "the Regulatory Staff and the Committee should review details of design, fabrication procedures, plans for in-service inspection and the analyses pertaining to the emergency core cooling systems, as soon as this information is available and prior to *irrevocable* construction commitments pertaining thereto." Prior to that time, construction permit approvals had been based largely on a commitment to meet rather general criteria, and the plant, as built, might or might not prove satisfactory to the regulatory groups. The requirement imposed by recommending ACRS review before irrevocable commitments were made by the utility was, in a sense, the forerunner of requiring an increased knowledge of most design aspects at the construction permit stage.

The draft general letter, after recommending further measures to help prevent a LOCA and a very conservative approach to the design of ECCS, discussed the potential for large-scale core melt. It went on to make a controversial recommendation, as follows:

"...because experience with emergency core cooling systems is limited, and because systems using current concepts necessarily are subject to certain low-probability modes of failure related to primary system rupture, the Committee believes it prudent to provide still greater protection of the public by some independent means, particularly for reactor sites nearer to population centers. Progress toward this objective will require an evolutionary process of design and a vigorous program of research, both of which should begin immediately and be aimed at reaching a high state of development in approximately two years. Future reactors relying solely on currently employed types of emergency core cooling systems to cope with the unlikely accident involving primary system rupture will be considered suitable only for rural or remote sites.

In connection with the issuance of its general letter on pressure vessels, the ACRS had received rather adverse comment from the Atomic Energy Commission to the effect that the AEC had not been consulted in advance and not been notified that the committee planned to issue such a general letter to the AEC. Hence, this time the committee decided to forward the draft general letter to Mr. Price so that he could see it and comment prior to its formal issuance by the ACRS.

The general letter was a way of asking that, as rapidly as possible, some alternate method, over and above adequate emergency core-cooling systems of the general type being used or proposed, be developed, so that the problem of the China Syndrome would be attacked in depth with diverse approaches. And the plan to issue such a report was part of the overall package agreed to by the ACRS at its August 1966 meeting. In fact, the basis on which some members agreed to the issuance of letters favorable to the construction permits of the Dresden 3 and Indian Point 2 reactors was that there would be such a general letter.

As was then the custom, all the discussions between the ACRS and the regulatory staff, or those with Commonwealth Edison and with Consolidated Edison, had been in closed session. The only thing on the public record which indicated that some new provision for full-scale core melt had been considered in any way was the amendment to the Preliminary Safety Analysis Report submitted for the Indian Point 2 reactor in late June, in which Consolidated Edison proposed to put a core-retention structure under the reactor vessel. It is interesting to look at the Public Safety Evaluation published for Dresden Nuclear Power Station Unit 3 on August 31, 1966, by the regulatory staff. There is no hint anywhere in this report that the China Syndrome and the inability of the containment vessel to withstand core melt had been a major concern. It is stated that the ACRS met with the applicant on June 10, to discuss the overall design of the station and particular features of safety significance, that the ACRS met with the applicant and the regulatory staff on August 5 and August 12, for further discussion on the emergency core-cooling systems for the proposed plant, and that following this meeting, the ACRS reported its views by a letter report dated August 16, 1966.

Depending on one's point of view, one might consider this a less than candid review of what had transpired. One might equally well say that the ACRS Dresden 3 letter was not completely candid, since it did not directly address the interrelationship between core meltdown and containment failure. However, that was a point which was to be made explicit in the general letter on problems arising from primary system rupture, (see appendix C to this chapter). This report was never issued, as we shall now discuss.

The AEC Intervenes

The seventy-seventh meeting of the ACRS was held on September 8–10, 1966. Members Thompson and Kouts, who had not attended the July and August meetings, were present, as was a new member, J. Hendrie. There was a meeting between the committee and the full Atomic Energy Commission, which included Chairman Seaborg and Commissioners W. E. Johnson, S. M. Nabrit, J. T. Ramey, and G. F. Tape, at which the proposed letter mentioned above was discussed. Some of the minutes of this meeting follow below.

Chairman Seaborg referred to the proposed letter on primary system failure. He said that the impacts on the industry might be serious, and he felt that any letter should await more study. To Dr. Seaborg, the letter failed to recognize the current efforts to meet this problem with large reactors. Mr. Price said that the core melting problem is one for the industry to pursue rather than for the Regulatory Staff. Dr. Kavanaugh, the Assistant General Manager for Reactors to the AEC, objected to the tone of the letter and, in particular, to its lack of recognition of the efforts under way. He saw the close relation of the time schedule for reactor designs and the site problems as a difficulty with any steps toward protection against primary system failure. Making public such a letter without the much related correspondence might lead to misunderstanding by the public. Commissioner Johnson noted that parts of the proposed letter might be considered decision-making which is the prerogative of the Commissioners.

To Commissioner Johnson, more facts are needed before outlining measures to avoid or cope with this primary system hazard. Assembling a task force group of experts from the AEC laboratory to assist the ACRS with this problem has been considered by Dr. Kavanaugh; no AEC conclusions on such an approach had been reached. After more discussion, a task force to develop the problem of primary system failure was again suggested by Chairman Seaborg, prior to any such letter from the Committee. Mr. Shaw of the AEC reported being impressed by the response of applicants to suggestions of the Committee and the Regulatory Staff; consequently he preferred no letter now.

After its discussions with the commission and the members of the AEC staff, the committee discussed at length in executive session the draft letter of August 16, 1966, concerning primary system rupture. Where agreement had existed in August, divided opinion was now present, and the presence of members who had not been there in August, particularly Kouts and Thompson, added to the division of opinion.

The General Letter Is Withdrawn

Although several members felt that they had agreed to the letters issued on the Indian Point 2 and Dresden 3 reactors only on the basis that a general letter of the type under discussion would also be issued, and, in fact, this exact statement is attributed to one member in the minutes, the final conclusion of the committee was not to send the general letter, but rather to endorse the recommendation of the Atomic Energy Commission that a task force be established. Thus a possible major change in safety for light water reactors, namely, the conscious development and provision of steps to mitigate core meltdown, was not undertaken . . . then or in the next 13 years.

We shall see in Chapter 11 ("The China Syndrome—Part 2") that there was an insignificant effort expended by the AEC (and NRC) in the decade following 1966 on research and development on phenomena important to understanding the course of a core melt accident, or on measures intended to ameliorate the consequences of core melt.

However, a major program on LOCA–ECCS and on primary system integrity was initiated as a result of the ACRS action in 1966. A new regulatory approach with great emphasis on preventing core meltdown, from any cause, rapidly evolved. And the MCA lost much of its meaning, although the prescription of Part 100 continued to be used in site evaluation.

The next month, at the October 1966 meeting, the ACRS wrote a letter on safety research, in which it placed great emphasis both on studying phenomena related to large masses of molten core, and also on an improved ECCS. However, unlike the draft general letter, this letter did not include a specific recommendation that some new engineered safety feature be developed for use a few years hence in light water reactors. About a year later the task force came out with a report which gave only weak support to a research and development program on core retention measures. The end result of the report by the task force was to provide a mechanism whereby any further work aimed toward the development of a means to mitigate the effects of core meltdown could be put aside as unnecessary and impractical by those who wished to argue against such efforts.

Chapter 8: Appendix A:
ACRS Report to the AEC on Dresden Nuclear Power Station—Unit 3

At its seventy-fourth meeting, on June 8–11, seventy-fifth meeting, on July 14–16, a special meeting on August 4–5, and its seventy-sixth meeting on August 11–13, 1966, the Advisory Committee on Reactor Safeguards reviewed the proposal of the Commonwealth Edison Company to construct a third nuclear power plant on the Dresden site, near Morris, Illinois. Unit 3 will include a boiling water reactor to be operated at 2256 MW(t) power level with pressure suppression containment. Unit 3 would be similar to Unit 2. The Committee had the benefit of discussions with representatives of the applicant, the General Electric Company, Sargent & Lundy, the Babcock & Wilcox Company, and the AEC Staff, and of the documents listed. A Subcommittee of the ACRS met to review this project at the Dresden site on June 2, 1966, and in Washington on July 7, 1966.

In its report on Dresden Unit 2, dated November 24, 1965, the Committee recommended that the AEC Staff follow development work by GE to resolve particular design problems. The Committee recommends that the Staff continue to follow the development work in connection with both Units 2 and 3, particularly with respect to operation with jet pumps, testing of emergency cooling methods, and studies of reactivity transients to assure no impairment of emergency cooling effectiveness as a consequence thereof.

The Committee also urged that the designers pay particular attention to the design of the pressure vessel, and of the high pressure steam lines with their isolation valves and fittings. The Committee reiterates its opinion on this matter in connection with Unit 3.

The Committee notes that the applicant has made improvements in the requirements for pressure vessel inspection during fabrication and urges that the applicant pursue vigorously the implementation of adequate in-service inspection techniques.

The effectiveness of emergency core cooling systems is a matter of particular importance in the unlikely event of a pipe rupture in the primary system. The applicant proposes the following improved complex of emergency cooling systems:

1. a high pressure coolant injection (HPCI) system,

2. a high-volume flooding system to permit rapid injection of water into the reactor vessel following blowdown to a low pressure,

3. two core spray systems,

4. a system that will make river water available to the feedwater pump for emergency cooling.

The applicant advised the Committee that equivalent changes in the emergency core cooling systems of the Dresden 2 unit would be made. Three diesel-driven generators will be installed to serve Units 2 and 3.

The Committee concurs that the proposed system should increase the reliability and effectiveness of emergency core cooling. Complete details of the systems are not available, but the Committee believes that these matters can be resolved during construction of this facility. The Committee believes that the Regulatory Staff and the Committee should review details of design, fabrication procedures, plans for in-service inspection and the analyses pertaining to the emergency core cooling systems, as soon as this information is available and prior to irrevocable construction commitments pertaining thereto.

Careful examination of the forces during blowdown on various structural and functional members within the pressure vessel is necessary to assure sufficient conservatism in the design. The Committee recommends that the AEC Staff satisfy itself fully in this respect.

The Committee believes that the combination of emergency cooling systems has a high probability of guarding against core meltdown in the unlikely accident involving rupture of a primary system pipe. In view of the present state of development of such emergency cooling systems, however, and since the cooling systems may be subject to certain low-probability inter-related modes of failure, the Committee believes that the already small probability of primary system rupture should be still further reduced by taking additional measures as noted below. The Committee would like to review the results of studies by the applicant in this connection, and the consequent proposals, as soon as these are available.

1. Design and fabrication techniques for the entire primary system should be reviewed thoroughly to assure adequate conservatism throughout and to make full use of practical, existing inspection techniques which can provide still greater assurance of highest quality.

2. Great attention should be given to design for in-service inspection possibilities and the detection of incipient problems in the entire primary system during reactor operation. Methods of leak detection should be employed which provide a maximum of protection against serious incidents.

The Advisory Committee on Reactor Safeguards believes that the various items mentioned can be resolved during construction and that the proposed reactor can be constructed at the Dresden site with reasonable assurance that it can be operated without undue risk to the health and safety of the public.

Chapter 8: Appendix B:
ACRS Report to the AEC on Indian
Point Nuclear Generating—Unit No. 2

At its seventy-fifth meeting, July 14–16, 1966, and its special meeting on August 4–5, 1966, the Advisory Committee on Reactor Safeguards completed its review of the application of Consolidated Edison Company of New York, Inc. for authorization to construct Indian Point Nuclear Generating Unit No. 2. This project had previously been considered at the seventy-second and seventy-third meetings of the Committee, and at Subcommittee meetings on March 30, May 3, and June 23, 1966. During its review, the Committee had the benefit of discussions with representatives of the Consolidated Edison Company and their contractors and consultants and with representatives of the AEC Regulatory Staff and their consultants. The Committee also had the benefit of the documents listed.

The Indian Point 2 plant is to be a pressurized water reactor system utilizing a core fueled with slightly enriched uranium dioxide pellets contained in Zircaloy fuel rods; it is to be controlled by a combination of rod cluster-type control rods and boron dissolved in the primary coolant system. The plant is rated at 2758 MW(t); the gross electrical output is estimated to be 916 MW(e). Although the turbine has an additional calculated gross capacity of about 10%., the applicant has stated that there are no plans for power stretch in this plant.

The Indian Point 2 facility is the largest reactor that has been considered for licensing to date. Furthermore, it will be located in a region of relatively high population density. For these reasons, particular attention has been given to improving and supplementing the protective features previously provided in other plants of this type.

The proposed design has a reinforced concrete containment with an internal steel liner which is provided with facilities for pressurization of weld areas to reduce the possibility of leakage in these areas. The containment design also includes an internal recirculation containment spray system and an air recirculation system consisting of five air handling units to provide long-term cooling of the containment without having to pump radioactive liquids outside the containment in the event of an accident. Even though the applicant anticipates negligible leakage from the containment, two independent means of iodine removal within the containment have been provided. These are an air filtration system using activated charcoal filters, and a containment spray system which uses sodium thiosulfate in the spray water as a reagent to aid removal of elemental iodine.

The reactor vessel and various other components of the system are surrounded by concrete shielding which provides protection to the containment against missiles that might be generated if structural failure of such components were to occur during operation at pressure. This includes missile protection against the highly unlikely failure of the reactor vessel by longitudinal splitting or by various modes of circumferential cracking. The Committee favors such protection for large reactors in regions of relatively high population density.

The Indian Point 2 plant is provided with two safety injection systems for flooding the core with borated water in the event of a pipe rupture in the primary system. The emergency core cooling systems are of particular importance, and the ACRS believes that an increase in the flow capacity of these systems is needed; improvements of other characteristics such as pump discharge pressure may be appropriate. The forces imposed on various structural members within the pressure vessel during blowdown in a loss-of-coolant accident should be reviewed to assure adequate design conservatism. The Committee believes that these matters can be resolved during construction of these facilities. However, it believes that the AEC Regulatory Staff and the Committee should review the final design of the emergency core cooling systems and the pertinent structural members within the pressure vessel, prior to irrevocable commitments relative to construction of these items.

The applicant stated that, even if a significant fraction of the core were to melt during a loss-of-coolant accident, the melted portion would not penetrate the bottom of the reactor pressure vessel owing to contact of the vessel with water in the sump beneath it.

The applicant also proposes to install a backup to the emergency core cooling systems, in the form of a water-cooled refractory-lined stainless steel tank beneath the reactor pressure vessel. The Committee would like to be advised of design details and their theoretical and experimental bases when the design is completed.

In order to reduce still further the low probability of primary system rupture, the applicant should take the additional measures noted below. The Committee would like to review the results of studies made by the applicant in this connection, and consequent proposals, as soon as these are available.

1. Design and fabrication techniques for the entire primary system should be reviewed thoroughly to assure adequate conservatism throughout and to make full

use of practical, existing inspection techniques which can provide still greater assurance of highest quality.

2. Great attention should be placed in design on in-service inspection possibilities and the detection of incipient trouble in the entire primary system during reactor operation. Methods of leak detection should be employed which provide a maximum of protection against serious incidents.

Attention should also be given to quality control aspects, as well as stress analysis evaluation, of the containment and its liner. The Committee recommends that these items be resolved between the AEC Regulatory Staff and the applicant as adequate information is developed.

The applicant has made studies of reactivity excursions resulting from the improbable event that structural failure leads to expulsion of a control rod from the core. Such transients should be limited by design and operation so that they cannot result in gross primary-system rupture or disruption of the core, which could impair the effectiveness of emergency core cooling. The reactivity transient problem is complicated by the existence of sizeable positive reactivity effects associated with voiding the borated coolant water, particularly early in core life. In addition, the course of the transients is sensitive to various parameters, some of which remain to be fixed during the final design. Westinghouse representatives reported that the magnitude of such reactivity transients could be reduced by installation of solid burnable poisons in the core to permit reduction of the soluble boron content of the moderator, thereby reducing the positive moderator coefficient. The Committee agrees with the applicant's plans to be prepared to install the burnable poison if necessary. The Committee wishes to review the question of reactivity transients as soon as the core design is set.

The Advisory Committee on Reactor Safeguards believes that the various items mentioned can be resolved during construction and that the proposed reactor can be constructed at the Indian Point site with reasonable assurance that it can be operated without undue risk to the health and safety of the public.

Chapter 8: Appendix C:
Draft of an ACRS Report to the AEC on
Problems arising from Primary System Rupture

In its continuing review of the safety of large nuclear power reactors, the Advisory Committee on Reactor Safeguards has paid considerable attention in recent months to the general problem of a primary system pipe rupture (or other cause of significant primary coolant leakage) followed by a functional failure of the emergency core cooling system. The probability of either of these types of failure occurring is considered to be very low, and the chance that both types of failure might occur within the same event is even more remote. Nevertheless, as more and more reactors come into existence, particularly reactors of larger size and higher power density, the consequences of failure of emergency core cooling systems take on increased importance. Accordingly, the Committee believes that action should be taken along the lines indicated below.

1. Additional precautions should be taken to reduce still further the low probabilities of failure of the primary coolant system and the emergency core cooling systems.

    a.   In the primary coolant system, careful and systematic analysis of all elements of the system should be made from the standpoint of possible contribution to leakage or rupture. All elements, active and passive, should be designed, fabricated, installed and tested with such care as to assure an operational system not marginal in any respect related to reliability. Deliberate surveys of new technological developments which might enable greater assurance of quality in fabrication and installation should be conducted, and use of such new or additional techniques as are appropriate should be instituted promptly. As one example, for pressure vessel welds on which ultrasonic inspection techniques can be used effectively to supplement radiographic inspection, both should be used.

        Current practice for assessing probable operability and reliability of the primary system in service should be reviewed, and design features incorporated to improve inspectability and testability. Efforts should be increased to develop improved means of leak detection and to assure appropriate operator response to signs of leaks. Operating procedures in the circumstance of incipient difficulty should be especially conservative.

    b.   Added conservatism should be used in evaluating possible effects of a primary system rupture with respect to rendering the core unamenable to emergency core cooling. Such effects should be recognized early during the design stage and be adequately protected against. For example, all vital structures within the reactor pressure vessel must be conservatively analyzed and designed to withstand the maximum forces resulting from blowdown associated with a major rupture.

    c.   For emergency core cooling, a very conservative design basis and high system quality are needed. Two or more independent systems should be provided. Preferably these should employ two different design approaches to provide maximum assurance that this engineered safeguard will function adequately under all conditions. Redundancy of all vital active components is required in each system, as are abundant capacity and adequate speed of response. Again, all elements should be designed, fabricated, installed, and inspected and tested with especial care, utilizing advanced techniques. The maximum practicable inspectability and testability during reactor operation should be provided. Experimental confirmation of the efficacy of this important safeguard under accident conditions should be pursued vigorously, in detail.

    d.   Appropriate design measures should be taken so that no possible mechanism exists by which a primary system rupture, through the effects of blowdown or by other means, might render the emergency core cooling system incapable of performing its function.

2. Design of water-cooled power reactors should develop in such a manner as to provide in the near future even greater certainty of adequate protection in the event of primary system rupture.

In the remote circumstance that failure of both the primary system and all emergency core cooling systems of a reactor were to occur, large portions of the reactor core could be expected to melt and the assurance of containment integrity would thereby be made substantially more difficult.

The ACRS believes that, with proper attention to the considerations of 1., above, a well-engineered set of emergency core cooling systems using the concepts employed in current applications for construction permits can exhibit very high probability of coping effectively with a primary system pipe rupture. Nevertheless, because experience with emergency cooling systems is limited, and because systems using current concepts necessarily are subject to certain low-probability modes of failure related to primary system rupture, the Committee believes it prudent to provide still greater protection of the public by some independent means, particularly for reactor sites nearer to population centers. Progress toward this objective will require an evolutionary process of design and a vigorous program of research both of which should begin immediately and be aimed at reaching a high state of development in approximately two years. Future reactors relying solely on currently employed types of emergency core cooling systems to cope with the unlikely accident involving primary system rupture will be considered suitable only for rural or remote sites.

As discussed in the Committee's report to you dated November 24, 1965, the problem of providing adequate emergency core cooling is considerably complicated in the unlikely event of reactor pressure vessel rupture, and the Committee suggests that attention be given to this aspect of the core cooling problem.

The Committee believes that, in view of the large number of reactors to be built in the future, it is prudent to undertake the above measures to assure continued protection to the public, and recommends that they be implemented as rapidly as possible.

9

Reactor Siting: 1966–1968

Had the pressure vessel issue not arisen in 1965, it is likely that the Brookwood and Millstone Point 1 construction permit reviews would have emphasized 10 CFR Part 100 in the manner previously developed in 1963–1964, namely, the substitution of engineered safeguards for distance. However, the ACRS letter of November 24, 1965, on pressure vessels, which recommended further measures to reduce failure probability and identified a potential need for measures to cope with vessel failure, changed this, and the question of how to implement the letter became a major part of the Brookwood and Millstone Point 1 reviews.

A difference in philosophic approach between the regulatory staff and the ACRS quickly emerged during the sixty-ninth ACRS meeting, January 6–8, 1966, when Harold Price, the Director of Regulation, said that if any applicant were forced to protect against pressure vessel failure, all other commercial reactors would have to comply, regardless of location. In contrast, ACRS thinking clearly differentiated among sites with regard to the possible requirements for such protection or other additional safety features (besides those dictated by the ritual of meeting Part 100). Then, the emergence of the China Syndrome problem and its resolution as adopted for the Dresden 3 and Indian Point 2 reactors produced a revolutionary change in regulatory review practices, with an ever-increasing stress on measures to prevent core melt. It also made much less likely a favorable

134

recommendation for a large LWR at a site substantially more populated than Indian Point. And, within the ACRS, it built up an increasing emphasis on still greater measures to prevent core melt for reactors at "borderline" sites like Indian Point, although the Indian Point 2 reactor itself had been accepted. The Indian Point 2 site was never a strong candidate for rejection based on considerations of site characteristics, despite its relatively large surrounding population density and the fact it was about 25 miles from New York City. Probably, the existence of the smaller Indian Point 1 reactor set so strong a precedent that the only matter receiving serious consideration was what safety features were needed to make Indian Point 2 acceptable.

From bits and pieces of discussion in the minutes of various ACRS meetings, one can deduce that during the 1960s the likelihood of a serious accident was estimated to be in the neighborhood of 10^{-5} per reactor-year. ACRS member Etherington, back around 1965 or 1966, estimated the possibility of pressure vessel failure to be in the vicinity of 10^{-6} per vessel-year, an estimate which was remarkably close to that published after considerable study some eight years later. In any event, the thinking was that the probability of a serious accident was likely to be small, especially when one included weather and population distribution into an estimation of probability.

Looking back at the situation some dozen years later, it is not obvious why more time was not taken in trying to fully develop a new regulatory approach in 1966. Nevertheless, an approach was developed during the summer of 1966 which, in essence, created a major change in the engineered safety requirements for light water reactors and really set light water reactor safety on a new path. It had become important to make the probability of core melt much lower than it was, whatever it had been. The first two major steps, which were taken in connection with the Indian Point 2 and Dresden 3 reactors, were: (1) to require much improved quality in the primary system, more inspection, and much more leak detection in order to reduce the probability of a loss-of-coolant accident: and (2) to require a much improved emergency core-cooling system in order to reduce the probability that a LOCA would lead to core melting. This was the beginning of a continuing series of efforts that looked in ever-expanding directions for possible causes of intiating events that could lead to core melt, and sought measures to reduce the probability of such events. Pressure vessel failure had become one of many possible sources of containment failure, and it had to take its place with other possible sources of large radioactivity release.

The three major LWR site reviews in the two years following that of Indian Point 2 were Browns Ferry, Burlington, and Zion. Although

Browns Ferry was a relatively remote site in Alabama, the Tennessee Valley Authority (TVA) was proposing to construct two boiling water reactors at Browns Ferry, which had a power level and peak power density (power output per unit volume) much larger than any BWRs previously approved for construction. Previous experience told one that construction permit applications would soon be submitted by other utilities for the same reactors at much more populated sites.

Burlington lay in New Jersey between Trenton and Philadelphia and was more heavily populated than the Indian Point site. Zion was near Waukegan, Illinois, not too far from Chicago, and was a site that had a similar average surrounding population to the Indian Point site, although this differed in detail.

The reviews of the Burlington and Zion sites were complicated by the proposed Bolsa Island reactors that were to be built at a highly populated site off Huntington Beach in southern California. Bolsa Island was to be the site of a proposed dual-purpose, nuclear power-desalination plant, and it represented a project that had strong support from the reactor development side of the AEC, as well as from some of the commissioners themselves. It was to have heavy federal funding support. The site was selected with the knowledge of the AEC commissioners; hence, when the Bolsa Island site came up for review, the Director of Regulation, Price, may have had much less freedom of decision than he would have had for a strictly commercial nuclear power plant at an equivalently poor site. Since Bolsa Island had a higher surrounding population density than the Burlington site, after the first two miles of water, if Bolsa Island were approved, it would be difficult to reject Burlington.

Browns Ferry Review

The ACRS held its first review of the proposed Browns Ferry BWRs at its eighty-first meeting, January 12–14, 1967. At 3,300 MWt these reactors had a power level about 45% higher than the Quad Cities (Illinois) or Dresden 3 reactors, and about 5 times larger than any operating BWR, thereby continuing the very rapid trend toward still larger reactors.

The ACRS review of the Browns Ferry site was difficult and controversial, and occupied a major portion of four full committee meetings in addition to numerous subcommittee meetings. At its eighty-third meeting, March 9–11, 1967, the ACRS finally arrived at a decision and wrote a letter report, signed by the ACRS Chairman, N. J. Palladino, to AEC Chairman Glenn T. Seaborg, which included may committee reservations,* as well as

---

*The Browns Ferry report represented the birth of the so-called "asterisked items," later to become the ACRS generic items.

dissenting remarks by ACRS member Hanauer. This Browns Ferry letter is reproduced in the appendix to this chapter.

During the intense discussion which led to the formulation of a committee position, ACRS member Hanauer told the committee that his feelings about the proposal were largely related to his experience with the ACRS within the past year. First, on the Dresden 3 and Indian Point 2 reactors, the committee had written letters approving construction and had intended to write a general letter on core cooling. Instead the task force was convened. Hanauer found the committee was already treating those reactors as proven types and others of the general class were getting approval with little difficulty. Hanauer objected that he found himself "stuck" with such systems regardless of the prudence of the initial approvals.

Secondly, Hanauer said that the first of the bigger power reactors were being reviewed for operating licenses, and for none of these cases were the problems which were of concern during the construction permit review even close to being solved. He had therefore concluded that it was no longer appropriate to assume that serious problems could be resolved between the construction and operating license reviews.

In view of the continued lack of resolution of problems identified on reactors like Dresden 3, and in view of the departure represented by the Browns Ferry proposal, and in view of the apparent reluctance of the ACRS to put teeth into its construction permit recommendations at the operating license stage, Hanauer did not feel that he could approve the Browns Ferry proposal on the basis of existing technology. At least two other ACRS members besides Hanauer had serious reservations about the evolving committee position.

ACRS member Zabel felt that there had been a gradual escalation in reactor systems and that there was some point beyond which it would not be prudent to go. Also, there were many questions about the Browns Ferry proposal, for which there would not be answers for a long time, even assuming an extensive research program. He could foresee other systems of the same type being proposed at much worse sites. Zabel felt that at times in the past the committee had been willing to accept some uncertainty on the basis that enough of a safety margin had been provided, but he had the uncomfortable feeling that such was not the case for this particular machine. In addition, attempts to alleviate the situation at the Browns Ferry site had only made it worse in Zabel's opinion, by forcing the use of unproven systems such as the large diesel-generators. Simply stated, Zabel felt that this was the place at which he would like to call a halt. However, he did not append written remarks to the letter.

ACRS member Bush submitted the following paragraph to the committee and said that it might form the basis of a dissent on his part to the letter.

Dr. S. H. Bush expresses a general concern that the acceptance of a double-ended pipe break, common to all water reactors, may lead to inherently less safe Emergency Core Cooling Systems. Such a pipe break is an admittedly incredible event in large austenitic steel piping when the normal failure is by limited circumferential or longitudinal cracking. The acceptance of the double-ended pipe break determines the time of blowdown and the sizing of pumps, valves and, more significantly, the diesel-driven emergency power generators. This break leads to severe requirements for short startup times of very large diesel-generators, representing a yet-undeveloped technology. A more realistic failure model, based on smaller break sizes, should permit a more rational sizing of ECCS equipment.

The committee decided to study the matter of the double-ended pipe break on a generic basis, and member Bush did not attach the remarks to the Browns Ferry letter.

Many other members who had strong reservations about recommending approval of the Browns Ferry proposal decided they could support the ACRS letter in view of the technical issues which the letter raised, coupled with a penultimate paragraph which explicitly noted the possibility that restrictions in operations, e.g., a reduction in power level, might be required.

Site-Dependent Design Criteria

In the months following the Indian Point 2 review, and prior to, during, and following the Browns Ferry review, the ACRS had been devoting considerable effort to reactor siting considerations. At its seventy-seventh meeting in September 1966, the ACRS requested one of its subcommittees to prepare a draft of possible design requirement criteria for use in future construction permit reviews, and a first draft was discussed at the seventy-eighth meeting, October 6–8, 1966. As member J. N. Hendrie put it in the discussion on design criteria, the ACRS was fundamentally enlarging the scope of credible accidents. This enlargement appeared during the Browns Ferry review.

A special ACRS meeting was held on December 2–3, 1966, particularly to provide time for discussion of reactor siting and reactor design criteria. At this meeting the committee adoped the following motion:

That it be recognized that there are differences in reactor sites which justify differences in safeguards provisions. In view of this, for the Committees' use in delineating reactor sites, three types should be recognized:

City—roughly characterized by Edgar and Ravenswood; Rural—roughly characterized by Dresden and San Onofre; Intermediate—roughly characterized by Indian Point and Millstone.

The committee decided to try to develop design requirements as a function of site-type for several safety concerns, including emergency power, tornado and hurricane protection, earthquake protection, decay heat removal, reactor scram (a fast insertion of all safety rods), turbine orientation, primary system integrity, instrumentation, emergency core-cooling systems, and containment requirements. Although the committee never issued a report in which it recommended such differences in design and engineered safeguards as a function of site types, the act of working on such possible requirements generated ideas as to different levels of safety, and led to a deeper examination of the adequacy of measures which had previously been accepted.

The ACRS also gave increased attention to the question of emergency planning, and the minutes of the eighty-third meeting, March 9–11, 1967, record the following discussion with the AEC commissioners.

ACRS Chairman Palladino observed that, at present, emergency plans were usually geared to the individual needs of existing facilities and were not adequate for wide application. Several factors had brought this to the Committee's attention, including the recent trend of power reactors toward population centers, the tendency toward multi-reactor siting, the recent large increases in design power, the increase in the handling of radioactive material which would result from the growth of the nuclear industry, and actions with local officials because of the possibility of public relations problems. Chairman Palladino concluded by saying that the ACRS was not suggesting that off-site drills be held, but rather that the situation be studied; the study perhaps going beyond evacuation capabilities to include the need for development of a wide-range stack monitor, cloud tracing techniques, training programs for local fire and police departments and the development of widespread hospital capability for treating irradiated patients. The Committee's feeling was that in some sense evacuation is being depended upon and the AEC should be made aware that a study might be indicated.

A Proposed Moratorium on Metropolitan Siting

At its eighty-fifth meeting, May 11–13, 1967, the ACRS changed its position drastically from that taken in March 1965 when it had opposed a moratorium on metropolitan siting; in May 1967 the ACRS adopted the position that a limit be placed on acceptable population density which, in effect, closely enveloped the existing approved sites for LWRs and that this be recommended to the AEC for use for the next five years. This would place a moratorium on sites worse than Indian Point, and was in contrast to the ACRS opinion against such a moratorium in the spring of 1965. This was not only an ACRS having a changed membership; the China Syndrome had removed containment as an independent bulwark.

The ACRS advised Mr. Price it was considering such a limit in the following mathematical form:

$$P_{TOT}(R) \leq 4000(R^2) \text{ using current population}$$
$$P_{TOT}(R) \leq 5000(R^2) \text{ using population pro-}$$
jected 25 years hence

where $5 \leq R \leq 25$ in miles and where $P_{TOT}(R)$ is the total population within a distance R miles from the reactor.

The regulatory staff appeared to be favorable toward proposing some such interim criterion to be applicable for about five years, by which time some operating experience should have been available from the larger reactors under construction. Also, this would have provided time for safety research and the development of improved safety systems.

However, prior to the June 1967 ACRS meeting, AEC Chairman Seaborg discussed this proposal with ACRS Chairman Palladino. Chairman Seaborg expressed concern over the use of a simple formula to establish a criterion for siting of reactors for several reasons, one of which was apparently its effect on the proposed Metropolitan Water District reactor on Bolsa Island. Dr. Seaborg indicated that any criterion must consider situations in which a site has particularly favorable meteorology or a relatively large, unpopulated area close to the site, together with a large population center further out (like Bolsa Island). In short, Dr. Seaborg encouraged the committee to proceed slowly.

The ACRS discussed the matter at considerable length at the June 1967 meeting without arriving at any agreement. It met with Mr. Price, who reported that he had met with the commissioners and had gotten the idea that they were generally unhappy with the idea of a fixed limit and with an arbitrary formula. They would prefer a more flexible approach and were influenced by several considerations. First, of course, the Metropolitan Water District site at Bolsa Island was beyond any of the threshold limits considered. There had already been a great deal of publicity given to this project in obtaining congressional approval. In addition, the proposed criterion did not permit a site with a very low population density within the first few miles to have a larger population density beyond five miles (as Bolsa Island needed). Mr. Price felt that, based on these meetings, it would be difficult to convince the commissioners to promulgate such a criterion any time soon.

The ACRS and the regulatory staff concluded that they would drop the proposal and continue to proceed as before, namely, on a case-by-case basis. During the discussion one point made was that recent testimony before the Joint Committee on Atomic Energy (JCAE) had, in effect, suggested a moratorium on metropolitan siting. In the ACRS testimony, which was presented by Palladino and Okrent to the JCAE in April 1967, the committee stated:

The ACRS believes that placing large nuclear reactors close to population centers will require considerable further improvements in safety, and that none of the large power reactors now under construction is considered suitable for location in metropolitan areas. The Committee believes that, in addition to favorable experience with reactor construction and with operation of reactor systems, components and safeguards in these reactors now under construction, further improvements in design are required to make accidents, large and small, still more unlikely; and the consequences limiting safeguards must be made more fool-proof, and provide protection from the consequences of accidents of still lower probability.

Testimony by the AEC at the JCAE hearings also indicated that the time was not ripe for metropolitan siting. But no quantitative definition was given of a metropolitan site, and the commissioners had avoided, at least temporarily, a siting problem for Bolsa Island (which had not yet received any review by the ACRS).

Burlington Review

The application for a construction permit for the proposed Burlington Nuclear Generating Station Unit 1 was submitted to the AEC on December 13, 1966, by the Public Service Electric and Gas Company of New Jersey. The proposed site was located on the east bank of the Delaware River estuary, adjacent to the city of Burlington, New Jersey, approximately 17 miles northeast of downtown Philadelphia and 11 miles southwest of Trenton, New Jersey. The population surrounding the Burlington site represented a significant increase compared to that around previously licensed sites for power reactors. It was proposed to build a 3,083 MWt PWR, which represented an increase of 12% over the Indian Point 2 reactor. The containment system and engineered safety systems were basically those of the Indian Point 2 reactor. The surrounding population distribution for several reactor sites including Burlington is given in table 2.

The application for the Burlington reactor site, coupled with the knowledge that plans were still under way for proposing a reactor for the heavily populated Bolsa Island site, made it clear that the pressure was still on from the industry and from the developmental side of the Atomic Energy Commission to move reactors into more-populated areas.

TABLE 2
Population surrounding four reactor sites

	Number of People Around Site			
Distance in Miles	Burlington	Indian Point 2	Turkey Point	Oyster Creek
0–1	4,700	1,080	0	200
0–2	18,600	10,800	0	1,600
0–5	119,400	53,000	0	4,600
0–10	536,200	155,500	42,000	32,800
0–20	3,904,000		232,000	136,000
0–25	4,502,000	1,393,000		

In July 1967, the ACRS actively began reviewing the proposed Burlington site. Excerpts from the minutes of the eighty-seventh meeting, July 6–8, 1967, give some insight into the course of events and the thinking at that time.

ACRS member Monson noted that the question being put to the ACRS was actually one of the acceptability of the site on the basis of population distribution, alone. The applicant had stated that due to the cost of transmission, there are no other suitable sites. In addition, air pollution problems have ruled out the construction of a fossil-fuel plant at the Burlington site.

Monson then quoted from a series of newspaper articles and public statements by Commissioner Ramey, Chairman Seaborg, etc., which indicated that the Bolsa Island project was very highly favored. In Monson's view, the Bolsa Island site, which essentially duplicated Burlington with respect to total population density, was even worse than Burlington because of the concentration of people to the landward side and the continuous, daytime on-shore winds. Since the wind direction frequency was more randomly distributed with respect to the people at the Burlington site, it had a lower effective population density.

Monson also recalled for the Committee many recent statements by the Director of Regulation and the AEC Chairman to the effect that AEC policy is to not site power reactors in metropolitan areas and not to allow them to encroach on such areas until there have been significant offsetting improvements in reactor technology. Monson thought that the inconsistency of this statement with the apparent predetermined approval of the Bolsa Island site made it difficult to reject the Burlington site on a population basis. He suggested that this matter be discussed with the Commissioners.

ACRS member Isbin felt that there were significant differences between the two projects. The Burlington proposal included conventional reactors under private ownership. Since the Bolsa Island facility would be heavily supported by the AEC, it might be a good starting point for establishing requirements which would allow the use of high population density sites.

Dr. Okrent thought that the ACRS should formulate its opinions on the Burlington project before having any discussion with the Commission. Since the Bolsa

Island design had not been completed there might be some willingness to go quite far on other features to allow the use of the site. Dr. Okrent did not agree with the conclusion that since a decision by the AEC had already been made on Bolsa Island, if indeed it had, that the Committee must accept it and therefore accept the equally undesirable Burlington site.

The ACRS decided to finalize its position on the Burlington site without discussing the question with the commission.

The governor of New Jersey had written a letter dated July 3, 1967, to Chairman Seaborg of the AEC, expressing concern about the prolonged schedule for the regulatory review of the Burlington site. The governor in his letter pointed out the need for power in his state from this facility and that this had a bearing on the public health and safety. He endorsed construction of nuclear plants as opposed to fossil-fuel burning plants as a step towards the solution of the problem of air pollution in the area.

The ACRS subcommittee on Burlington visited the site and held a meeting on August 9, 1967. Public Service Electric and Gas Company of New Jersey stated that they had no site better than Burlington, and that no sites with, say, half the population density of Burlington existed except in the pine woods of central New Jersey where no cooling water was available. Public Service also stated they were proposing no special design or operating features to compensate for the high population density.

The regulatory staff had submitted a report to the ACRS in late July, which estimated the off-site doses in terms of the classical Part 100 recipe, namely, assuming 100% of the noble gases and 50% of the iodine released to an intact containment vessel, with various assumptions on iodine removal within the containment, etc. The report said that unless significant removal of radioiodine was achieved in the containment vessel, doses could exceed Part 100 guidelines within a matter of a few hours for several thousand people. The report did not discuss accidents involving a loss of containment integrity and contained no conclusions concerning the acceptability of the site.

However, at the eighty-eighth meeting, August 10–12, 1967, Mr. Price told the ACRS that the regulatory staff thought Burlington was a poor site. The staff had told this to Public Service in 1965, but in December of 1966, without further discussion, the company publicly announced its intent to construct a reactor on the Burlington site and followed this announcement with a formal application.

Mr. Price noted that he had had a telephone call from AEC Commissioner James T. Ramey, who was concerned as to how the application had found itself in this condition and why the situation had not been handled in the same way as the Boston Edison application for the highly populated Edgar Station site. Mr. Price had informed the commissioner that the situation had arisen because the company elected not to follow the approach taken by Boston Edison.

Dr. Beck of the staff stated that the Public Service proposal represented a long step toward higher population densities without substantial operating experience in any of the larger power reactors. Dr. Beck concluded that this site should not be approved. Mr. Price also noted that with respect to the argument that no appropriate sites were available within the Public Service Company service area, there was nothing to prevent them from choosing a site outside their service area.

ACRS member Isbin asked, with respect to the lack of operating experience, if negative results were to come in during the next four years on some of the items identified as potential problems in recent committee letters, were not the companies who had engaged in the projects taking the risk. Dr. Peter A. Morris of the staff replied that they would be taking the risk but predicted that considerable pressure would be put on the AEC to approve operation of the plants in any case. Mr. Price added that after 15, 20, or 30 such large power reactors came into operation the country would be heavily dependent on their power production capability.

ACRS member Manglesdorf asked if the Burlington case bore any relation to the Bolsa Island project. Mr. Price replied that at least the Bolsa Island site was a good one for approximately 1½ miles from the reactor, while the Burlington site was bad after 500 feet.

After discussion of the case with the applicant, the ACRS* met in executive session and adopted the following position:

> The Committee believes that it has now received essentially all of the information necessary to evaluate the Burlington site and has given careful consideration to this information. It is the unanimous opinion of the Committee that it does not see how the site can be approved.

The committee met with the applicant and advised him orally of its position. Mr. Price noted that a public response from the ACRS and the regulatory staff could be avoided.† The spokesman for Public Service replied that he preferred that there would be no ACRS letter and noted that the company would have to reconsider its position.

---

*The ACRS members at the eighty-eighth meeting were the following: N. J. Palladino, chairman; S. H. Bush; H. Etherington; W. L. Faith; F. A. Gifford; S. H. Hanauer; J. N. Hendrie; H. S. Isbin; H. G. Mangelsdorf; H. O. Monson; A. A. O'Kelly; D. Okrent; W. R. Stratton; and C. W. Zabel.

†Prior to the Federal Advisory Committee Act and the Government in the Sunshine Act, ACRS meetings were not open to the public.

The ACRS recorded its position, as stated to Public Service of New Jersey, in its monthly summary letter to the chairman of the AEC. At that time, however, the summary letters were not routinely made public by the AEC. Nevertheless, the fact that the Burlington site had been rejected became generally known.

Public Service withdrew its application for the Burlington site. The Bolsa Island site, which was similar to that at Burlington regarding population distribution, except for the first 1½ to 2 miles, still remained for consideration soon. It is interesting to note that in testimony both by AEC Commissioner Ramey, and also by Mr. Price, to the JCAE on April 4, 1967, no mention was made of Bolsa Island as a site which presented problems arising from the large surrounding population density, although the seismic design aspects of the proposed project were discussed. The one exception arose in a comment by Congressman Hosmer of California.

Back in my mind I have the question of the Bolsa Island reactors, two reactors, which will be in the same general location (the Los Angeles basin). I would believe that every effort would be made to make these compatible with their intended locality.

To which Mr. Price replied, "That is right."

At the same hearings, Congressman Hosmer asked, "Do you think you could really justify building reactors with lower safety standards in remote areas than would be required in metropolitan areas anyway?" Dr. Beck answered,

We are not, in fact, suggesting that reactors be built to any different standards at one place or another. We are building reactors to the best standards we know, at any location. We say that there are still some residual uncertainties in reactors.

The actual experience with reactors in general is still quite limited and with large reactors of the type now being considered, it is non-existent. Therefore, because there would be a large number of people close by and because of lack of experience, it is not a matter of difference of standards; it is a matter of judgment and prudence at present to locate reactors where the protection of distance will be present.

Industry Pressure for Metropolitan Siting

During part 2 of the 1967 hearings by the JCAE on Licensing and Regulation of Nuclear Reactors (September 12–14, 1967), statements supporting metropolitan siting of reactors were made by several representatives of the nuclear industry. Mr. Jack Horton, speaking on behalf of the Edison Electric Institute testified:

The siting of nuclear power plants in metropolitan areas is important to the electric utility industry and the public which it serves. Utilities serving metropolitan areas must have their sources of generation close to the load they serve if they are to continue giving reliable service. Therefore, siting of nuclear power plants in metropolitan areas must be a key factor in the design of our future electric power systems. We believe the AEC is moving toward this goal.

A.E. Schubert of General Electric said:

We agree that requirements for metropolitan siting need to be defined. To this end, we believe it would be helpful for the AEC to establish a government-industry task force, which would include appropriate ACRS and Regulatory Staff membership, to make recommendations for metropolitan siting criteria.

J. C. Rengel of Westinghouse testified:

We believe that a change is called for in AEC policy on siting of nuclear power reactors in or near metropolitan areas. We believe that utility groups can make an economic case for such locations and plants can be designed which can be constructed and operated safely in metropolitan areas.

Thus the nuclear industry continued to press for metropolitan siting. At the ninetieth ACRS meeting, October 5–7, 1967, Mr. Price informed the committee of the next try by the industry, namely, the proposal by the Jersey Central Power and Light Company for the location of a reactor at its Union Beach site. Jersey Central was so much in favor of this site that they had suggested approaches such as double containment, etc. They had also requested some informal reaction from the regulatory staff. Mr. Price observed that the Union Beach site seemed better than Burlington for the first few miles out, but then it became much worse. Later during the ninetieth meeting the ACRS voted to have its chairman inform Mr. Price that "if the proposed Union Beach facility is similar in design to those now being reviewed, the Committee would have great difficulty reaching a favorable conclusion."

The year following the eighty-eighth ACRS meeting in August 1967 at which the Burlington site was turned down, involved a variety of complex matters relating to the siting of reactors in populated areas. The ACRS attempted to provide guidance on what might make sites that were more populated than Indian Point acceptable for reactors. The probable need to review the Bolsa Island site was always present. Construction of two large PWRs at the Zion site north of Chicago was proposed. Consolidated Edison proposed three sites near Indian Point, which were somewhat more populated, as possible locations for large boiling water reactors.

Further down the road Public Service of New Jersey would propose the Newbold Island site as its alternate to Burlington.

Chapter 9: Appendix A:
ACRS Report to the AEC on Browns Ferry Nuclear Power Station

At its eighty-third meeting, March 9–11, 1967, the Advisory Committee on Reactor Safeguards completed its review of the application of the Tennessee Valley Authority for authorization to construct Browns Ferry Nuclear Power Station Units No. 1 and No. 2. This project was previously considered at the eighty-first and eighty-second meetings of the Committee, January 12–14, 1967 and February 9–11, 1967, respectively, at a special meeting on February 28, 1967, and at subcommittee meetings on November 25, 1966, January 4–5, and January 28, 1967. Representatives of the Committee visited the site on February 27, 1967. During its review, the Committee had the benefit of discussions with representatives of the Tennessee Valley Authority, General Electric Company, and the AEC Regulatory Staff. The Committee also had the benefit of the documents listed.

The Browns Ferry Units are to be located in Limestone County, Alabama, on the shore of Wheeler Lake approximately 30 miles west of Huntsville. Each Unit includes a boiling water reactor to be operated at a maximum power level of 3293 MWt, the highest power level for any reactor reviewed for a construction permit to date. The average core power density is about 40 percent higher than for the previously reviewed Quad-Cities boiling water reactors. The increase is achieved by flattening the power density distribution and employing an approximately 20 percent higher fuel element maximum linear heat rate. The margins between thermal operating limits and fuel element damage limits are thereby reduced. In relation to margin on critical heat flux, the applicant uses new heat transfer correlations developed from recent experimental data.

The complex of emergency core cooling systems for Browns Ferry is similar to that proposed for the Quad-Cities reactors. Each reactor is provided with a high pressure coolant injection system; a low pressure coolant injection, or flooding, system; and two core spray systems. Because of the higher core power density and power level, substantial increases have been made in the flooding system and core spray system capacities. The Committee feels that the emergency core cooling systems proposed have a high probability of preventing core meltdown in the unlikely event of a loss-of-coolant accident. It notes, however, that although calculated peak fuel temperatures in such an accident are similar to those for the Quad-Cities reactors, the calculated number of fuel elements reaching undesirably high temperatures is greater. Also, the time margin available for actuation of the systems is less. Because of these factors and the importance of the effective functioning of emergency core cooling systems, the Committee believes the adequacy of these systems should be further corroborated by the following two measures:

1. Analysis indicates that a large fraction of the reactor fuel elements may be expected to fail in certain loss-of-coolant accidents. The applicant states that the principal mode of failure is expected to be by localized perforation of the clad, and that damage within the fuel assembly of such nature or extent as to interfere with heat removal sufficiently to cause clad melting would not occur. The Committee believes that additional evidence, both analytical and experimental, is needed and

should be obtained to demonstrate that this model is adequately conservative for the power density and fuel burnup proposed.*

2. In a loss-of-coolant accident, the core spray systems are required to function effectively under circumstances in which some areas of fuel clad have attained temperatures considerably higher than the maximum at which such sprays have been tested experimentally to date. The Committee understands that the applicant is conducting additional experiments, and urges that these be extended to temperatures as high as practicable. Use of stainless steel in these tests for simulation of the Zircaloy clad appears suitable, but some corroborating tests employing Zircaloy should be included.

The applicant stated that the control systems for emergency power will be designed and tested in accordance with standards for reactor protection systems. Also, he will explore further possibilities for improvement, particularly by diversification, of the instrumentation that initiates emergency core cooling, to provide additional assurance against delay of this vital function.

Steam line isolation valves are provided which constitute an important safeguard in the event of failure of a steam line external to the containment. One or more valves identical to these will be tested under simulated accident conditions prior to a request for an operating license.

Operation with a fuel assembly having an improper angular orientation could result in local thermal conditions that exceed by a substantial margin the design thermal operating limits. The applicant stated that he is continuing to investigate more positive means for precluding possible misorientation of fuel assemblies.

The applicant considers the possibility of melting and subsequent disintegration of a portion of a fuel assembly by inlet coolant orifice blockage or by other means, to be remote. However, the resulting effects in terms of fission product release, local high pressure production, and possible initiation of failure in adjacent fuel elements are not well known. Information should be developed to show that such an accident will not lead to unacceptable conditions.*

A linear heat generation rate of 28 KW/ft is used by the applicant as a fuel element damage limit. Experimental verification of this criterion is incomplete, and the applicant plans to conduct additional tests. The Committee recommends that such tests include heat generation rates in excess of those calculated for the worst anticipated transient and fuel burnups comparable to the maximum expected in the reactor.*

The Rod Block Monitor system should be designed so that if bypassing is employed for purposes other than brief testing no single failure will impair the safety funtion.

The diesel-generator sets for emergency power appear to be fully loaded with little or no margin (on the design basis of one of three failing to start). They are required to start, synchronize, and carry load within less than thirty seconds. The applicant stated that tests will be conducted by the diesel manufacturer to demon-

*The Committee believes that these matters are of significance for all large water-cooled power reactors, and warrant careful attention.

strate capability of meeting these requirements. Any previously untried features, such as the method of synchronization, will be included in the tests. The results should be evaluated carefully by the AEC Regulatory Staff. In addition, the installed emergency generating system should be tested thoroughly under simulated emergency conditions prior to a request for an operating license.

The Committee continues to emphasize the importance of quality assurance in fabrication of the primary system and of inspection during service life. Because of the higher power level and advanced thermal conditions in the Browns Ferry Units, these matters assume even greater importance. The Committee recommends that the applicant implement those improvements in primary system quality which are practical with current technology.*

The Browns Ferry Units have been designed to provide the same accessability for inspection of the primary system as for the Quad-Cities plants. A detailed inspection program has not yet been formulated by the applicant. The Committee will wish to review the detailed in-service inspection program at the time of request for an operating license.

Considerable information should be available from operation of previously reviewed large boiling water reactors prior to operating of the Browns Ferry reactors. However, because the Browns Ferry Units are to operate at substantially higher power level and power density than those on which such experience will be obtained, an especially extensive and careful start-up program will be required. If the start-up program or the additional information on fuel behavior referred to earlier should fail to confirm adequately the designer's expectations, system modifications or restrictions on operation may be appropriate.

The Advisory Committee on Reactor Safeguards believes that the items mentioned above can be resolved during construction of the reactors. On the basis of the foregoing comments, and in view of the favorable characteristics of the proposed site, the Committee believes that the proposed reactors can be constructed at the Browns Ferry site with reasonable assurance that they can be operated without undue risk to the health and safety of the public.

The following are additional remarks by Dr. Stephen H. Hanauer. "It is my belief that the substantial increase in power and power density of the Browns Ferry reactors over boiling water reactors previously approved should be accompanied by increased safeguard system margins for the unexpected. The emergency core cooling system proposed should in my opinion be redesigned to provide additional time margin and to reduce the severe requirements for starting of large equipment in a few seconds. The dependence on immediate availability of a large amount of emergency electrical power, using diesel generators operating fully loaded in a previously untried starting mode, is of special concern, as are the high temperatures and numerous fuel-element failures predicted even for successful operation of the emergency core cooling system in a large loss-of-coolant accident."

---

*The Committee believes that these matters are of significance for all large water-cooled power reactors, and warrant careful attention.

10
Zion

In the summer of 1967, Commonwealth Edison filed an application for construction of two 3,250 MWt PWRs on Lake Michigan at Zion, Illinois (population 14,000), between Chicago and Milwaukee, six miles north-north-west of Waukegan, Illinois (population 55,719 at the time). The preliminary report on the Zion site by the regulatory staff to the ACRS, dated September 13, 1967, notes "Our review of the site will emphasize population distribution in the vicinity. A preliminary comparison reveals a definite similarity to the distribution around the Indian Point site."

The review of the construction permit application for the Zion reactors is of interest in several ways. Zion was the first highly populated reactor site to receive a complete review after the Burlington site had been rejected. The population density around Zion was sufficiently large that the site was not automatically accepted with regard to demographic considerations. The eventual acceptance of the Zion site was based on an estimated equivalence of its population distribution to that at Indian Point, thereby reinforcing the importance of the precedent set at Indian Point.

There were also many other interesting issues associated with the ACRS review. As a site on the borderline of acceptability, it raised questions for the ACRS concerning the possible need for additional safety features. Also, during specific construction permit cases over the previous one or two or three years, a growing list of unresolved safety issues had been

150

identified as applicable or generic to all LWRs, or to all PWRs, or BWRs. These including the asterisked items in the ACRS letter on the Browns Ferry site (see appendix to Chapter 9). The resolution of these safety issues appeared to be too slow to the ACRS, and at least some of its members felt disquieted. The Zion reactor represented a difficult case and hence a good focal point on which to try to force more rapid resolution of the outstanding generic issues.

Furthermore, because the Zion reactor was like that at Indian Point 2, the question of a "core catcher," which was still a matter of controversy for Indian Point 2, was discussed somewhat in connection with the Zion review.

The regulatory staff basically completed its review of Zion and issued a report to the ACRS, dated March 18, 1968, which was favorable to construction with only a few (typical) minor reservations, including one requiring charcoal filters in the containment vessel to further reduce the iodine inventory during a postulated MCA. Although the first ACRS subcommittee meeting on Zion was not held until March 21, 1968, the proposed reactors had been the focus of discussion within the committee prior to that time. For example, at the ninety-third meeting, January 11–13, 1968, the ACRS discussed the lack of progress on the list of items which had been asterisked in the Browns Ferry review, a year earlier. ACRS member Okrent urged positive committee action to obtain resolution of these items and moved that this be accomplished (if possible) prior to completion of review of the next bad site similar to Indian Point 2, namely, Zion. However, ACRS member Mangelsdorf opposed this and moved instead that a subcommittee be set up to review the progress on the asterisked items. The Mangelsdorf position was adopted by the committee.

At the first meeting of the Zion subcommittee, a major interest of the ACRS members was about additional measures that might be recommended for the Zion reactors, which were essentially a replica of the recently approved Diablo Canyon reactor near San Luis Obispo, California; the latter, however, was a relatively remote site. Various potential improvements were identified including partial protection against pressure vessel failure, improved protection against sabotage, increased margins in engineered safeguards, and pressure vessel cavity flooding (to provide cooling water should reactor vessel failure result from the thermal stresses due to the injection of cold water following a LOCA). Member Hanauer questioned whether it was appropriate to use Part 100 as a basis for designing such a plant, as the regulatory staff were doing.

According to the analysis of the regulatory staff, there was no meaningful difference in population distribution between the Zion and Indian Point sites. Zion was on a lakefront about forty-miles north of Chicago and its average population distribution was similar to that at Indian Point 2; however, in some directions it had significantly higher population densities, e.g., four times, if one considered the sector which encompassed Waukegan. The ACRS asked its metropolitan siting subcommittee to provide an independent assessment of the relative population characteristics of the Zion and Indian Point sites.

The Core-Catcher Controversy

The Zion site received an initial hearing by the full ACRS at the ninety-sixth meeting, April 4–6, 1968. The general consensus of the committee was toward approval of the site if it was like Indian Point, but requiring safety improvements on the reactor. However, the seemingly strong trend toward improvements in safety for Zion was undercut by the frequent disparaging references made by some ACRS members, especially Mangelsdorf, to the Consolidated Edison core-catcher proposal for Indian Point 2. These members stated that the committee's desire for "something more" might force the applicant into something which was ill considered. The core-catcher issue was somewhat of a sore point within the committee, since other members felt let down by the ACRS decision of September 1966 to accept the task force on core melt problems recommended by the AEC, instead of issuing the general letter than had been drafted. Now this feeling was reinforced by the task force report which came out in early 1968, and (1) endorsed the existing approach to safety as adequate, and (2) very weakly supported any serious research and development effort on means to cope with core melt.

The core melt controversy within the ACRS was further aggravated because Consolidated Edison discussed the status of the core catcher they had proposed in 1966 for the Indian Point 2 reactor with the ACRS during the spring of 1968. In March 1968 Consolidated Edison nominally provided a status report to the ACRS on Indian Point 2, but the question implicit in the ensuing discussion was whether the applicant could remove the core catcher. It is important to realize here that there had been no further technical development by Westinghouse of a core-catcher system, and there was no more assurance in 1968 than there had been in 1966 that the design proposed could cope with core melt and prevent containment failure. In effect, it was a take-it or leave-it kind of core catcher that was still in the construction permit application. Consolidated Edison was not going to show it would work.

At its next meeting in April 1968 the ACRS again reviewed the question of the core catcher for the Indian Point 2 reactor. The ACRS was not inclined to recommend a core catcher for the Zion reactor, especially since a technically feasible approach did not exist. However, in his discussions with the ACRS, Mr. Price took the approach that, if a core catcher was going to be required for Indian Point 2, then he would require core catchers for Zion 1 and 2. After discussing this difficult subject in considerable detail, the ACRS decided to defer making a decision on the Indian Point 2 reactor until after the committee heard from its metropolitan siting subcommittee which was to provide a comparison of the Zion and Indian Point 2 sites. The matter is reviewed in some detail in the appendix to this chapter.

Comparison of the Zion and Indian Point Sites

During March, April, and May 1968, the metropolitan siting subcommittee held three meetings to develope a basis for comparing the population characteristics of sites, and specifically, to determine the relative population characteristics of the Zion and Indian Point 2 sites. A range of serious accidents involving lethal doses beyond the site boundary was postulated, and alternative methodologies for estimating the consequences were evaluated.

There were a wide range of approaches considered and a wide range of opinions expressed. For example, ACRS member Hendrie argued that the concern regarding metropolitan siting was that a large number of people were subjected to a small probability of a severe accident. He believed that this concern led to the need to consider the population distribution in angular sections. It was his opinion that a site with a certain population density per square mile equally distributed around the site is not significantly different from one with a 45° sector with the same population density per square mile and few or no persons located in the 315° sector. On the other hand, member Monson thought the site with persons located only in a 45° sector would be eight times better than a site with the same population density all around the site.

There was considerable discussion on whether to include meteorology into the comparative site evaluation, and the consensus was not to.

When comparisons were made of the Indian Point and Zion sites on an integrated population basis, Zion usually appeared slightly better. If only the worst angular sector was used, Zion was "better" than Indian Point for distances from one to four miles, but worse between four and twenty miles.

By the May 8, 1968, subcommittee meeting, a methodology which weighted persons close to the site more importantly, and looked at both sectors and integrated population, had been chosen for interim purposes. The subcommittee concluded that the Zion site was slightly better than Indian Point from the population point-of-view (and hence acceptable).

The strong precedent set by the original acceptance of the Indian Point 1 reactor in 1956, and of Indian Point 2 in 1966, was clear. Only now and then do the minutes record an opinion by a member expressing unhappiness with the earlier acceptance of the Indian Point 2 reactor. And the idea that there should be a reexamination of the acceptability of the Zion site, even if it is equivalent to or "better" than Indian Point 2, is rarely found.

Although major accidents were postulated for use in making comparisons of different sites, no quantitative estimate of the probability of such large accidents or of the risk (a summation of the product of probability and consequences over all accidents) was available. Nor did there exist a quantitative risk acceptance criterion* (except the indirect one that might be derived from the AEC commissioners opinion on the Malibu site, that even though a fault had not moved in the last 10,000 or 14,000 years, displacement along the fault should be considered in the design of that proposed reactor).

At the ninety-eighth meeting, June 5–8, 1968, the siting subcommittee reported its findings to the full committee, and the ACRS concluded that the Zion site was comparable to Indian Point and acceptable. The ACRS also decided that the Zion reactor did not need a core catcher and that the core catcher could be removed from the Indian Point 2 design without objection from the ACRS. Considerable discussion was held with Commonwealth Edison concerning various safety matters, with emphasis on the possible need for a system to cope with vessel failure from thermal shock. Commonwealth Edison agreed to include provisions in the design which would make it feasible, at a later date, to add a slow-flooding capability for the vessel cavity. However, the Zion reactor did not include really substantive additional safety features compared to Diablo Canyon.

A Dissent on Zion

The ACRS agreed at the June 1968 meeting that it would prepare a letter favorable to construction of the Zion station, with member Okrent indicating that he would include additional remarks which were aimed primarily at future reactors proposed for populated sites similar to or worse than Zion.

Extensive and controversial discussion ensued within the committee. This included (a) the propriety of adding remarks concerning future reactors, and (b) the possibility that the committee would adopt a position similar to that formulated in the additional remarks with regard to more-populated sites than Zion and Indian Point, and might write a general

---

*For example, that a serious accident leading to tens of hundreds of fatalities should occur less frequently than, say, one in a million reactor-years of operation.

(non-case-specific) letter on the subject.

Discussion of the matter was carried over the ninety-ninth meeting, July 11-13 and 21, 1968, at which time the committee decided to issue only a report on the Zion reactor with dissenting remarks. The letter to AEC Chairman Glenn T. Seaborg from ACRS Chairman Carroll W. Zabel is reproduced below, together with a personal letter from ACRS member Joseph M. Hendrie to ACRS member David Okrent.

ACRS Report to the AEC on Zion Station Units 1 and 2

At its ninety-ninth meeting, July 11-13 and 21, 1968, the Advisory Committee on Reactor Safeguards completed its review of the application by the Commonwealth Edison Company for authorization to construct nuclear generating Units 1 and 2 and its Zion Station in Zion, Illinois. This application was considered also at the ninety-sixth, ninety-seventh, and ninety-eighth meetings, on April 2-6, 1968, May 9-11, 1968, and June 5-8, 1968, respectively. Members of the ACRS visited the site on June 6, 1967, and Subcommittee meetings were held at the Argonne National Laboratory on March 21, 1968, and in Washington, D. C., on April 17 and May 29, 1968. During its review, the Committee had the benefit of discussions with representatives of the Commonwealth Edison Company and their consultants, with the Westinghouse Electric Corporation, and with the AEC Regulatory Staff and their consultants. The Committee also had the benefit of the documents referenced in this report.

The Zion Station is located on the west shore of Lake Michigan in Zion, Illinois. Zion has a population of 14,000 and Waukegan, Illinois, with a population of 65,000, has its nearest boundary 3.6 miles from the site. The site comprises 250 acres.

Each of the two 3250 MWt pressurized water reactors is similar in design to the Diablo Canyon reactor. The containment for each reactor is a prestressed concrete vessel similar to previously approved designs (e.g., Turkey Point, Palisades, and Point Beach). The reactors to be built at the Zion Station are the largest reactors reviewed to date for construction in a region of relatively high population density.

The applicant has considered the possibility of reactor vessel failure as a result of thermal shock caused by emergency core cooling system action in the unlikely event of a loss-of-coolant accident during the later portions of vessel life. He has conducted engineering studies which have established the feasibility of a cavity flooding system that could flood to a level above the top of the core and thereby provide additional protection in the event of such failure. He stated that this system would be installed at a future time if studies now under way indicated that vessel failure as a result of thermal shock could occur. The present design provides for reactor cavity flooding to about two feet above the bottom of the core. Additionally, the reactor cavity has been designed, as at Indian Point 2, to limit vessel movement in the highly unlikely event of failure of the reactor vessel by longitudinal splitting during operation. The Committee continues to favor such protection for large reactors at sites of relatively high population density.

The applicant has proposed using signals from the protection system for control and override purposes. The Committee reiterates its belief that control and protection instrumentation should be as nearly independent of common failure modes as possible, so that the protection will not be impaired by the same fault that initiates a transient requiring protection. The applicant and the AEC Regulatory Staff should review the proposed design for common failure modes, taking into account the possibility of systematic, non-random, concurrent failures of redundant devices, not considered in the single-failure criterion. In cases where hypothesized control or override failure could lead to the need for action by interconnected protection instrumentation, separate protection instrumentation channels should be provided or some other design approach be used to provide equivalent safety.

The applicant described programs for development and utilization of instrumentation for prompt detection of gross fuel failure and for detection of primary coolant leakage.

The Committee continues to emphasize the need for quality in the manufacture, storage, and installation of the reactor and primary system components. The applicant described the quality assurance program that he and his contractors intend to carry out for this purpose. In this connection, the applicant described the testing program for engineered safety features, including a full flow test of the emergency core cooling system delivering water to the reactor vessel. The Committee recommends that the applicant give further consideration to testing the containment spray systems with full flow to the spray nozzles at least once at an appropriate time during construction.

The applicant described his emergency plans for the Zion Station, which are based partly on experience acquired in developing plans for the Dresden Nuclear Station.

The Committee continues to call attention to matters that warrant careful consideration with regard to reactors of high power density and other matters of significance for all large, water-cooled power reactors. In addition, attention is called to safety-related questions specifically identified for the Diablo Canyon reactor class. The applicant reviewed his research and development program designed to resolve safety-related problems and stated that he expects resolution of these problems before operation of the reactors. System modifications or restrictions on operation may be appropriate if the startup program, additional operating experience, or the research and development should fail to confirm adequately the proposed safety margins.

The Committee believes that the items mentioned can be resolved during construction and that, if due consideration is given to the foregoing, the nuclear Units 1 and 2 proposed for the Zion Station can be constructed with reasonable assurance that they can be operated without undue risk to the health and safety of the public.

Additional remarks by Dr. David Okrent are appended. The matters discussed by him were considered by the Committee during its meetings. The Committee believes that the status of these matters, as they pertain to the Zion units, is satisfactory.

Additional Remarks of Member David Okrent

While I am not objecting to a construction permit for the Zion reactors, I am suggesting that in connection with its issuance there are certain matters that warrant consideration and resolution before construction is completed.

In its report of November 24, 1965, on reactor pressure vessels, the ACRS recommended that further attention be given "to methods and details of stress analysis, to the development and implementation of improved methods of inspection during fabrication and vessel service life, and to the improvement of means for evaluating the factors that may affect the nil ductility transition temperature and the propagation of flaws during vessel life." The ACRS also recommended that "means be developed to ameliorate the consequences of a major pressure vessel rupture" and suggested as a possible approach the provision of "adequate core cooling or flooding which will function reliably in spite of vessel movement and rupture". The ACRS went on to state that "the orderly growth of the industry, with concomitant increase in number, size, power level and proximity of nuclear power reactors to large population centers will in the future make desirable, even prudent, incorporating in many reactors the design approaches whose development is recommended above."

Since November, 1965, considerable additional emphasis has been placed by the nuclear industry and the AEC on providing still greater quality in pressure vessel fabrication. An important research program is under way by the AEC to provide a better understanding of the behavior of thick-walled, steel pressure vessels. Our reactor vessel operating experience, although limited, has been good.

On the other hand, some questions have arisen in connection with specific design and fabrication aspects of pressure vessels. Resolution is required concerning the potentially adverse effect on vessel integrity of thermal shock arising from operation of the emergency core cooling system in the unlikely event of a sizable primary system loss, and questions exist with regard to the behavior of highly irradiated, thick section, pressure vessel walls in the presence of flaws and at significant vessel pressure.

Increasing attention has been given to the development of in-service inspection techniques and to the provision during reactor design of the necessary accessibility for thorough in-service inspection. Both industry and AEC regulatory groups are currently working on access and periodic inspection requirements for water reactor primary systems, including the pressure vessel. Means of remote, volumetric inspection of pressure vessels in service are under development by the nuclear industry, as are other flaw detection devices.

I believe that, with regard to water reactors of current design to be sited in less populated areas, the efforts under way to provide improved vessel quality and adequate, thorough, in-service inspection, in conjunction with satisfactory resolution of the thermal shock matter, with acceptable results from safety research programs on irradiation effects, subcritical flow growth, etc., in thick-walled vessels, and with deliberate conservatism and thoroughness in pressure vessel design and fabrication practice, should provide an acceptable basis for dealing with safety questions arising from pressure vessel integrity.

The Zion site has a relatively large surrounding population density. For large water reactors proposed for such a site, I believe that, in addition to the above steps, careful consideration should be given in the initial engineering design to provision of the capability to cope with a loss in primary system integrity arising from a leak or split in the pressure vessel wall. Such provisions should include necessary steps to maintain the containment integrity. It appears likely that means to maintain the general core geometry and to provide the necessary emergency cooling water would be required. It is important that such provisions, if they are to be implemented, provide a significant degree of additional protection, albeit not perfect or complete, and that they should not, of themselves, provide a means of detracting from the integrity of the pressure vessel. It is to be expected that the development of means to deal with a loss of primary system integrity arising with the pressure vessel will be a process of evolution. Careful and thorough study should lead to a definition of these potential areas of degradation in pressure vessel integrity for which protective measures are practical and appropriate. In view of the very low probability of a pressure vessel rupture, the design of these protective features could be based on fairly realistic rather than highly conservative analyses. A reliance on off-site power sources in connection with these protective features may be acceptable, if the capability of the external power system to withstand sudden, unexpected shutdown of the reactor can be clearly demonstrated and periodically verified.

For the Zion reactors, where the engineering design is now well along and could not be readily modified without major delays and significant additional costs, I believe that the applicant should study what provisions could be made, within the limitations of the existing design, to provide further protection against a loss in primary system integrity arising from a limited size leak or split in the pressure vessel wall, particularly in the region that receives the highest neutron irradiation dose during reactor lifetime.

I also believe that, at this time, additional conservatism in design, construction and operation is desirable for the Zion reactors, as compared to similar reactors at less populated sites. To be most effective, this additional conservatism should be part of the applicant's basic philosophic approach. The following aspects might be included.

1. Both for the primary coolant system and for other features of vital importance to the protection of the health and safety of the public, additional conservatism in design and further steps to assure quality of construction and continued integrity and reliability during operation should be used, where practical.

2. Safety issues remaining to be resolved between the start of construction and the initiation of operation at power should be minimized; well-defined research and development programs, adequate to clearly resolve the issues in timely fashion, should be committed. Where questions remain to be resolved, and where complete resolution may not be accomplished by the time of reactor operation, the reactor design should proceed on the basis of incorporating the appropriate safety provisions.

3. Since it is highly unlikely that a clear demonstration of the efficacy of the several engineered safety systems and other protective features under representative accident conditions will occur as a consequence of actual accident experience in the reasonably near future, it is desirable that extra margins be provided in the design of the usual engineered safety systems, particularly those for which some degree of uncertainty or some problem requiring resolution remains.

4. Additional, detailed examination of potential accidents leading to moderate releases of radioactivity to the environment (small accidents) should be made, and steps be taken to reduce still further the probability of occurrence of such accidents.

In my opinion, additional steps such as these, which are taken to protect the health and safety of the public with regard to reactors to be sited close to population centers, need not necessarily be applied to reactors in less populated sites.

Letter from Joseph M. Hendrie to David Okrent

Since everybody else is giving you advice on the Zion matter, I don't see why you should be spared my two cent's worth. I've been trying to answer, for myself, three questions about your additional remarks on Zion. I think I have, and the answers may be of interest to you.

I ask myself:

a) Do I now approve of, or object to your additional remarks?

b) What is the reason for the answer to a), and why, since I strongly support your position in general, do I refuse to join you in these additional remarks:

c) What position do I take on Zabel's move to prevent your (and Mangeldorf's) additional remarks from being attached to the Zion letter?

The answer to a) is that I object, in a very mild way, to your additional remarks as they now stand. The answer to b), the reasons for objecting, and for not joining you in the remarks go as follows. First of all, the remarks as written seem to have a very limited applicability to the Zion application itself. I know you don't agree, and the fact that we differ on this matter is a principal reason for this lengthy discourse.

As I read the additional remarks, they say for Zion,

1) do what is practicable (and reasonable?) within the limits of the existing design to deal with vessel leaks and ruptures, and

2) take the steps #1 through #5 listed on pages 6–7 of Draft 2 (attached to letter Draft 7).

But since you do *not* oppose the Zion application, which your first paragraph specifically says, and since the application does not now include these items, I must conclude that these items do *not* really apply to Zion in your view. (As a side note, the steps 1–5 you list are only inferred to be appropriate for Zion through the reference, middle of page 6, to sites with population densities equal to, etc., Zion. I think some, maybe most, people would conclude the steps 1–5 do *not* apply specifically to Zion.)

Now I know the syllogism above will be offensive to you (note Webster's alternate definition of syllogism as a " . . . specious or crafty argument."), but it is a difficulty for me, and for others too, I am sure. From my standpoint, the additional remarks would be much clearer and would squarely meet the test of applicability to Zion if they said something like:

"I believe that the applicant should, in addition to the matters cited in the ACRS report, 1) do what is practicable within the limits of the existing design to deal with pressure vessel leaks and ruptures, (etc.), 2) take the steps enumerated below (steps 1-5, etc.)."

In this form the additional remarks would be a clear dissent on the Zion application approval. They might be tempered by a remark that these matters are resolvable during construction, in your view.

I know the above introduces an "applicability" criterion for additional remarks with which you may disagree. But I do think some of the Committee reaction to your remarks is brought on by such feelings about applicability. (Later on, in discussing the answer to my third question, I will tell you what I think of such a criterion.)

There remains the question of how to treat the matters in your additional remarks pertaining to future reactors, and this raises the second reason for my objection to the remarks as they now stand. Basically this is that you are preempting the Committee in two important areas: first in the area of requiring protection against vessel failure, and second, in the area of Metropolitan Siting, where you suggest that steps along the lines you suggest might open the way to sites in high population areas. I don't think I am being completely rational with this objection, because you have given the Committee ample time to consider your point of view, and if they will not agree with it, you have a right and a duty to express it individually. But I am frustrated by a) agreeing with you, b) being personally unwilling to join in these remarks on the *Zion letter*, and therefore, c) being unable to participate in the expression of views on vessel failure protection. So, rational or not, I wish you were not going to put the "non-Zion" sections of your remarks into the additional remarks on Zion. (I think Palladino and some others share this feeling.) Furthermore, I have the feeling that the presence of these remarks, appended to the Zion letter over the protests of many Committee members, may really hamper future acceptance by the majority of the position that protection against vessel failure should be provided at poor sites.

If you were to ask what I suggest doing (obviously I am going to tell you whether you want to know or not), it is as follows:

A) Confine additional remarks for the Zion letter to those areas listed as 1) and 2) above, and consider whether they should not have the more direct dissent form suggested above.

B) Work with the Committee toward incorporating the "non-Zion" portions of the present remarks in a general letter. Since this is unlikely to produce the desired general letter in the near future, I suggest it *only* as a prelude to an individual letter to Seaborg on vessel failure in which I would join you, and in which some others might join, notably Palladino. Your current drafts of such a letter would suit me just fine, and we could allow the rest of the Committee to make "additional remarks" if they wanted to.

Now as to the answer to my third question to myself, about Zabel's attempt to deny your additional remarks a place with the Zion letter. I have been trying to think of a reasonable basis for limiting the scope of individual remarks on case letters. I conclude there is none. Even the test of applicability of the remarks to the case at hand, which I do apply for myself and which is the central reason I do not join with you on the Zion remarks, I would not want as a general rule for the Committee. The problem with such a general rule, or any general rules related to individual remarks, is that the majority will have to decide how to apply the rule in each specific case, and the nature of the situation is that the majority is automatically prejudiced against the individual remarks. So I am against Zabel's move, and I will try hard to prevent anything along that line.

I have tried to think what I might do, if I were in your position and were denied the Zion letter as a vehicle for my individual views. Being unfortunately inclined to emotional outbursts when frustrated, I expect I would resign with a fiery letter to Seaborg, and would subsequently regret it. I judge you to be considerably more self-possessed than I am, and if Sunday's meeting goes badly, I will count on you to fall back on the general letter approach and on a joint individual letter on the vessel failure subject if necessary.

Chapter 10: Appendix A
Removal of the Core Catcher from the Indian Point 2 Reactor

A very detailed description of the difficult and very controversial discussion within the ACRS concerning core catchers in general, and their potential applicability to the Indian Point 2 and Zion 1 and 2 reactors, in particular, can be found in the minutes of the ninety-fifth meeting, March 7–9, 1969; ninety-sixth meeting, April 4–6, 1968; ninety-seventh meeting, May 9–11, 1968; and ninety-eighth meeting, June 5–8, 1968.

At the ninety-fifth meeting, AEC Chairman Seaborg expressed his concern to the ACRS about piling one safeguard on to back up another, referring to the core catcher at the Indian Point 2 reactor. At the same meeting, Consolidated Edison came to see the committee, nominally to report on its ECCS in accordance with the August 16, 1966, ACRS letter, but more specifically to get ACRS approval of removal of the core catcher which was formally included in the plant in June-July, 1966.

A very complex discussion ensued. The applicant stated that they had done only one to two man-years of work on the core catcher, and some members expressed disappointment that so little effort had been made to examine its possibilities, despite the admittedly difficult problems involved. Some members felt that a decision on the core catcher should be related to how effective the new ECCS really was, a matter not yet evaluated. Others were in favor of dropping the core catcher as having no value, independent of other considerations.

The applicant first said he planned to drop the core catcher, but then said it was not his intent at that time to request ACRS approval for such action.

The regulatory staff basically treated the core catcher as if it did not exist, even on paper.

The ACRS finally took no action at the March meeting.

At the April meeting, both the Zion and Indian Point 2 reactors were discussed. Mr. Price said that the regulatory staff did not know how to design a core catcher and so he did not see any need for it. Dr. Morris of the staff stated that the Zion site was acceptable without a core catcher; however, if a core catcher was required for the Indian Point 2 site, then it would have to be included for Zion.

In preliminary vote in executive session, one member favored a core catcher for the Zion site, eight did not, while six abstained.

At the same meeting, Mr. Price reported that he had received a telephone call from Consolidated Edison reporting that they had decided to include a core catcher for the Indian Point 2 reactor. The reasons for the decision were complex. Apparently, the applicant was reluctant to initiate a formal request for its removal. Some ACRS members thought that the tone of the questions in the March meeting may have suggested an ACRS position in favor of the proposed device.

The discussion and arguments over a core catcher for the Zion and Indian Point sites occupied much of the May meeting. Three members (Bush, Mangelsdorf, and Stratton) felt so strongly about the issue that they implied they would write their own letters that month against a core catcher for the Indian Point 2 site, if the full committee failed to act to remove it. Their deadline for action was later withdrawn when it was agreed that the matter would be given priority at the June meeting and resolved for both the Indian Point and Zion sites.

At the June meeting, the issues discussed on the Zion site were not related to a core catcher, which the ACRS agreed they would not recommend. Instead discussion centered on the measures included in the design to enable a later addition of a cavity flooding device intended to assist in keeping the core cool, should thermal shock lead to a reactor vessel leak following a LOCA, and on additional remarks by one member dealing with other matters. In effect, the applicant, by provision of the capability for cavity flooding, had added a special site-related safety feature, albeit nothing approaching a core catcher in its objective.

With regard to the Indian Point 2 site, the ACRS reached the position (with one negative vote) that "the core catcher was not an essential engineered safety feature for this reactor," but did not request that it be removed from the design. The regulatory staff and the applicant were advised orally of this decision. The applicant indicated that construction of the core catcher would be terminated and that it did not need a letter report from the ACRS on the matters discussed.

During the series of meetings, there had been indications to the applicant that, were the ACRS to write a report, it might have to include consideration of the asterisked (or generic) items which had arisen during the Browns Ferry review and elsewhere since its August 1966, report on the Indian Point 2 site. Hence, for this and other reasons, the applicant's final decision to ask that no ACRS letter be written at that time was understandable.

The ACRS asked that the final safety analysis report be submitted as soon as possible, so that the applicant's final decision to remove the core catcher would be made public as soon as possible, since there was to be no letter on the matter, and at that time ACRS meetings were not public.

11

The China Syndrome—Part 2

Chapter 8, "The China Syndrome—Part 1," concluded with a discussion of the meeting between the ACRS and the AEC commissioners in September 1966 concerning the ACRS draft letter entitled "Problems Arising from Primary System Rupture" in which the ACRS had recommended the development and implementation, in about two years, of safety features to protect against a loss-of-coolant accident in which the emergency core-cooling system did not work. The ACRS had decided to submit the draft letter for comment, rather than just sending it, in large part because of the strong reaction by the AEC to receiving the ACRS pressure vessel letter in November 1965 without warning.

The AEC commissioners recommended delay in the issuance of an ACRS report, proposing instead that the commission would establish a task force to study and report in a few months on questions arising from the China Syndrome matter. The ACRS discussed the AEC recommendation in a very difficult executive session during the September ACRS meeting, a session which left the committee very much divided. The ACRS members present had been unanimous a month earlier that the committee would send a letter to the AEC recommending the development and implementation of measures to deal with core melt. This position was part of the treaty that allowed committee members to agree to the writing of letters favorable to construction of the Indian Point 2 and Dresden 3 reactors. The stand

163

was changed, in September, however, when a majority of the members at the seventy-seventh ACRS meeting decided to accept the AEC proposal of a task force. The ACRS did take the somewhat unusual step of documenting the details of its discussion with the commissioners in its monthly summary letter to the AEC dated September 14, 1966, as follows:

The Committee met with the Commission and discussed several items related to ACRS activities. The Committee devoted considerable attention during this meeting to problems associated with the low probability accidents involving primary system rupture followed by functional failure of the emergency core cooling system. During the discussion between the Committee and the Commission and members of the commission staff on this subject, the Commission suggested that a technically competent task force, including personnel from the AEC staff, Commission laboratories, industry, universities, etc., be formed to gather pertinent information.

The Committee endorsed this suggestion, urged rapid convening of such task force, and recommended that the topics to be assigned to the task force include the following: "The degree to which core cooling systems could be augmented for additional assurance that substantial meltdown does not take place; the potential history of large molten masses of fuel following a hypothetical accident; the engineering problems associated with possible 'core catcher' systems; and the build-up of excessive pressure or an explosive atmosphere in the containment."

At the meeting, the Commission suggested to the Committee that a review of existing information bearing on these problems might be available within approximately two months. The Committee has stated that early completion of such a review would be of considerable use in connection with current license applications. The Committee also suggested that the task force propose a course of action to assure development of any additional information needed. The Committee has expressed its willingness to cooperate with this task force.

Failure of the committee to maintain the agreement made in August remained a sore point within the ACRS for many, many months. Several members, including Hanauer, Okrent, and Palladino, felt let down because they had agreed to the position taken in August strictly as a package. What is more important, however, is that for the next 12½ years, up to the time of the Three Mile Island 2 accident on March 28, 1979, there was no serious effort in the United States to develop systems to cope with an accident involving core melt or severely damaged cores.

The nuclear industry, the development side of the AEC, and the AEC commissioners themselves were opposed to any consideration within the regulatory process of accidents involving core melt or even severely degraded core cooling, such as later occurred at Three Mile Island, when the raising of the China Syndrome issue had removed the concept that containment was an independent bulwark. After formation of the NRC, the aversion to considering core melt accidents as a safety matter in the licensing process persisted, even though accidents involving core melt were, of course, studied and constituted the major aspect of the reactor safety study, WASH–1400 (frequently called the Rasmussen report), which was issued in draft form by the AEC in 1974 (AEC 1974a) and in final form by the NRC in 1975 (NRC 1975a).

The lack of a substantive AEC program on the core melt problem was not because there were no such recommendations, however. At its seventy-fifth meeting, October 6–8, 1966, the ACRS prepared a report on the AEC safety research program. The ACRS letter of October 12, 1966, to AEC Chairman Seaborg gave highest priority to research on the China Syndrome as follows:

A vigorous research program should be initiated promptly on the potential modes of interaction between sizeable masses of molten mixtures of fuel, clad and other materials with water and steam, particularly with respect to steam explosions, hydrogen generation, and possible explosive atmospheres. Work should be directed to understanding the mechanisms of heat transfer connected with such molten masses of material, the kinds of layers formed at cooled surfaces, the nature and consequences of any boiling of the fuel, and the manner and forms in which fission products escape from molten fuel mixtures. Further studies should be initiated by industry to develop nuclear reactor design concepts with additional inherent safety features of new safeguards to deal with low probability accidents involving primary system rupture followed by a functional failure of the emergency core cooling system.

Second on the ACRS list of emphasis was increased work on emergency core-cooling systems. Third was the development of practical methods for extensive periodic inspection of pressure vessels. And fourth, out of about fifteen points, was a strong research program on the behavior of thick-walled pressure vessels.

This was the inception of a major AEC program of research on emergency core cooling. And the AEC and industry did a lot of research on pressure vessel behavior and inspection. But very little work was done on the China Syndrome despite repeated ACRS recommendations.

Establishment of the Task Force

The Atomic Energy Commission announced establishment of a task force on October 27, 1966. The eleven-man task force had six members from the nuclear industry and five from national laboratories or nonprofit research institutions under contract to the AEC. The task force chairman was William Ergen of Oak Ridge National Laboratory. The AEC press release stated that the task force would consider the following topics:

1. The degree to which core cooling systems could be augmented, by way of design modifications and/or new design concepts, for additional assurance that a substantial meltdown is prevented.
2. The potential history of large molten masses of fuel following a hypothetical accident.
3. The possible interactions of molten fuel with materials or atmospheres in containments, and phenomena associated with such interactions.
4. The design and development problems associated with systems whose objective is to cope with large molten masses of fuel.

At the special ACRS meeting, December 2–3, 1966, task force Chairman Ergen gave the ACRS a brief review of the status of the effort of the task force. He noted that although the ACRS concern was that the core might melt even with improved primary system integrity and improved emergency core-cooling systems (ECCS), one of the task force members, H. Mandil, had been urging that primary system quality should be improved to match that of the pressure vessel. In Mandil's opinion this would make the probability of a loss-of-coolant accident so low that the newly improved emergency core-cooling systems would not be necessary, let alone provisions to cope with core melt. The rest of the task force did not agree that piping failures could be completely ruled out, however.

The task force concluded that containment integrity could be assured up to the point at which the molten core melts through the reactor pressure vessel, but after that the uncertainties became so significant that the containment vessel could not be considered a barrier. The task force was concerned about a "steam explosion" that could occur if the core melted through the reactor vessel allowing the molten fuel to fall into a pool of water below which might lead to an extremely rapid transfer of heat from the fuel to the water.*

Dr. Ergen advised the ACRS not to expect answers concerning how to cope with core meltdown from the task force. He went on to say:

Most Task Force members are agreed that containment is required in the event of lesser accidents. However, there is a body of opinion within the Task Force which says that the containment design pressure should be reduced to reflect the reduction in probability of major accidents due to recent improvements in cooling system design. If a major accident should occur, the containment will not offer protection

---

*This is at odds with the conclusion in the draft WASH–1400 report in 1974 (AEC 1974a) that if a steam explosion was a threat to the containment vessel of a PWR, it would be from an explosion in the vessel. It is not known conclusively if so violent a steam explosion can occur.

anyway. A second group within the Task Force is of the opinion that, since containment cannot be considered sure protection against very large accidents, one need not be concerned with assuring the very low leak rates required following such accidents.

Dr. Ergen remarked that he had been concerned that the conclusions of the task force might be overly influenced by its more optimistic members. ACRS member Okrent commented that, "If the Task Force were to make judgments concerning the suitability of present design approaches, the effect would be that of the industry judging itself."

Dr. Ergen replied that at the last task force meeting there were signs that such judgments might be attempted in the final report. He felt that some effort would be required to keep such findings from appearing under the aegis of the task force.

On February 8, 1967, there was a meeting of an ACRS subcommittee with the AEC task force. The subcommittee was clearly dissatisfied with the directions the task force was taking, and not only its failure to focus on possible methods of coping with core melt. When the subcommittee asked about task force efforts on improved ECCS design approaches, Dr. Ergen said the task force had come up with no new ideas on this and had not given it much attention. He indicated that, by choice of a majority of the task force and "their bosses," the task force had redefined their assignment to considerations of existing emergency core-cooling systems and the primary system. Task force member E. Beckjord commented that when the task force had looked at the China Syndrome and found it could not be sure of maintaining containment integrity after a core meltdown, instead of charging about in many directions, it elected to try to do something about primary system integrity and the ECCS.

ACRS member Hanauer observed that in the past people had thought they understood things like reactivity excursions, and then there had been some surprises. He wondered where the acknowledgment by the task force was that there might be some things we had not thought about for real loss-of-coolant accidents in real, large power reactors. Task force member Mandil stated that you can always postulate things from "blue heaven" and attach a lot of safety features which are not really helpful. Dr. Hanauer expressed his personal dissent that all the problems had already been identified.

Task Force Conclusions

The work of the task force took considerably longer than the two months originally proposed by the Atomic Energy Commission. In late 1967, an un-numbered report became available under the auspices of the Atomic Energy Commission entitled "Report of the Advisory Task Force on Power Reactor Emergency Cooling." The major finding of the task force report, which contained 12 principal conclusions, was as follows:

The Task Force has concluded that within the framework of existing types of systems, sufficient reliance can be placed on emergency core cooling following the loss-of-coolant, and additional steps can be taken to provide additional assurance that substantial meltdown is prevented.

The Task Force reached the above conclusion on the basis of the findings in the report that the evnts associated with blowdown and core heatup are definable within existing technology; that core structural response can be evaluated within conservative bounds; that appropriate requirements can be placed on core-cooling system design; and that the phenomena associated with the currently incorporated concepts represent satisfactory approaches to emergency core cooling.

The task force went on to recommend improvements in primary system integrity, as the ACRS had already done in its letters on the Indian Point 2 and Dresden 3 sites.

With regard to core meltdown the task force said:

While with present technology the integrity of the containment cannot be assured in the event of a postulated core meltdown, there is likelihood that a length of time will elapse before breachment of the containment might occur. It may be possible to develop preventive measures which are effective during this period and which could significantly reduce the hazards resulting from subsequent failure of the containment.*

The desirability of utilizing such systems and the merits of requiring containments to be designed to assure such time availability should be evaluated after the effectiveness of these systems has been established through necessary development work. The use of such safeguards will depend on weighing their merits with those of other safety features to obtain the desired objectives in overall reactor safety.

The Task Force considered also the design and development problems associated with systems whose objective is to cope with the consequences of core meltdown, such as large molten masses of fuel, and releases of energy and fission products.

We recommend for the near future a small-scale, tempered effort on these problems. The reasons for this are as follows:

(a) if such systems could be developed and their reliability established, they would have certain advantages. They would probably not have to be connected to the primary system. In that case, the likelihood that they would be incapacitated

*This refers primarily to providing the ability to vent the containment vessel through a large filter capable of retaining essentially all the radioactivity except the noble gases, an idea proposed by ACRS consultant Ergen during the hot summer of 1966.

coincident with the primary-system break would be still smaller. Any increase in confidence obtained from these systems could be used to reduce emphasis on other safety related features.

(b) to produce effective designs, if indeed feasible, might require both considerable fundamental research and practical engineering application. Both laboratory investigations and large-scale meltdown tests might be required, as scale-model tests or single-fuel-assembly test might not be adequate. Core meltdown tests would require a large expenditure of funds, manpower, and an extended schedule to complete the design, fabrication, testing and evaluation. Important aspects which could be included in the scope of such a basic development program are discussed in this report, but before any large effort is started, the necessary contents of the program would have to be defined.

(c) for the time being, assurance can be placed on existing types of reactor safeguards, principally emergency core-cooling.

The purpose of the small-scale effort would be an improved understanding of the related phenomena and possibly a definition of the content of a larger program. A larger program should be undertaken only if it can be shown to have adequate prospect of success.

The ACRS Disagrees

The ACRS was unhappy with the task force report. On February 26, 1968, the ACRS sent AEC Chairman Seaborg a letter signed by ACRS Chairman Zabel, which is reproduced in appendix A of this chapter, commenting on the task force report. The committee noted its agreement with the need for improved primary system integrity and a good ECCS. And it pointed out some deficiencies in the task force statements concerning loss-of-coolant accidents.

Of particular interest, however, are the last three paragraphs of the ACRS letter, in which the committee strongly recommended that a positive approach be adopted towards studying the workability of protective measures to cope with core meltdown. The ACRS noted that, while the task force report presented considerable information on phenomena associated with large-scale core meltdown, there was little examination or discussion of the degree to which the efficacy of core-cooling systems might be augmented by way of design modifications; and similarly the report did not provide recommendations on design approaches to cope with large molten masses of fuel, or on the particular research and development problems related to these approaches. Finally, the ACRS recommended that additional design and development effort be aimed at means of providing protection against the extremely low-probability loss-of-coolant accident, in which emergency core-cooling systems of current design might not be effective.

To be more blunt, the ACRS was saying that the task force had not provided any answers to the issue of ameliorating the effects of core meltdown or coping with core meltdown. Also the task force had not defined a research and development program which could provide answers in this regard; and the committee was reiterating its previously stated recommendations that this be done.

In effect, 18 months had passed with no new effort toward coping with core meltdown; and there had been a negation of the major recommendation in the draft ACRS letter of August–September 1966 namely, "that still greater protection of the public by some independent means be provided, particularly for reactor sites nearer to population centers . . . that progress for this objective would require an evolutionary process of design and a vigorous program of research, both of which should begin immediately and should be aimed at reaching a high state of development in approximately two years."

What is unfortunate is that lack of study of the problem left the nuclear industry, the regulatory groups, and society in general, in a poor position to judge whether significant improvements in safety were feasible, and hence, in no position to make an educated decision on whether such additional measures should be considered. On the other hand, there did exist a considerable school of thought that LWRs were already adequately safe, if not more safe than necessary when compared to other existing societal risks.

The report by the task force on power reactor emergency cooling was used for policy decisions by the Atomic Energy Commission during the ensuing years, in that the AEC placed its emphasis on improvements in quality control and improvements in emergency core-cooling systems, both of which had been recommended by the ACRS in its letters of August 16, 1966, on the Dresden 3 and Indian Point 2 reactors. The AEC never pursued a serious study of the filtered vented containment idea of task force chairman Ergen, and it did very little on core melt, in general.

That the judgments reached by the task force were subject to technical flaws was already clear by early 1968. For example, the task force had missed the fact that the Zircaloy clad might embrittle at temperatures far below its melting point, thus requiring that the highest clad temperature be held far below the melting point during a LOCA, if core geometry was to be maintained. With time other flaws came to light. For example, the first task force conclusion stated that current technology was sufficient to enable prediction, with reasonable assurance, of the key phenomena associated with the loss of coolant, and to provide quantitative understanding of the accident. Actually, our knowledge of LOCA–ECCS in 1967 did not include very important effects in PWRs such as steam binding, which could sharply reduce the rate at which ECCS water could reflood the core, and flow bypass effects which could cause water to go out through the break without cooling the core. There were equally important omissions in our knowledge of the behavior of ECCS for BWRs. Similarly, problems in the prediction of dynamic forces during a LOCA were not foreseen by the task force.

The ACRS Reiterates Its Recommendation

In various ways over the next few years, the ACRS attempted to get research done on the phenomena associated with large-scale meltdown and the possible design measures that could be taken to ameliorate large core meltdown. For example, in a letter dated April 14, 1967, to Milton Shaw, Director of the AEC Division of Reactor Development, who was responsible for all AEC reactor safety research, the ACRS said:

It is not clear that substantial early effort will be devoted to gaining an understanding of the various mechanisms of potential importance in describing the course of events following large-scale core melting, including steam explosions and hydrogen generation. Information should be gained which would provide a better foundation for assessing the possibilities of coping with large-scale core melting.

On April 24, 1968, in a memorandum from Beck to Shaw, the regulatory staff reiterated strong support of a research and development program on the means of handling large masses of molten fuel. However, in a report from the Director of Regulation, Harold Price, to the commissioners dated February 20, 1969, it is specifically noted, "there are no current plans to study events following large-scale core meltdown." And for the next 10 years, the regulatory staff did not really support such research.

In a report on the reactor safety research program, dated March 20, 1969, from ACRS Acting Chairman Hendrie to AEC Chairman Seaborg, the ACRS said, "With regard to containment of molten cores, no AEC work is currently planned. Interest in this work continues, however, and the problem may be a more critical concern for larger reactors and much more populated locations than are used at present. Some work in this area in the nature of scoping studies and possible solutions is appropriate. Also, research aimed at providing a better understanding of the more important phenomena involved should be undertaken."*

In a letter dated November 22, 1969, from ACRS Chairman Hendrie to AEC Chairman Seaborg, the committee reiterated its statement made in

---

*Interestingly, when in 1977, Hendrie became chairman of the Nuclear Regulatory Commission, the NRC still did not initiate a major research program on the containment of molten cores.

the previous letter of March 20, 1969, and noted that the ACRS had strongly recommended safety research of this kind several times during the previous three years. However, only very modest efforts had been initiated by the AEC.

The Study by du Pont

In 1970, a draft report, concerning the various phenomena important in consideration of core meltdown and the possible design of features to prevent loss of containment integrity after core meltdown, was prepared by Battelle Memorial Institute, Columbus, Ohio, under the auspices of the AEC. This report represented a very good collection of the information available up to that time. It did not include much in the way of design approaches, although there were some considerations of that sort. Shortly after this draft became available, the ACRS arranged, through the AEC Division of Production, (and not through Mr. Shaw, who was hostile to such work) to have a short study done on its behalf by members of the Savannah River Research Laboratory, which was operated for the AEC by the du Pont Company. On September 16, 1970, a presentation entitled "Concepts for Mitigation of Postulated Power Reactor Core Meltdown Accidents," was given by representatives of du Pont at a meeting of the reactor safety research subcommittee of the ACRS. The summary of this meeting states that:

Du Pont believes it is feasible to control the situation involving molten uranium. All of the tentative concepts du Pont has devised depend upon active components and require both power and a cooling water supply. Du Pont believes that further design studies would require an intimate knowledge of modern power reactors. Because of this and a reduction in engineering staff, du Pont appeared reluctant to perform additional studies themselves but recommended that as the next (or second) step, three years of research and development be carried out on 1) properties of molten material, 2) reaction of molten material with water, and 3) methods of dose reduction with vented gases. Du Pont believes that these studies would cost one to two million dollars. They recommend that, if a third step were to be carried out later, it should be an engineering effort to provide a specific design concept for a specific reactor. They believe that this would cost a factor of ten more than the above research and development effort.

It is very likely that this study generated friction within the Atomic Energy Commission. Milton Shaw was positively against any such efforts; and the arrangement to have du Pont do the study had, in effect, been a short end run around his position. In any event there was great difficulty in getting any such further work performed for the ACRS, and du Pont fairly abruptly "lost interest."

The ACRS met with Shaw and members of his staff, and with representatives of the Battelle Memorial Institute and du Pont during its one hundred twenty-ninth meeting, January 7–9, 1971, to discuss the matter of postulated core meltdown accidents. Following this meeting, the ACRS presented its recommendations in a memorandum from R. F. Fraley, the ACRS executive secretary, to Shaw, dated January 11, 1971. The memorandum is reproduced in appendix B of this chapter.

In this memorandum the ACRS observed that both Battelle and du Pont had separately concluded that it appeared technically feasible to mitigate the consequences of a core meltdown accident. The committee stated that it believed that, even though a core-retention system might not be effective for all causes and modes of core meltdown, it could, as an independent backup, decrease the probability of an untenable fission product release to the environment by at least an order of magnitude, a result that became increasingly difficult to achieve by refinement of systems designed to preserve core integrity within the reactor vessel.

Shaw Opposes Any Such Research

Shaw responded a year later in a lengthy letter dated February 3, 1972, to Fraley (Okrent 1979). He disagreed that the committee's safety concerns about design, construction and operating errors could be compensated for, or that protection could be significantly increased, by the AEC's undertaking a research program on molten core retention. He refused once again to follow ACRS recommendations in this regard, urging instead greater emphasis on quality assurance.

The ACRS chose not to respond directly to this letter from Mr. Shaw. However, in a letter dated February 10, 1972, from ACRS Chairman C. P. Siess to the new AEC Chairman, James Schlesinger, the committee noted that "although the ACRS has recommended that research and design studies be undertaken on systems which might be capable of coping with a largely molten core, little such work appears to be underway." But there was no change in the reactor safety program with regard to core melt under AEC Chairman Schlesinger.

In a letter dated November 20, 1974, from ACRS Chairman W. R. Stratton to the Honorable Dixy Lee Ray, the new chairman of the Atomic Energy Commission, the committee again reiterated, "its previous recommendations for research into phenomena involved in core meltdown, including the mechanisms, rate and magnitude of radioactive releases, and the study of means of retaining molten cores or ameliorating the consequences. In this connection more knowledge of the possibility and extent of steam explosions in the presence of large quantities of molten fuel and steel is of particular importance." By the time this letter was issued, the draft version of the reactor safety study, WASH–1400 (AEC 1974a), was available with its interesting analyses of various possible accident paths that could lead to core melt, and its estimates of the probabilities and consequences of such accidents in light water reactors. The reactor safety study estimated the probability of core melt as 1 in 20,000 per reactor per year, with an uncertainty of about a factor of 5. This probability (or recurrence frequency) was much larger than the regulatory staff had been estimating previously, and has, in turn, been questioned by many, as possibly giving too small a probability.

Finally, in a letter dated June 12, 1975, from ACRS Chairman William Kerr to Mr. Ralph V. Carlone, Assistant Director of the United States General Accounting Office, the committee reiterated once again its recommendation for research into phenomena involving core meltdown.

In summary, by the time of the Three Mile Island accident more than a decade had passed since 1966 when the ACRS almost made very strong recommendations, in the ACRS letter that was never issued, that a means to ameliorate or cope with core melt be developed and *implemented*. There was essentially no response from the nuclear industry or the Atomic Energy Commission to the continuing ACRS recommendations for research on core melt, and it was not until the very latter part of this era that the Nuclear Regulatory Commission initiated a modest amount of research into the phenomena. Even then, no conceptual design studies were included until about the time of Three Mile Island.

Chapter 11: Appendix A:
ACRS Comments on Report of Advisory Task Force on
Power Reactor Emergency Cooling

The Advisory Committee on Reactor Safeguards offers the following comments on the recently issued Report of the Advisory Task Force on Power Reactor Emergency Cooling.

The Committee believes that the Task Force has performed a valuable service by assembling in a single document discussions covering many of the problems associated with postulated loss-of-coolant accidents and the phenomena important to proper functioning of emergency core cooling systems. Also, the Task Force has reviewed in a useful manner the many phenomena involved in the course of a postulated large-scale core meltdown.

Certain of the report's conclusions and recommendations appear to constitute expressions of judgment as to the adequacy or sufficiency of particular reactor safety provisions in respect to their capability for providing assurance against undue risk to the health and safety of the public. No attempt is made to comment on these. There are, however, a number of other conclusions in the report concerning which the Committee wishes to recommend emphasis, supplementation, or a differing viewpoint. Comments on these are given below.

In Conclusion 1, the report states in connection with the loss-of-coolant accident: "... for quantitative understanding of the accident, the analysis of such an event requires that the core be maintained in place and essentially intact to preserve the heat-transfer area and coolant-flow geometry. Without preservation of heat-transfer area and coolant-flow geometry, fuel-element melting and core disassembly would be expected. With the start of core disassembly there would be a major increase in the uncertainty of prediction of core behavior, and degeneration of the core to a meltdown situation could not be ruled out." The ACRS is in substantial agreement with this observation.

With respect to assuring that the core remains essentially intact during a loss-of-coolant accident, the report emphasizes the importance of properly assessing and designing for the hydraulic effects incurred, and lists several important specific aspects of the problem that must be recognized and dealt with in designing to cope with such effects. The ACRS agrees with this emphasis.

The possibility of fuel element failure from high internal pressure and high clad temperature during a loss-of-coolant accident is mentioned. In this connection, the Committee notes that present license applications show that a large fraction of fuel rods may fail in such accidents even though the emergency core cooling system works as designed. The Committee believes that, in addition to the work proposed by the Task Force, further research is needed to ascertain the modes of fuel rod failure and to determine that failures will not propagate or tend to block coolant flow excessively.

Conclusion 2 discusses further the importance of controlled and acceptable structural deformation during reactor blowdown in a loss-of-coolant accident. The ACRS agrees with this and calls attention to the need for considering deterioration during the life of the reactor and the role that periodic inspection could play in alleviating this potential difficulty. Also, more conservatism in design and fabrication may be needed where structural member response to accident-induced hydraulic forces is not testable. Further, the Committee continues to be concerned with the possibility of thermal shock effects on the pressure vessel, or other parts of the primary system, as a consequence of the rapid introduction of emergency cooling water in the unlikely event of a large primary system rupture.

The Committee endorses Conclusion 4 which recommends further testing of emergency core cooling at higher temperatures and for degenerated conditions such as core distortion.

The systematized approach to the design and evolution of emergency core cooling systems described in Conclusion 5 appears potentially useful. However, deliberate allowance should be made for the possibility of aggravated accident conditions introduced by possible design errors, by weaknesses common to redundant components, or by other unexpected conditions, and full attention should be given to the potential advantage of diverse approaches to the design of emergency core cooling subsystems. It should be recognized, also, that new design features may introduce new potential safety issues in specific reactor designs.

The Committee endorses the recommendation of Conclusion 7 for improvements in primary system integrity to reduce still further the already low probability of primary system boundary failure.

The ACRS agrees with the statement of Conclusion 10 that: "If emergency core-cooling systems do not function... the current state of knowledge regarding the sequence of events and the consequences of the meltdown is insufficient to conclude with certainty that integrity of containments of present design, with their cooling systems, will be maintained." Recognizing that absolute certainty cannot exist concerning any facet of safety, the Committee strongly recommends that a positive approach be adopted toward studying the workability of protective measures to cope with core meltdown. Basic safety research experiments would provide valuable insight and, possibly, direct attention to potentially profitable avenues of design which eventually could lead to substantial additional protection in this area. The proposal in Conclusion 11 for study of preventive measures to be made effective prior to loss of containment integrity to minimize the ultimate hazard is a helpful step in this direction.

In summary, the Task Force Report presents considerable information of interest on primary system integrity, key phenomena effective during loss of coolant and core heatup, functional considerations for emergency core cooling systems, and phenomena and effects associated with large-scale core meltdown. However, there is provided little examination or discussion of the degree to which the efficiency of core cooling systems might be augmented by way of design modifications or new design concepts. Similarly, the report does not provide recommendations on design approaches to cope with large molten masses of fuel, or on the particular research and development problems related to these approaches.

The ACRS endorses the Task Force recommendations for improvement in primary system boundary integrity and for additional research and development work on emergency core cooling systems. The Committee further recommends, as it did in its 1966 report on safety research, that a vigorous program be aimed at gaining better understanding of the phenomena and mechanisms important to the course of large-scale core meltdown. It also recommends that additional design and development effort be aimed at means for providing protection against the extremely low probability type of loss-of-coolant accident in which emergency core cooling systems of current design may not be effective. The ACRS urges that these matters be pursued vigorously by manufacturers of nuclear equipment, the electric utilities, and the AEC, as appropriate.

Chapter 11: Appendix B:
ACRS Comments to Milton Shaw on a Core Retention System to Mitigate the Consequences of a Core Meltdown

The Advisory Committee on Reactor Safeguards appreciates the meeting with you, members of your staff, and representatives of Battelle Memorial Institute and

E. I. du Pont de Nemours and Company, on January 8, 1971, to discuss the matter of postulated core meltdown accidents. As you know the Committee has had a continuing interest in this matter.

Following its meeting with you, the ACRS further reviewed the usefulness and feasibility of a core retention system to mitigate the consequences of a core meltdown. The Committee agrees with you that quality assurance, including assurance of proper functional performance, of present systems having safety functions is of primary importance. The Committee believes that the probability of meltdown with present systems is very low, and that more stringent application of principles of quality assurance will make the probability still lower. However, improvements to systems and system quality cannot lead to continued significant increase of safety without limit; external phenomena, unforeseen events of very low probability, common mode failures, and human error will set a practical limit to system reliability.

The Committee believes that even though a core retention system may not be effective for all causes and mode of core meltdown, it could, as an independent backup, decrease the probability of an untenable fission-product release to the environment by at least an order of magnitude, a result that becomes increasingly difficult to achieve by refinement of systems designed to preserve core integrity within the reactor vessel.

The Committee has found the work by Battelle and du Pont to be very helpful in its considerations. Both groups have separately concluded that it appears technically feasible to mitigate the consequences of a core meltdown accident. Both groups have recommended that, if work in this area is to be continued, the logical next step is to choose one or possibly two design approaches which appear to have the best potential of success, to evaluate this design in greater depth, and to pursue an associated research and development program organized so as to obtain information vital to the success or failure of the particular design approach. The ACRS believes that the results of the studies thus far are encouraging.

The Committee recognizes that physical and physiochemical properties of the molten fuel and structural materials would be required before a good conceptual design could be made, but it is believed that present knowledge of these properties may be sufficient to establish basic feasibility. However, there appear to be other major uncertainties that do affect basic feasibility. For example, sudden admission of a stream of molten fuel into water (even hot water), especially in a manner that can trap water under the fuel, as in steel retaining channels, could lead to a steam explosion of such violence as to make the entire scheme impractical. The Committee believes that, early in the program, such problems should be explored qualitatively, for example, with material that can be readily melted in conventional furnaces, but using quantities that are large enough to give confidence in the results.

The Committee believes that a program of conceptual design studies and analyses for a core retention system, coupled with the kind of exploratory experiments cited above, should be undertaken. One to two million dollars over a period of two to three years might be a reasonable estimate of the effort and time scale to accomplish this step. The Committee recommends that a program of this type be undertaken with a completion goal of 1973–1974 for this phase. The Committee believes it important that the group undertaking the task have considerable background and resources in practical engineering and metallurgy, as well as a strong research and development capability.

12

Some Papers and Speeches on Siting and Safety: 1966–1969

In this chapter we shall take a brief look at some thinking by the British and others on reactor siting, acceptable risk, and philosophic approaches to reactor safety. In addition, we shall look at two rather different congressional opinions on how safe nuclear power reactors should be, and we shall get some measure of the disfavor with which the ACRS was viewed from some important quarters in the United States. The material dealt with represents only a small sample of the overall discussion which was ensuing nationally and internationally at this time.

The International Atomic Energy Agency (IAEA), which is located in Vienna and is an international organization funded and governed by a large number of countries, frequently holds symposia on subjects relevant to atomic energy. In the spring of 1967, the IAEA held one of its symposium series on reactor safety entitled "Containment and Siting of Nuclear Power Plants" (IAEA 1967). We shall review briefly some of the papers in this symposium that dealt with engineered safeguards and also a trio of papers from the United Kingdom of special interest.

In a paper from the United States, S. Levy of the General Electric Company (Levy 1967) argued for a systems approach to containment design. He recognized that containment is only part of a total system which prevents and limits the consequences of a release of fission products. Levy argued that containment performance requirements should be determined by looking at the overall system as a single entity rather than by making arbitrary assumptions about the source of the accident and the effectiveness of some of the provided features. In effect he was criticizing the Part 100 approach which postulated that the core melted and much of the fission product inventory was released into an intact containment vessel without looking at whether there was consistency concerning the phenomena postulated to occur in the reactor vessel with those postulated for the containment.

Levy called attention to the previously oversimplified interpretation of the role of the containment vessel as a last-ditch barrier. Interestingly, he suggested consideration of the capability for controlled venting of the containment, through a system which filtered out all but the noble gas fission products, to prevent overpressurization. This idea had earlier been proposed by Ergen and was later reemphasized by the American Physical Society study group report on light water reactor safety (Lewis 1975) and is currently receiving considerable attention.

Levy did not explicitly mention the China Syndrome, but as a member of the Ergen task force on emergency core cooling, he was very knowledgeable about the subject. Levy's argument for a systems approach to safety could be used in two ways. One was to advance the case for dealing explicitly with core melt in reactor design, and the other was to argue for reduced requirements on containment and engineered safeguard systems on the basis that the probability of core melt was very low. Levy's company pushed the latter approach and in his paper he estimated the reliability of the emergency core-cooling system (ECCS) as 0.99999, a very high reliability that would be considered overly optimistic today.

There were several European papers at the same IAEA symposium, for example by Vinck and Maurer (1967) and by Kellerman and Seipel (1967), which illustrated that these authors had not yet recognized the relationship between core melt and containment failure and still viewed the containment as an independent bulwark.

The British View on Reactor Safety

Three interesting papers from the United Kingdom dealing with siting were presented at the IAEA symposium. Charlesworth and Gronow of the British Inspectorate for Nuclear Installations, the equivalent of the Nuclear Regulatory Commission in the United States, gave the paper "A Summary of Experience in the Practical Application of Siting Policy in the United Kingdom" (Charlesworth and Gronow 1967). They described the system of site classification that was used in the United Kingdom at that time, which arbitrarily compared the effects of a postulated iodine release on the most populated 10° sector and on the entire surrounding population density, using weighting factors for the expected dispersion with distance. (The United Kingdom was maintaining the requirement of a zone with limited population next to the plant, but was tending at that time to move from a policy of relatively remote siting, to sites more populated than Indian Point for their gas-cooled power reactors.)

Charlesworth and Gronow (1967) came to the following conclusions:

The time has not yet come when it can be pretended that decisions on the safety of a nuclear plant can be based more firmly on objective assessment than on judgment.

When the consequences of a fault or accident cannot be securely limited or contained, then the chance of the fault or accident occurring must be sufficiently remote to justify taking the risk. To attempt to quantify this statement, it might be stated that the probability of such an accident must be such as to give an acceptable margin of safety over it occurring during the lifetime of the plant, and of course the margin must be wide enough to take into account the uncertainties of the probability figure. Before any decision can be taken as to the wisdom of expressing policy in such terms, due regard must be given to the possible consequences which follow from enshrining figures of this nature often to the detriment of the original intention. One point is that if a certain probability figure is approved, any improvement required will be seen as illogical. Over the whole range of releases, effort must still be made to reduce the uncertainties which must enshroud the figures, while in the range of possible releases where results are very serious, and too serious to be allowed to occur, broader logic demands that improvements should continue to be made in the light of an assessment of what is worthwhile in terms of cost and advantage. Another point is that a number of aspects of design are not amenable to a statistical assessment of failure, and due regard must be paid to these, otherwise attention will be focused on matters which can more readily be treated statistically to the detriment of a balanced judgement.

The analysis of a number of existing and prospective sites for nuclear power reactors in terms of potential population risk reveals that there is no unique classification of the sites in order of merit. Orders of merit will vary depending on whether the criterion is the risk to the population as a whole or only that part resident in the most densely populated sector. If the exposure of individuals living close to the site is considered then a further merit order would be obtained. Some sites can be distinguished that are low in the merit order on all counts and clearly a special case would have to be made to justify their use in the immediate future. Moreover, there are too few practical sites high in the merit order on all three counts to form an adequate basis for the substantial development programme which is envisaged for nuclear energy in the UK.

An examination of the practical sites demonstrates that siting alone is an inadequate means of providing proper safeguards for the public. Therefore greater emphasis must be placed on the safety aspects of the design, construction and operation of a nuclear plant. Criteria proposed for doing this have been mentioned in this paper and are further discussed elsewhere in the Symposium. At present no conclusions have been reached, although work is in hand to define standards which will meet the immediately foreseeable siting requirements.

Charlesworth and Gronow were referring to two other British papers which proposed numerical risk criteria for nuclear reactors.

Proposals for Risk Criteria

Adams and Stone (1967) of the national British utility, the Central Electricity Generating Board, proposed that the parameter determining acceptable siting be taken as risk to the individual, and presented arguments as to why a constant, incremental, annual risk of 10^{-5} would be acceptable. In particular, they calculated an average decrement in life of only 3 to 6 days from such a risk (although, of course, a few would suffer much larger decrements and almost all others none).

Adams and Stone suggested a siting policy which would require an exclusion area, i.e., a controlled area in which development that would prevent emergency action would not be allowed, and then an area of unrestricted population.

This paper drew considerable comment. For example, O. Ilari of Italy suggested that "community risk" must also be considered. Adams countered in two ways. First, he argued that it would be difficult to tell thousands of people they were being exposed to a risk that would not be permitted, if they were a much larger group. Perhaps more importantly, he said that, if a criterion is based on the total potential number of casualties, the uncertainty which would exist in calculating that number because of uncertainties as to the conditions and magnitude of the radioactive release is far greater than the difference that the choice of any site could make.

Adams did concede that quantitative risk assessment would not be possible when construction of a plant began, and would be subject to large uncertainties in any event.

F. R. Farmer of the United Kingdom Atomic Energy Authority presented a paper that would become much quoted, "Siting Criteria—A New Approach" (Farmer 1967). In it he proposed that a probabilistic approach be employed in reactor safety assessment. He also suggested a risk acceptance criterion in the form of a limit line in which the acceptable frequency of occurrence of an accident fell off as the consequences increased at a rate such that the expected contribution to risk (frequency times consequences) was less for very large accidents than for smaller ones (a negative slope of 1.5 on a log-log plot).

Farmer proposed his limit line as dividing acceptable and nonacceptable events, and suggested that only a relatively few accidents would be near the line for a reactor; these would lead to the principal contribution to risk, for which he suggested a criterion of less than 0.01 premature deaths per reactor year.

Farmer assumed that the release of fission products which might cause one casualty on a typical reactor site was about 10^4 curies of radioactive iodine (^{131}I.) He set a target objective that the release of 10^3 curies of ^{131}I should have a probability of occurrence of less than 10^{-3} per reactor year. He further proposed that an accident having 10 times the consequences (i.e. 10^4 curies of ^{131}I) should have a lower probability of occurrence by at least a factor of 10.*

Farmer for many years had argued against the maximum credible accident approach which dominated the regulatory process in the United States prior to the advent of the China Syndrome. His argument for use of a probabilistic approach has slowly gained increasing support over the years. Two major difficulties that such an approach faces are the following:

(1) There is insufficient experience and data to estimate accident probabilities and consequences with confidence for accidents expected to have a very low frequency of occurrence such as one in ten thousand or one in a million reactor years.

(2) A probabilistic approach ultimately acknowledges that there is some low probability of an accident leading to large numbers of casualties for a nuclear reactor meltdown or failure of a dam. When one uses the probabilistic approach, one identifies and quantifies the probability, however small, that an accident can occur. In this fashion, the installation, quite correctly, is not defined as being without risk. However, the statement that a low probability exists of some large accident is taken by many as being incompatible with being "safe," although this very situation is actually the typical, unadvertised, and frequently unrecognized fact for essentially all parts of society which are considered "safe."

We turn next to some examples of congressional attitudes toward the ACRS and reactor safety in the late 1960s.

Congressman Aspinall Supports the ACRS

In 1966 Congressman Aspinall of Colorado made the following remarks urging a conservative approach toward reactor safety in an address to the American Nuclear Society in Denver, Colorado.

---

*On November 10, 1977, Farmer told the NRC risk assessment review group, or Lewis panel (NRC 1978a) that he thought that a probability of occurrence of 10^{-4} per year per large industrial installation of an accident leading to serious casualties was on the borderline of acceptability; he didn't have much confidence that 10^{-4} was being met (universally). He stated that chemical plants were not within that value now, but efforts to improve were ongoing.

The large, so-called "conventional" reactors being sold today involve extrapolations from operational reactors that are three to five times smaller in size, and these larger reactors are new and in fact pioneering efforts. It is, therefore, necessary that we remain ever vigilant of the safety aspects of nuclear plants. If we should relax our efforts to maintain the nuclear industry's remarkable safety record and a significant nuclear incident were to result, the consequences to the public—not to mention the industry itself—could be most unfortunate.

I therefore say, "Make haste slowly." If the Atomic Energy Commission's regulatory program is marked by "unparalleled conservatism," as some pundits have observed, then I say: So be it. Another commentator has remarked, perhaps with some exaggeration, that the safety features required by the Commission to be built into a reactor are somewhat akin to a man wearing three belts and two pairs of suspenders. My retort to this comment is simply this: Caution is the parent of safety. Safety consciousness is and should remain the number one criterion in the atomic energy field. Therefore, I *want* the AEC to continue to rigorously apply its high safety standards to the construction and operation of nuclear reactors. I *want* the Advisory Committee on Reactor Safeguards to continue to insist that multiple engineered safeguards be built into each reactor.

However, the ACRS actions, first in writing the pressure vessel letter, and then in raising the China Syndrome issue and forcing major changes in the AEC approach to reactor safety, created much antagonism in the nuclear industry and among some other important members of Congress.

Congressman Hosmer Attacks the ACRS

Congressman Craig Hosmer of California, who for many years was the ranking Republican member of the Joint Committee on Atomic Energy, attacked the ACRS in three speeches during 1968 and 1969.

On May 23, 1968, at the Conference on Nuclear Fuel in Oklahoma City, Oklahoma, before a largely nuclear industry audience, Congressman Hosmer said:

The Joint Committee might take a good hard look at the propostion of eliminating the Advisory Committee on Reactor Safeguards. AEC already has asked relief from the mandatory ACRS review of every license application. That could provide a good opportunity to evaluate the organization's future.

ACRS has no responsibiity for the economics of the nuclear business and apparently could care less. The reactor manufacturers are afraid to approach it with many new safety improvements, particularly in the area of an integrated, systems approach to safety. With good reason, they are afraid ACRS will act as an Advisory Committee on Reactor Redundancy, and simply order them to add the new features to the existing safeguards, thus increasing costs further. I cannot help but wonder if ACRS had outlived its usefulness—if it now serves less as a protective boon than it does as an anachronistic burden.

At the joint winter meeting of the American Nuclear Society and Atomic Industrial Forum in 1969, Congressman Hosmer gave a speech which included the following comments:

Six months ago I called for burning the Advisory Committee on Reactor Safeguards at the stake. Since then it has moved slightly off cloud #9 and we have had a couple of interesting off-the-record chats. So tonight I'm going to recommend that we temporarily forego the burning, but instead of letting ACRS itself pick and choose what it wants to create fear and trepidation about, that in the future the Commission designate to it the matters which AEC believes important enough for specific ACRS review. After all, ACRS is an advisory setup. So it seems appropriate that the Commission tell it what it wants to be advised about instead of ACRS telling AEC what advice it is going to get.

Frankly, I am concerned that if ACRS cannot somehow speed up doing its homework at the operating license stage, this country may be in for some serious brownouts. Some 72,000 megawatts of nuclear electricity is scheduled to come on the line in the period 1971–1973. It is required for the utilities' basic load growth. If ACRS cannot do its job with considerable dispatch as to the operating licenses involved, the delay will run us into drastic power shortages, and severe public censure of the whole idea of nuclear kilowatts would be inevitable.

Thus, in this speech Congressman Hosmer was proposing the imposition of strict limitations on the ACRS. Rather than permit the ACRS to set its own agenda and make recommendations on whichever subjects it believed relevant, Congressman Hosmer suggested that the ACRS be able to provide advice only on those subjects for which it received specific requests from the AEC.

Hosmer's allusion to delays in the regulatory process caused by the ACRS reflected a commonly stated position of the nuclear industry, one which was incorrect. In a talk at a meeting of the Atomic Industrial Forum not too long afterwards, ACRS member Okrent set the record straight, showing that the ACRS generally concluded its review within a month of the time that the regulatory staff had completed and submitted its safety evaluation report on each project.

Finally, at the nuclear safety program information meeting at Oak Ridge, Tennessee, on February 17, 1969, Congressman Hosmer said the following about the ACRS.

Recently, I conducted a review among some of the knowledgeable people on the problems in reactor licensing. And if I could summarize their comments, I would say they are concerned primarily about the leadtime, in the name of safety and licensing, that it takes to receive a construction permit.

They are also very concerned about the continual changing of regulatory requirements. Maybe this is necessary in a rapidly changing field, but every application seems to bring new requirements, and more often than not, the requirements seem to change during the processing of a particular application. This, they contend, makes it impossible to file a satisfactory construction permit application, and the long lead-times make orderly load-growth planning unreasonably difficult.

Additionally, many complained, rightly or wrongly, about what they regard as overly strict requirements for design detail at the construction permit stage of the licensing process. This results, they say, in a premature freeze on design, and a trade-off of progress for safety objectives which they believe may be ill-defined or even non-existent.

One suggestion made was to revamp the functions of the ACRS. Half in jest, I suggested last summer during a speech out in Oklahoma City that the ACRS be abolished and got a standing ovation. I certainly don't expect that kind of a reaction here tonight. I didn't even expect it then. My purpose simply was to get some discussion going about ACRS roles and missions, and it seems to have succeeded.

There is no question in my mind that these men perform a valuable service to the nation by providing some independent expertise on reactor safety. The questions I have are two. One is whether or not ACRS members worry more about keeping their personal records clean against all contingencies by striving for absolute safety rather than the high relative safety society and progress require. My second concern is whether the functions we have presently assigned to the Committee are the proper ones.

As you are probably aware, a bill to relieve ACRS of its responsibility for reviewing each and every construction permit application was introduced in Congress last year at AEC's request, although no action was taken. I am hopeful that when the Joint Committee gets around to taking up this matter, we might broaden our vision and consider some alternatives.

One particular frequent suggestion has been to re-establish ACRS as a general advisory committee on reactor safety. I have a feeling that such an approach would be more consistent with the original objectives when ACRS was established. The idea here, obviously, is to remove ACRS from the regulatory mainstream and let it devote its time to the major safety questions affecting all reactors. The nuts-and-bolts review of specific applications would then be left to the Regulatory Staff with the continued safeguard of a public hearing. It seems to me that the important question is whether today's reactors would be any less safe without a specific ACRS review. I'm not prepared to answer that tonight. I only suggest that we take a good look at it.

As Congressman Hosmer said, at the initiative of the AEC, a bill had been introduced into the Congress to change the role of the ACRS. In the early draft form of the bill, the ACRS would have been relieved of the mandatory responsibility of reviewing each reactor case, and would only have made case reviews at the request of the AEC. The ACRS did not agree with such a proposal and the draft congressional bill was changed so that if the ACRS said, in writing, that it did not think ACRS review of a specific case was warranted, such review would not be required unless the AEC asked the ACRS for such a review.

This bill was acceptable to the ACRS since it permitted the ACRS to choose not to participate in cases which introduced only minor safety issues, thereby giving it additional time for more important matters. To have been completely removed from the review of specific projects would have removed what was perhaps the committee's greatest power, namely, to choose not to provide a recommendation until it believed it had sufficient information. Since the licensing process could not proceed without an ACRS recommendation, even if it was ignored, this was a non-trivial power which was not exercised but was always present.

No change in the mandatory responsibilities of the ACRS was made in the decade following Congressman Hosmer's speech.

13

Reactor Siting: 1968–1970

At the ACRS meetings in June and July of 1968, the committee decided to write a letter favorable to the construction of two large reactors at the relatively highly populated Zion, Illinois, site. One ACRS member, D. Okrent, appended additional remarks to the Zion letter in which he recommended that for such a site additional measures to protect the public safety were appropriate. Furthermore, he stated that such improvements in safety should not automatically be required for reactors at relatively remote sites.

Actually, most, if not all, members of the ACRS supported this general point of view; however, the majority were not then ready to recommend significantly more stringent safety measures for the Zion site. The regulatory staff, on the other hand, continued with the position that all sites required the same level of safety, except as the ritual of meeting the Part 100 maximum credible accident imposed requirements on the acceptable containment leak rate and on systems to remove radioactivity from the containment atmosphere, features which would not reduce the probability of a bad accident and would not be expected to reduce the consequences significantly for many serious accidents which led to breach of the containment.

The regulatory staff and the ACRS were both opposed to a truly metropolitan siting at that time (except that Bolsa Island, which had strong AEC commissioner support, seemed to be left in a special category by the staff). However, the staff appeared to emphasize the lack of operating experience as a principal reason for not approving sites having much higher population densities than Indian Point, with the implication that favorable operating experience would lead to their acceptability. The ACRS, on the other hand, continued to believe that substantially greater safety protection was appropriate for sites much worse than Indian Point. They had some qualms that if new criteria for poorer sites were not prepared and enunciated, the time would soon come when the same reactors that were currently being approved for good sites would be accepted for poor sites.

The Bolsa Island Project

At the ninety-ninth ACRS meeting, July 11–13 and 21, 1968, there was some discussion between the ACRS and Mr. Price concerning the Bolsa Island review. In addition to population density questions, seismic design was a major question for this proposed facility, and the group responsible for the Bolsa Island project had taken the somewhat unusual step of appointing their own "blue ribbon" panel of experts to make recommendations as to what constituted adequate seismic design. To some ACRS members it appeared that an attempt was being made to preempt the regulatory groups with regard to the decision on appropriate seismic design bases for this site. The regulatory staff and the ACRS had been working on the development of seismic and geological siting criteria, and a draft had been agreed to by the ACRS some months earlier. Its release in the Federal Register for public comment had been held up by Mr. Price, however, while he awaited comments from Milton Shaw, Director of the AEC Division of Reactor Development.

At the July 1968 ACRS meeting, Mr. Price was asked if the seismic criteria would be put out for public comment in the near future. ACRS member Okrent pointed out that these criteria could be relevant for the Bolsa Island site, although its review could probably be carried out without the criteria. Mr. Price felt that it would not be wise to judge this or any other facility on the basis of criteria which had not yet been issued. He stated frankly that the delay in issuing the criteria was partly due to the fact that he wanted to know whether or not the draft criteria would rule out the Bolsa Island site.

The regulatory staff had submitted a noncommital report on Bolsa Island to the ACRS dated July 8, 1968, just before the ninety-ninth ACRS meeting. It summarized the recommendations of the "blue ribbon" panel and gave some population comparisons with the Indian Point, Zion, and Burlington sites.

There was no discussion of this report at the July 1968 ACRS meeting. At the next committee meeting, August 8–10, 1968, ACRS Chairman Zabel said that Mr. Price had telephoned him to advise the committee that the Bolsa Island project had been terminated because of rising project costs. And so ACRS review was never made of that site.

It is interesting that a year earlier, Dr. Donald Hornig, the science assistant to the President, had written a letter to AEC Chairman Seaborg in which he mentioned that the proximity of Bolsa Island to a large population appeared to present a more serious problem than that introduced by high-population density at the Burlington site, which had been turned down. He suggested early resolution of the matter.

However, no request was made by the AEC to the ACRS for an early review of the suitability of the Bolsa Island site, although the population information existed. And the regulatory staff refrained from completing any position on suitabilty of the site with regard to population.

The Trap Rock Site

At the one hundredth meeting, August 8–10, 1968, the ACRS reviewed Consolidated Edison's proposed Trap Rock, Montrose, and Bowline sites, all in New York in the vicinity of Indian Point, but each having its own demographic characteristics. The ACRS metropolitan siting subcommittee had reviewed the three sites for population considerations at a meeting held on August 7, 1968, and concluded that the Trap Rock site was fairly similar to Indian Point. They used a comparative approach which was based on estimating off-site casualties and latent cancer effects and employed an average, rather than worst sector, basis. The subcommittee recommended to the full committee that the Trap Rock site was not unacceptable on the basis of population alone, but that the ACRS should not take a position on the Montrose or Bowline sites, both of which were less desirable than Trap Rock.

Consolidated Edison had proposed a secondary containment vessel with a design pressure of 10 or 15 psi, instead of the normal, very low design pressure containment building placed around the pressure suppression containment system of a BWR. A special, very low temperature (cryogenic) off-gas system to trap and store noble gas fission products which would normally be released as routine effluents was also proposed.

In its report to the ACRS on May 31, 1968, the regulatory staff stated, "These sites are somewhat worse demographically than Indian Point 2; however, we believe they are in the Indian Point 2 class of sites. We do not consider them in the Burlington class of sites." The staff went on to say, "We anticipate the applicant would need to put more emphasis on design and performance adequacy of the containment and engineered safety features than, for example, was required at Peach Bottom." Presumably, this staff emphasis was once again to try and meet the guidelines of Part 100 for the MCA and other design basis accidents.

The committee discussion at the one hundredth meeting showed a very considerable divergence of opinion. Some members felt that the three sites were relatively indistinguishable, in contrast to the subcommittee conclusion. Some felt all might be acceptable for a PWR like Indian Point 2, although there was concern about a BWR with the primary system extending outside the containment building. At least one member stated that he did not like the Indian Point 2 site, felt he had been taken in when he approved it, and felt the ACRS should be more demanding now.

The committee had available to it a long list of safety matters related to BWRs and these, rather than the site characteristics, served as the principal focus for the discussion with the applicant.

From the minutes of the committee meeting, it is not clear that all members agreed that the Trap Rock site was acceptable. However, an ACRS position was adopted without dissent, and subsequent to the committee caucus, the ACRS chairman read the following statement to the representatives of Consolidated Edison.

The ACRS feels that the Trap Rock site is not unacceptable on the basis of population alone for a reactor of the approximate size proposed. The Montrose and Bowline sites seem less desirable than Trap Rock, but the Committee makes no statement now as to their acceptability. The Committee feels that considerable attention should be given to emergency plans, including evacuation, particularly in view of the high close-in population. With regard to the reactor design matters discussed thus far, the Committee continues to be concerned with the fact that a part of the primary system is outside containment, and will watch the detailing of the steam line break accident and associated matters with great interest. The Committee emphasizes the importance of in-service primary system inspection and of the provisions for adequate access for such inspection. The Committee feels that at difficult sites such as these, more careful study and evaluation of the problems associated with possible reactor vessel splitting are needed. Other items noted in previous Committee letters will be of even greater interest in this case, in view of the difficulties of the site. The Committee feels that a more conservative approach on the part of the applicant, leading to fewer items to be resolved after the construction permit stage, is appropriate for a plant at this kind of site. The Committee believes that the double containment and the off-gas systems are useful approaches.

Not too long after the one hundredth ACRS meeting, Consolidated Edison advised the regulatory staff that their schedule for these two proposed BWRs had become somewhat indefinite, and, in fact a construction permit review was never completed.

However, at the Trap Rock site, as at Indian Point and Zion before, the ACRS was continuing its emphasis on trying to get a higher level of safety at poorer sites, either by decreasing the probability of accidents or by improving safety features that were intended to prevent an accident leading to core meltdown. This was in contrast with the regulatory staff which continued to make adherence to Part 100 the only aspect of safety which was dependent on population density for an accepted site. To a considerable extent, the ACRS position on the Trap Rock site had adopted most of the points made by ACRS member Okrent in his additional remarks on the Zion site.

As it turned out, by its one hundredth meeting the ACRS had dealt with all but two of the most-populated sites it would review through the end of 1978, namely, Newbold Island and Limerick.

Criteria for Bad Sites

In the months following the Trap Rock site review, the ACRS continued to work on possible design criteria for reactors at sites worse than Indian Point and Zion. At its one hundred eleventh meeting, July 10–12, 1969, the committee decided to forward a proposed ACRS letter on this subject to Mr. Price and the AEC commissioners. The draft letter was written to AEC Chairman Glenn T. Seaborg, would have been signed by ACRS Acting Chairman Joseph M. Hendrie, and included a dissent by ACRS member Hanauer. However, it was never formally sent to the AEC. The letter is reproduced in the appendix to this chapter. In large part, it extends the approach the ACRS took to the Trap Rock site. Hanauer said that while he supported the ideas, he disagreed that the necessary knowledge and experience was available.

The draft ACRS letter was discussed with Mr. Price and with the AEC commissioners at the one hundred thirteenth ACRS meeting, September 4–6, 1969. Mr. Price noted that the requirement for 10 reactor-years of satisfactory operation associated with the issuance of an operating license would be difficult to implement if operation were not completely satisfactory at each reactor but a large number of reactors were already built. He suggested that perhaps 10 years of operation should be required for a construction permit, rather than an operating license. This was a more conservative position than that proposed by the ACRS.

On the other hand, Mr. Price noted that there was a public relations as well as a political problem of convincing people near the presently approved sites that the nuclear units were adequately safe compared to those which might be approved at more densely populated sites, if *more* engineered safety features were to be provided at the more-populated sites.

ACRS member Hanauer noted there seemed to be an apparent inconsistency in this point of view. He said that, if industry came forward with improved designs for more-populous sites and it were required to backfit (or retrofit after completion of construction) such requirements to presently approved sites, the industry would shy away from coming forth with improvements. Price recommended that the ACRS letter should not be issued. He thought that releasing such a report at that time would aggravate and worsen the public relations problem. He felt, also, that the nuclear industry was not placing any great pressure on the AEC to approve more densely populated sites, and he did not consider the upcoming Newbold Island site worse than the Indian Point and Zion sites.

The AEC commissioners also indicated strongly that they would prefer not to receive the ACRS report at that time. Individual commissioners noted that most of the items mentioned were already under active consideration by the commission and/or the staff and suggested that several, particularly those relating to reduction in the acceptable levels for routine release of radioactive wastes and the increased level of safety required for "city" reactors, could create difficult public relations problems.

The commissioners suggested that ACRS representatives work with the nuclear industry and the staff in establishing an acceptable level of safety for those items mentioned in the draft ACRS report as requiring improvement for sites worse than Zion. The ACRS decided to accept the suggestion of the commissioners to set up joint meetings of the regulatory staff and the ACRS with representatives of industry.

Mr. Price issued letters of invitation to various representatives of the nuclear industry to present their views on a range of matters related to metropolitan siting, and on March 4, 1970, the ACRS site evaluation subcommittee and the regulatory staff met with representatives of the Atomic Industrial Forum and 10 utilities in the first of several such meetings. In his letter of invitation Mr. Price said, "the Commission believes that before the siting of power reactors in closer proximity to metropolitan centers can be favorably considered, further advances are needed in reactor plant design and the capability of safety systems and engineered safety features in adapting critical components and systems to accomodate reactor inspection and testability, and in the practical demonstration of dependability of performance of vital systems." And a list of specific items for discussion was included.

At the March 4 meeting, ACRS member Monson said that he would like to secure the opinions of the industry representatives present about what improvements they believed could be made to allow the use of worse sites. He stated that the ACRS was not interested in promoting nuclear power, but was interested in assuring adequate safety for whatever sites were used.

He indicated that the committee was under the impression that new techniques could be proposed, but might not be because of fear that the regulatory staff and the ACRS would require that any new improvements be incorporated for all reactors, no matter what site was involved. Monson said that, judging on a technical basis, improvements developed for reactors to be located at worse sites need not necessarily be required for reactors at better sites. He thought, however, that a public policy question arose as to whether it was advisable to provide a greater level of safety at one site than at another solely because of population density differences.

Industry Viewpoints

Mr. H. Brush of the Nuclear Reactor Safety Committee of the Atomic Industrial Forum presented a statement in which he said that there were at least two large utilities, one on the East coast and one on the West coast, where there was a pressing need to use worse sites than Indian Point-Zion. He stated that the use of metropolitan sites would alleviate the problem of obtaining rights-of-way for overhead transmission lines, might mitigate problems with environmentalists arising from industrial plants in non-urban areas, and would provide additional taxes in metropolitan areas where such money was badly needed. Brush said that a record of trouble-free operation over the short-term would have little or no relevancy to reactor safety, per se. Brush went on to say that, although his group would readily agree that a number of outstanding issues should be resolved, there was little evidence to suggest that urban siting must await the results of further research and development. Rather, he said that they believed what was needed was more comprehensive review and analysis of the data already on hand, a revision and updating of the outmoded assumptions used in such criteria as TID–14844 (the "appendix' to 10CFR Part 100; DiNunno et al. 1962), and a meaningful interpretation of how the AEC would permit new ideas to be applied to licensing problems. When Mr. E. Case of the regulatory staff asked Mr. Brush what he considered to be an acceptable level of safety, Brush replied that many in the nuclear industry believed that the plants under construction presented an acceptable risk for urban siting.

Mr. W. J. Cahill of Consolidated Edison stated that there was a precedent for additional safety features for reactors in more densely populated areas. He pointed out that Consolidated Edison reactor sites, which involved higher population densities than others, had led to the use of additional safeguards for their reactors. Mr. Cahill said that he thought that plants could be adequately designed for a site like Ravenswood. He stated that industry did not design plants using probabilistic methodology to establish the design requirements and the adequacy of the design, and that the reliability of equipment was not known. Mr. Cahill said that industry could design the plant to protect against accidents which were considered credible, and that it was ultimately the responsibility of the regulatory groups to make a judgment as to what was credible.

When members of the ACRS and the regulatory staff pointed out that operating experience could turn up unfavorable aspects of operation and that lessons could be learned and the mistakes rectified in the next reactors, the general response from the industry representatives was negative. The industry group stated that operation of large power reactors during the next few years would not be meaningful with regard to safety evaluation and design.

In summary, two utilities, Consolidated Edison and Southern California Edison, expressed a need to use worse sites than Indian Point and Zion. Consolidated Edison wanted to use sites having population densities equivalent to or not far different from the Ravenswood site, while Southern California Edison desired to use sites like Bolsa Island or Burlington.

In the next months similar meetings were held with each of the light water reactor vendors, Babcock and Wilcox, Combustion Engineering, General Electric, and Westinghouse. Only Mr. J. M. West of Combustion Engineering made specific recommendations for engineered safety improvements that he thought would be appropriate in connection with metropolitan siting. Some excerpts from the statement by West at the meeting of May 5, 1970, follow:

To make any substantial progress in resolving the dilemma which now exists, we must make a clear distinction between the safety requirements for nuclear plants in metropolitan areas versus the requirements for rural locations. These requirements cannot be the same. I do not believe for one minute that after a reasonable amount of satisfactory operating experience is obtained on the present generation of light water reactors, an identical reactor will be licensed for a site in the heart of New York City or Los Angeles. If that were to occur I would feel that the safety requirements for non-populated sites had been too stringent.

Starting with satisfactory operating experience from essentially identical reactors, what additional requirements are likely to be imposed for metropolitan siting? In my opinion the answer does not lie in further ratcheting toward the items listed under Item D of this agenda (increased attention to quality assurance, further protection against common failure modes, increased margins, etc.). Instead I believe there should be a fresh look at the whole safety matter with an objective of defining what is desired and then attempting to meet the criteria for metropolitan nuclear plants by radical changes in design if necessary. Let's face it—we arrived where we are today on safety requirements for non-metropolitan plants by a torturous path which in retrospect does not appear too logical. Rather than add safety pins to the belt and suspenders already required to support the trousers, perhaps we should change the basic garment to something which is inherently more suitable—say coveralls.

One utility company has recently estimated that just in the past five years the cost of an 800 MWe nuclear power plant has increased by 20 to 30 million dollars solely due to additional licensing requirements. Even so, nobody would conclude that the plants licensed five years ago were unsafe—nor conversely that the present plants are necessarily overdesigned. We still assume, as we did then, that the fuel melts, major fractions of the fission products mix with air and steam in the containment building, that some of these fission products leak out under adverse meteorological conditions, etc.

What is a metropolitan nuclear plant likely to need that a rural plant does not have? To catalyze discussion I will mention the following things which have occurred to me:

• Engineered safeguards must be much less dependent upon correct sequential operation of "active" devices such as detectors, switches, valves, and motors. Our safeguards systems should be passive to the maximum practical extent.

• A fracture of the reactor pressure vessel or any other part of the primary system must be accommodated.

• An additional zero leakage barrier must be imposed to prevent the escape of fission products from the containment building. This leakage barrier must be effective over long periods of time without operator attention and without power.

• Containment building leakage must be continuously monitored.

• Discharges of radioactive materials to the atmosphere or to waterways must be further reduced.

• A large heat sink must be provided such that residual and decay heat can be accommodated for many hours without a source of electrical power and without any operator action.

• The plant must be designed against any credible natural forces such as earthquakes and hurricanes, and against external missiles such as aircraft.

None of the safety criteria which I visualize for a metropolitan nuclear power plant appear to be impossible to satisfy. A fracture of the reactor pressure vessel could be accommodated by having a double pressure vessel or by having a restraining envelope outside the reactor pressure vessel. This would cost several million dollars. We have a conceptual design of a Passive Reactor Containment System (and associated auxiliaries) to control leakage of radioactivity to the environs following an accident. This would also cost several million dollars. Even the "China Syndrome" could probably be accommodated at a finite cost. Radioactive discharges during normal operation could be reduced by a large factor at additional cost.

No problem has occurred to me which could not be handled, if it must be, in order to locate a nuclear plant in a populated area. The cost of safety features to handle these problems would be high. Utilities could not afford the cost of these features added to the already high cost disadvantage of transmission lines from remote sites. However, we must remember that several tens of millions of dollars can be spent, if necessary, on safety items in order to qualify a 1000 MWe nuclear plant for a highly populated zone. There is at least a fair chance that safety requirements can be met at a cost which is tolerable. I recommend that we attempt to identify what will be required, develop these designs to the point where cost estimates can be made, and assess whether metropolitan siting appears to be economically and technologically feasible. If so, a strong Government-industry program should be initiated.

Mr. West mentioned a core catcher, but stated that the core melt-through accident was not on his list of items requiring resolution for metropolitan siting, and stated that he hoped it was not on the ACRS list. This was almost the only mention of the China Syndrome problem throughout the meetings with the representatives of industry.

Another ACRS Letter Is Not Sent

The ACRS decided it would prepare a written report to the Atomic Energy Commission on the outcome of the meetings which had been held with members of industry, a report that would review the opinions of the ACRS concerning the possible use of sites worse than Indian Point and Zion. A tentative final letter was adopted by the committee at its one hundred twenty-seventh meeting, November 12–14, 1970, with the under-standing that the committee would talk to the commission prior to trans-mission of the letter, and that the letter could be revised after the discussion, if the committee so chose. The letter was similar in its recom-mendations to the draft letter discussed with the commissioners in September, 1969.

The draft ACRS letter was discussed in fairly general terms between the ACRS and the commissioners at the one hundred twenty-eighth ACRS meeting, December 10–12, 1970, after which the committee approved the report for issuance. However, before the letter had actually been sent, the commissioners requested that the ACRS defer issuing the report until it could be discussed further. The committee agreed, and the letter was discussed during the January, 1971 ACRS meeting with commissioners W. E. Johnson, C. E. Larson and J. T. Ramey. Commissioner Ramey stated that there was some doubt in the AEC's mind whether the committee letter on the use of reactors at more densely populated sites than employed to date would serve any useful purpose at this time, e.g., the same matters were already covered in the ACRS letter on the Newbold Island site.

Commissioner Johnson added that the letter would hurt in that it would provide information which would lead to increased public intervention regarding nuclear plants at currently acceptable sites. He saw no difficulty in imposing specific requirements mentioned in the siting letter for a specific case.

Commissioner Ramey noted that it would probably be possible to prepare a letter which only advised the public that the AEC and the ACRS had had discussions with industry about siting of plants near dense populations.

The committee decided to "file" and not send the 1970–1971 version of its report of "Use of water-cooled power reactors at sites more densely populated than those employed to date." This was based primarily on the desire of the commission not to receive such a report.

This was the last effort by the ACRS to write a "general" letter on safety requirements for the use of sites more populated than Indian Point and Zion. The upcoming Newbold Island construction permit review first opened the door toward the use of a somewhat more populated site. However, the regulatory staff decision that, under NEPA (National Environment Policy Act) considerations, a less populated site should be used by the applicant drew a rather firm line, at least for the interim, at sites such as Indian Point and Zion.

Chapter 13: Appendix A
Draft of an ACRS Report to the AEC on Location of Power Reactors at Sites of Population Density Greater than Indian Point-Zion

As the use of nuclear power has grown, water-cooled power reactors of progressively larger size and reactor sites of increasingly higher population density have come to be employed. Simultaneously, progress has been made in improving the design and construction of such reactors so as to reduce the already low probability of occurrence of accidents and mitigate further any potential accident consequences. Although this progress has been reasonably commensurate with the increases in reactor size and population density, the Advisory Committee on Reactor Safeguards believes that additional steps are necessary to justify the use of sites more densely populated than the Indian Point-Zion type of site. The Committee believes that operation of large water-cooled power reactors at sites of somewhat greater population density may be appropriate if:

(a) Prior to the time of receiving a construction permit, at least one year of satisfactory operating experience has been obtained with a reactor of generally similar design, power density, and power rating;

(b) Prior to the time of starting power operation, at least ten reactor-years of satisfactory operating experience have been obtained with reactors of essentially the same design, power density, and power rating;
and if the measures described below, additional to those required for the Indian Point-Zion type of site, are adequately effected.

1. The containment system should be designed to reduce to substantially lower levels the off-site radiation doses in the unlikely event of a major reactor accident, and should be capable of maintaining this level of protection even with substantial

degradation of the system. These more stringent requirements for the containment systems are appropriate for higher population density sites, because the present guidelines assume that in an emergency members of the public in the low population zone can be evacuated or otherwise protected in timely fashion. For sites of higher population density, evacuation or other effective protective measures for the close-in population are less certain to be achieved in the short times required, and it is necessary that the containment system provide a greater degree of protection.

2. Increased emphasis should be placed on detailed consideration of possible accidents leading to small or moderate releases of radioactivity to the environment, and means should be provided to reduce still further the probability of occurrence of, and the consequences of, such accidents. In particular, increased attention should be given to potential radwaste accidents. Similarly, current practices related to fuel handling, storage, and shipping should be re-evaluated and changes implemented wherever found appropriate to increased safety.

3. Further reduction in the already small releases of radioactivity from routine plant operation should be effected.

4. The number of safety issues remaining to be resolved between the start of construction and initiation of operation at power should be minimized. Where it appears that resolution of a safety issue may not be accomplished by the time of start of reactor operation, the plant design should incorporate whatever alternative features are necessary to provide adequate protection. An example of such a feature is the provision of permanent in-core instrumentation for use in the event that out-of-core instrumenation should not prove adequate.

5. Because of the small likelihood that proof of the efficacy of engineered safety systems under accident conditions will be obtained as a consequence of actual accident experience, extra margin should be provided in the design of these systems wherever such provision is practical and will clearly improve safety. As an example of extra margin, additional emergency core flooding capability might be appropriate.

6. Steps should be taken during the design of the reactor plant to provide further protection in areas related to possible degradation of reactor vessel integrity, such as leaks and vessel wall ruptures. The protective features provided should be of such design as to preclude their jeopardizing vessel integrity.

7. Additional consideration should be given in the design of the plant to protection against damage by missiles.

8. Greater assurance of maintainence of integrity of any portions of the primary system outside the containment, and appropriate additional means for coping with possible loss of their integrity, should be provided. The ACRS emphasizes again the vital importance of quality assurance, and the necessity for adequate consideration of diverse and independent means of protection against common failure modes in safety systems.

The Committee believes that realization of item (a) and demonstration of reasonable assurance of realization of item (b), together with adequate implementation of items 1–8, could provide a basis for considering applications for construction permits for large water-cooled power reactors at sites of somewhat greater population density than that of the Indian Point—Zion type of site (e.g., approximating that of the Burlington site). The Committee also believes that the additional protective features eventually resulting from these measures need not necessarily be incorporated in reactor plants either existing or yet to be constructed at sites of population density equal to or less than that of Indian Point-Zion.

Additional remarks of Dr. Stephen H. Hanauer are attached.

Additional remarks by Dr. Stephen H. Hanauer

In my opinion, approval of sites more densely populated than Indian Point—Zion for large water-cooled reactors should be based on verified facts, rather than reasonable assurance regarding the outcome of work not yet completed, as is appropriate for present sites, where evacuation of the surrounding population is feasible in an unforeseen emergency. For this reason, I cannot agree with the conclusion of this report, even though I concur with many of its recommendations. I do not believe that the necessary knowledge and experience are now available to support such a conclusion. It is my hope and expectation that the needed knowledge and experience will be obtained; that would be the appropriate time to consider the use of more densely populated sites, and suitable criteria for such use.

14

The Newbold Island Controversy

By letter dated May 6, 1969, the Public Service Electric and Gas Company of New Jersey requested an informal review of their proposed Newbold Island site by the regulatory staff and the ACRS. The proposed Newbold Island site consisted of 530 acres located on Newbold Island in the Delaware River in Mansfield Township, New Jersey. It was 4½ miles south of Trenton, New Jersey, and 11 miles northeast of Philadelphia, Pennsylvania. Cumulative population within the first 4 miles of the Newbold Island site was significantly lower than that of Zion, Indian Point, or Burlington. The cumulative population 5 to 33 miles around the Newbold Island site exceeded that of Indian Point or Zion. Beyond 34 miles, the projected cumulative population for the Indian Point site was greater than that of Newbold Island.

The ACRS Newbold Island subcommittee met on July 1, 1969. The population distribution for the Newbold Island site was compared with that of Indian Point, Zion, Trap Rock, and Burlington, using a rather simple, arbitrary, site-population index (Okrent 1979). The subcommittee members concluded that the Newbold Island site had a similar population index to Trap Rock, which the committee had previously decided was "not unacceptable" at its one hundredth meeting on August, 1968. One of the subcommittee members was Dr. Hanauer, who had dissented from the draft ACRS letter of July 1969, on "Location of Power Reactors at sites of

201

population density greater than Indian Point-Zion". This reinforces the idea that the subcommittee members felt the Newbold Island site was not significantly worse than Indian Point.

Approval of the Site

At its one hundred thirteenth meeting, September 4–6, 1969, the ACRS considered the Newbold Island site proposal, and, as shown in the ACRS letter of September 10, 1969, the committee decided the site was not unacceptable. This letter to AEC Chairman Glenn T. Seaborg signed by ACRS Chairman Stephen H. Hanauer is reproduced in appendix A of this chapter. However, the ACRS called attention to several matters that required attention, including measures to cope with pressure vessel leaks and rupture, as practical. In effect, many of the comments in the Newbold Island report were similar to those previously included in the never-issued, draft letter of July 1969 on the use of sites worse than Indian Point and Zion (see Chapter 13). The ACRS consciously withheld approval of the proposed Newbold Island containment building at the site review stage.

On February 27, 1970, Public Service Electric and Gas Company filed a formal application for the licenses required for the construction and operation of the proposed Newbold Island nuclear generating station. The applicant proposed certain additional features for these plants compared to those being incorporated in similar boiling water reactors in much less-populated sites. These included an additional main-steam-line stop valve; enclosure of the main-steam-line and isolation valves within a tunnel chamber which would discharge into the reactor building in the event of an accident; charcoal adsorbers in the system which processes radioactive waste gas in order to permit a holdup of the release of the noble gases; and a system to reduce turbine seal releases of radioactivity. The proposed secondary containment was a low-leakage building (10% per day), held at negative pressure to provide controlled and filtered venting of minor radioactivity releases from the primary containment, and having a 2 psig design pressure capability. These additional features were aimed at reducing either releases from routine operation, or those related to the existing design basis accidents for BWRs. They did not cope with accidents more serious than the existing design basis, and they were not trying to make serious accidents less probable.

In May 15, 1970, the AEC regulatory staff issued a preliminary report to the ACRS which indicated no major problems with the Newbold Island site.

The ACRS began its review of the Newbold Island plant very early in the process, rather than waiting for the staff to have essentially completed its own review. A first subcommittee meeting was held June 3, 1970, and both the topics listed in the September 10, 1969, site letter (see appendix A of this chapter), and others, were discussed. The reactor vendor, General Electric, was very reluctant to discuss the capability of the reactor to withstand postulated pressure vessel ruptures of various sizes, including rupture of a large nozzle. However, under strong pressure by the subcomittee, they agreed to review these matters. This topic was then discussed at considerable length in ensuing subcommittee and full committee meetings. General Electric concluded that the structural members within the reactor vessel and the ECCS were such that a break area up to four square feet anywhere in the vessel could be withstood and the core cooled acceptably. They also thought that the concrete shield wall around the vessel could be strengthened so that it would maintain its basic integrity and not generate unacceptably large missiles.

One matter that the ACRS forced a partial decision on during the Newbold Island review was the issue of anticipated transients without scram (ATWS).* General Electric had previously reported in generic studies and during the review of other projects that an automatic shutoff (or trip) of the recirculating pumps in the primary system would resolve short-term concerns in ATWS (leaving the long-term cooling requirements still to be reviewed). However, they had never included the feature. The regulatory staff had taken no position on what, if anything, was required or acceptable. The ACRS advised General Electric that, for the Newbold Island site, a commitment on the pump trip (or some other method for dealing with ATWS) would be required prior to completion of the ACRS construction permit review. The committee took a similar position with regard to the need for a way to control hydrogen following a LOCA.

There was much discussion of the adequacy of the ECCS and the potential for increasing its functional capability or reliability. General Electric maintained that the function of the system was well understood and that the proposed ECCS was adequate (General Electric said it had an 800° F margin for the design basis pipe break) and could not be significantly improved from a functional or reliability viewpoint. The regulatory staff officially remained aloof from such discussions. At a subcommittee meeting on April 26, 1971, they indicated they were divided concerning the desirability of improved ECCS at the Newbold Island site. The staff

---

*These refer to expected interruptions (transients) in normal operation which would ordinarily be terminated with rapid insertion of the control rods (scram). In an ATWS event, it is postulated that the control rods fail to go in for some unspecified reason. In this event rapid trip of the recirculating pumps would quickly reduce water flow in the core, leading to an increase in steam formation, a loss of reactivity, and earlier termination of the power transient. See Chapter 16 for a detailed discussion of ATWS.

officially favored improvements only in safety features as required by compliance with Part 100.

There was also considerable discussion of the relative advantages of the proposed low-pressure secondary containment vessel versus the higher-pressure (10–14 psig) building included in the preliminary proposal by Consolidated Edison for the Trap Rock site.

Comparison of the Newbold Island and Limerick Sites

Actually, at about the same time that application was made for a construction permit for the Newbold Island site, Philadelphia Electric applied for a construction permit for a BWR at the Limerick site west of Philadelphia. This was also a highly populated site. According to the rough site-population index used by the ACRS, Newbold Island was worse than the average Indian Point-Zion site, while Limerick was somewhat better. This index did not, however, include meteorology. The local meteorology for the Limerick site was worse, and the prevailing winds were generally easterly, that is towards Philadelphia from Limerick. Also, the population around the Limerick site exceeded that at Newbold Island for the first four miles. In the eyes of the regulatory staff this made the "Limerick site equivalent to Newbold Island, or worse." The regulatory staff stated that whatever requirements were made for the Newbold Island site should also apply to Limerick. However, in practice, they did not require strengthening of the shield wall around the reactor vessel, as was done at Newbold Island, or any other steps which went beyond the design basis accidents they normally reviewed.

One small sidelight of the review of the Newbold Island reactor was that the regulatory staff was asked by the ACRS to take a new independent look at the possible sources and effects of fire. However, the ACRS subcommittee minutes indicate that the staff did no additional review in this regard, and the matter was not pursued. This was in 1971, roughly four years before the Browns Ferry fire.

At the June 3, 1971, subcommittee meeting there was an inconclusive discussion of the proposed seismic design basis for the Newbold Island site. ACRS consultant S. Philbrick had suggested that a much larger earthquake than the applicant had postulated could occur, and ACRS consultant J. T. Wilson had estimated that the recurrence interval for the proposed safe shutdown earthquake* (a maximum acceleration equal to 20% of gravity) was only about 2,500 years. The regulatory staff had no

*Each reactor is designed so that it could be shut down safely with no damage to vital components if the safe shutdown earthquake were to occur. The safe shutdown earthquake is generally chosen to be a very improbable event.

response concerning an acceptable recurrence interval. The applicant stated he had no probabilistic basis in proposing the seismic design basis. The proposed seismic design basis was left unchanged.

At the June 3, 1971, subcommittee meeting, there were also some inconclusive discussions about the reliability of various engineered safety systems. When asked how the applicant knew that his plant had the necessary safety-related reliability, Mr. R. Schneider of Public Service replied that their decision was based on experience and on making logical improvements. He noted that if the AEC went much further in requiring expenditure of time and money for extremely conservative designs, it might stop the construction of nuclear power plants.

Three Dissents on Newbold Island

In its report to the ACRS of June 30, 1971, the regulatory staff concluded that the Newbold Island facility could be built and operated at the proposed location without undue risk to the health and safety of the public. At its one hundred thirty-sixth meeting, August 5–7, 1971, the ACRS completed its reviews and reported favorably concerning the proposed construction permit applications for both the Newbold Island and Limerick sites. However, the recommendation for Newbold Island was not unanimous; there were additional comments by three members of the ACRS, H. O. Monson, D. Okrent and N. J. Palladino, to the effect that a construction permit should not be issued unless there were a major change in the containment design, and the high-pressure portion of the emergency core-cooling system (which was intended to deal with small breaks in the primary system) was made redundant.

The Newbold Island letter was sent to AEC Chairman Glenn T. Seaborg and signed by ACRS Chairman Spencer H. Bush. It is reproduced in appendix B of this chapter.

The ACRS again reported favorably on the Newbold Island site on July 17, 1973, with the same dissents, following a re-review of the significance of increased population estimates within the first several miles from the site.

After the favorable ACRS letter, the construction permit application for the Newbold Island reactor proceeded to a hearing before an Atomic Safety and Licensing Board (ASLB). The case had become very controversial, with some members of state government and some congressmen from the region expressing concern or outright opposition.

On October 3, 1973, Dr. Ralph E. Lapp, appearing before the ASLB on behalf of the Commonwealth of Pennsylvania, testified against issuance of a construction permit. He stated his belief that the ACRS request for extraconservatism in plant design, and its requirement for additional safeguards, was a reflection of the need for insuring a higher degree of public protection because of the high population at risk at the site. He stated that in his opinion, the ACRS letter of July 17, 1973, was a conditional approval of the station and that the Atomic Energy Commission should resolve this condition before the construction permit was approved. "When three highly qualified scientists on the ACRS express reservations about a reactor installation. I believe that their recommendation should be given the most serious consideration." Then he went on to note that the ACRS recognized an unusual population risk. He called attention to the fact that the committee believed that emergency plans should include appropriate measures several miles beyond the proposed low-population zone with its radius of only one mile. Dr. Lapp stated "to the best of my knowledge this is the first time that ACRS has expressed a specific requirement for emergency plans to treat a Class 9 accident situation." Dr. Lapp concluded "I believe that the Board is faced with a problem that is greater than that of a single nuclear power site. If Newbold Island is approved, then the way is open for utilities to site future plants in locations of equal or higher population-at-risk situations."

A Change in the Staff Position

In an unusual step, the regulatory staff on October 5, 1973, issued an opinion that, while the Newbold Island site was acceptable, they had applied considerations of the National Enfironmental Policy Act (NEPA) and concluded that the applicant had other significantly less populated sites available. They, therefore, had concluded that the reactors should not be built at the Newbold Island site but elsewhere. Shortly thereafter, Public Service announced it had decided to move th Newbold Island station near to the relatively rural Salem, New Jersey, site.

The regulatory staff did not take a similar position on the Limerick station, however, and these reactors received a construction permit and were constructed. The staff had previously argued to the ACRS that the Limerick site was "equivalent or worse" than Newbold Island when they combined population and meteorological considerations. Yet, the staff did not apply NEPA to the Limerick site, nor did they supply any rationale for the difference in actions taken on the Newbold Island and Limerick sites. Newbold Island was the site with the highest surrounding population density to be approved by the regulatory staff and the ACRS. The decision by the staff to apply NEPA considerations to the Newbold Island site came as a surprise to the ACRS.

A year earlier, the regulatory staff had submitted to the ACRS for discussion a reactor site index, which had been developed by the staff. This draft approach from the staff was discussed with an ACRS subcommittee on January 5, 1973. The reactor site index was the product of a site-population factor and a plant design factor. The site-population factor was proportional to population weighted by a factor [(distance $^{-1.5}$)] to allow for meteorological dispersion, using a sector approach in a fairly conventional way. The plant design factor proposed to take into account any differences in accidental radioactive releases by comparing doses at the low-population zone boundary, using the traditional approach of Part 100, which assumes containment integrity and gives credit for containment cleanup systems.

What is somewhat surprising is that, as late as October 1972, the regulatory staff was still completely ignoring considerations of class 9 accidents in the development of a reactor site index. In fact, the proposed method gave a high (bad) plant design factor to some plants having a large exclusion area and a large, surrounding low-population zone. The curious result was that on a scale where the Indian Point 2 site had a rating of unity, Newbold Island had a lower (substantially better) reactor site index while relatively remote sites like Peach Bottom and McGuire had higher reactor site indexes and appeared to be much worse than Indian Point 2.

The subcommittee seriously questioned the proposed staff reactor site index and it was dropped.

In 1975, in section 2.1.3 of the USNRC Standard Review Plan (NUREG–75/087) the staff defined a quantitative approach to acceptability of population distribution, as follows (NRC 1975b).

If, at the construction permit stage, the population density, including weighted transient population, projected at the time of initial plant operation exceeds 500 persons per square mile averaged over any radial distance out to 30 miles (cumulative population at a distance divided by the area at that distance), or the projected population density over the lifetime of the facility exceeds 1,000 persons per square mile averaged over any radial distance out to 30 miles, special attention should be given by the Staff to the consideration of alternative sites in the environmental review.

Transient population should be included for those sites where a significant number of people (other than those just passing through the area) work, reside part-time, or engage in recreational activities, and are not permanent residents of the area.

The specified low population zone is acceptable if (a) it is determined that appropriate protective measures could be taken in behalf of the enclosed populace in the event of a serious accident; (b) dose computations for the outer boundary of the LPZ, as discussed in the review plans for Section 15, are within Part 100 guidelines; and (c) the nearest boundary of the closest population center (as defined in Part 100) is at least one and one third times the distance from the reactor to the outer boundary of the low population zone.

The population center distance is acceptable if there are no likely concentrations of greater than 25,000 people over the plant lifetime closer than the distance designated by the applicant as the population center distance.

The Standard Review Plan was like the plan proposed by the ACRS and vetoed by the AEC commissioners in 1967. It effectively limited sites to those no worse than Indian Point and Zion. Under these guidelines which did not take into account meteorology, the Limerick site would have looked better than Newbold Island.

Since issuance of the Standard Review Plan, the regulatory staff have reported unfavorably on at least one site on the basis of population and other considerations, namely the Perryman site in Maryland.

Chapter 14: Appendix A:
ACRS Report to the AEC on
Public Service Electric and Gas Company—Newbold Island Site

At its 112th meeting, August 7-9, 1969, and its 113th meeting, September 4-6, 1969, the Advisory Committee on Reactor Safeguards considered the Newbold Island site, which the Public Service Electric and Gas Company proposes as the location for a nuclear power plant including two boiling water reactors of approximately 3400 MW(t) each. The site consists of approximately 500 acres located on Newbold Island in the Delaware River. A relatively high population density is associated with this site; it is 4½ miles south of Trenton, New Jersey (1960 population—114,000) and 11 miles northeast of Philadelphia, Pennsylvania (1960 population—2,000,000). The nearest population center is a grouping of suburbs in Bucks County, Pennsylvania, known collectively as Levittown (1960 population— 70,000), with its nearest boundary 3.4 miles from the site. An ACRS subcommittee visited the site on July 1, 1969. During its review, the Committee had the benefit of discussions with representatives of Public Service Electric and Gas Company and their consultants, and the AEC Regulatory Staff, and of the documents listed below.

Preliminary studies of the geology, seismology, hydrology, and meteorology of the site have been made and have revealed no significant problems. Natural draft cooling towers will be used in the plant.

The conventional dry well and suppression- chamber containment system will be enclosed in a low leakage reactor building with air recirculation and filtration to reduce the releases of radioactivity in the unlikely event of an accident. The Committee believes that the proposed containment system is a useful approach, but cannot comment at this time on its adequacy.

Special attention will be required with regard to the integrity of any portions of the primary system outside the containment and to the steamline isolation valves. Appropriate additional means for coping with possible valve leakage or a loss of integrity outside the containment should be provided.

Public Service Electric and Gas Company describes procedures involving additional hold-up of off-gas releases during routine plant operations. The Committee believes that special attention should be given to the control of liquid waste releases and to the prevention of radwaste accidents, as additional means of keeping radiological releases at a very low level.

The Committee believes that, for this site, additional study of the problem of possible degradation of reactor vessel integrity, such as leaks and vessel wall ruptures, is needed. Measures that will ameliorate these problems should be implemented to the extent that they are practical and significant to public safety. The features provided should be of such design as to prevent their interference with other engineered safety features.

Other matters noted in previous ACRS letters pertaining to large water reactors should receive appropriately greater attention in the design of the plant. The Committee believes a more conservative approach is appropriate in the design of a plant at this site, with regard to the margins in the engineered safety systems, protection against possible internally-generated missiles, and the number of items to be resolved after the construction permit review.

The Committee emphasizes again the vital importance of quality assurance, and the necessity for adequate consideration of diverse and independent means of protection against common failure modes in safety systems.

The conclusion reached by the Committee regarding this site has been influenced in part by its expectation that some satisfactory experience will have been obtained with reactors of this general type by the time a construction permit is issued, and some satisfactory experience will have been obtained with reactors of this type having the same power and power density as those proposed for this site by the time an operating liscense is issued.

The Advisory Committee on Reactor Safeguards believes that, subject to the above comments, the Newbold Island site is not unacceptable with respect to the health and safety of the public, for a plant having the general characteristics described above and designed with due attention to the other matters discussed.

Chapter 14: Appendix B
ACRS Report to the AEC on
Newbold Island Nuclear Generating Station Unit Nos. 1 and 2

At its 136th meeting, August 5–7, 1971, the Advisory Committee on Reactor Safeguards completed its review of the application by the Public Service Electric and Gas Company for a permit to construct the dual-unit Newbold Island Nuclear Generating Station. This project was also considered at the 130th, 133rd, 134th, and 135th meetings of the Committee on February 4–6, May 6–8, June 10–12, and July 8–10, 1971, respectively; and at Subcommittee meetings on June 3, 1970 at Argonne National Laboratory, and on February 3, March 29, April 26, June 3, July 7, and August 4, 1971 in Washington, D.C. During its review the Committee had the benefit of discussions with representatives and consultants of the applicant, the General Electric Company, and the AEC Regulatory Staff. The Committee also had the benefit of the documents listed below. The Committee reported the results of its pre-application site review to you in a letter dated September 10, 1969.

The station will be located in New Jersey on 530-acre Newbold Island which is near the east bank of the Delaware River about 4½ miles south of Trenton, New Jersey (1970 population—105,000) and 11 miles northeast of Philadelphia, Pennsylvania (1970 population—2,000,000). The nearest population center is a grouping of suburbs in Bucks County, Pennsylvania, known collectively as Levittown (1970 population—72,000), with its nearest boundary 3.4 miles from the site. The applicant has specified a radius of one mile for the low population zone, which had in 1969 a transient population associated with industry of approximately 1200, and a small resident population which is expected to be about 100 by 1985. The minimum exclusion distance is 700 meters, which extends to the west bank of the Delaware River. As pointed out in the Committee's report of September 10, 1969, a relatively high population density is associated with this site.

Each unit includes a boiling water reactor to be operated at 2393 MWt. With respect to core design, power level, and other features of the nuclear steam supply system, the Newbold Island units are essentially duplicates of the Browns Ferry Units 1, 2 and 3, and Peach Bottom Units 2 and 3. Waste heat from the station will be rejected to the atmosphere by natural draft cooling towers.

In its report of September 10, 1969, the Committee listed several matters which it believed warranted special attention in the design of a plant for the Newbold Island site. In response to those recommendations, the applicant has included in the Newbold Island design several features, in addition to those normally provided for boiling water reactor units, to reduce still further the potential for release of radioactivity to the environment. The principal additional features are described below:

Reactor Building. For each unit, the conventional steel drywell and suppression chamber primary containment, the fuel handling area and spent fuel pool, and the principal components of the engineered safety features are contained in an unlined reinforced concrete building of cylindrical shape with a domed roof. This building is designed to Class I seismic standards and to resist the standard tornado, and missiles from this or other sources. The building can resist an internal pressure of 2 psig, and inleakage at a differential pressure of ¼ inch of water will be limited to 10 percent of the building volume per day. A filtration, recirculation, and ventilation system (FRVS) is provided to recirculate and filter the reactor building atmosphere and maintain the building at a negative pressure relative to the outside environment.

Main Steam Lines. A low-leakage, slow-acting, stop valve has been added downstream of the two fast-acting valves in each main steam line, and a seal air system has been provided to further reduce leakage of radioactivity after main steam line isolation. The portion of the main steam lines containing the isolation valves is enclosed in a Seismic Class I tunnel chamber connected to the reactor building so that any out-leakage following the unlikely event of a design basis loss-of-coolant accident will be treated by the reactor building FRVS before release to the atmosphere. The entire length of the main steam lines up to and including the turbine stop valve will be designed to Class I seismic standards. The main steam lines from the third isolation valve to the turbine stop valve will be designed and fabricated in substantial accordance with the requirements of AEC quality assurance Classification Group B. In addition, selective inspection of critical areas of this piping will be performed during refueling outages.

Radioactive Waste Disposal. The radioactive waste disposal systems include several features beyond those normally provided in boiling water reactor plants. The liquid waste system permits the recycling of equipment and floor drain wastes and the evaporation of chemical and laundry wastes before discharge to the environment. The gaseous waste system provides for the recombining of hydrogen and oxygen, condensing the vapor, hold-up for decay of short-lived isotopes, and cryogenic separation of the noble gases. Krypton and xenon may be stored for periods sufficiently long that krypton-85 becomes the only significant remaining radioisotope. Provisions will be made to utilize nonradioactive steam in the turbine gland seals and to process containment purge gases when deinerting. The Committee believes that these waste management systems are capable of limiting releases of radioactivity to the environment to levels that are as low as practicable.

Reactor Vessel Integrity. The applicant has described improvements in the design and fabrication of the reactor vessel. These include redesign of the large nozzles to reduce stress concentrations; redesign of the bottom band to reduce the number of welds and improve the capability for inservice inspection; and improved procedures and standards for inspection during fabrication. The applicant has studied the problems related to possible degradation of reactor vessel integrity and has concluded that a nozzle failure or a small break would not impair the integrity of the biological shield, the primary containment, or the reactor internals, and would not affect the ability to cool the core. In addition, the biological shield has been redesigned to increase substantially its ability to withstand internal pressures, jet forces, or missiles.

Emergency Core Cooling System. The emergency core cooling system (ECCS) has been modified in two ways. The high-pressure coolant injection (HPCI) system has been changed to inject water directly to the core through the core spray sparger rather than into the downcomer region via the feedwater sparger. In addition, the applicant has stated that the steam-turbine driven HPCI pump will be modified to the extent feasible to increase the volume of water delivered to the core. The low-pressure coolant injection (LPCI) system has been changed to inject water inside the core shroud through four separate vessel penetrations, rather than through the recirculation lines. The applicant has stated that these changes provide increased reliability of these systems and reductions in the peak clad temperatures attained in the unlikely event of a loss-of-coolant accident.

The Committee believes that the design changes described above are suitably responsive to the concerns stated in its letter of September 10, 1969 regarding additional matters which should be considered for a plant at the Newbold Island site.

In the event of an unisolable break of an instrument line or a process line, reactor coolant will be discharged to the reactor building. Since the instrument lines will contain a ⅜-inch flow-restricting orifice inside the primary containment, failure of as many as eight such lines will not lead to pressures inside the reactor building greater than the 2 psig at which it relieves to the environment. However, failure of a process line, if not isolated in a very short time, could lead to pressures in excess of this relief pressure and significant amounts of reactor coolant would be discharged to the environment. Although the off-site doses from such an accident would be well within the 10 CFR Part 100 guidelines, they would be comparable to or greater than the doses calculated for other less probable accidents. The Committee believes, therefore, that the applicant should make design provisions for reducing the quantity of reactor coolant discharged to the reactor building in the event of a process line break.

The applicant has studied design features to make tolerable the consequences of failure to scram during anticipated transients, and has concluded that automatic tripping of the recirculation pumps and injection of boron could provide a suitable backup to the control rod system for this type of event. The Committee believes that this recirculation pump trip represents a substantial improvement and should be provided for the Newbold Island reactors. However, further evaluation of the sufficiency of this approach and the specific means of implementing the proposed pump trip should be made. This matter should be resolved in a manner satisfactory to the AEC Regulatory Staff and the ACRS during construction of the plant.

The applicant has stated that a system will be provided to control the concentration of hydrogen in the primary containment that might follow in the unlikely event of a loss-of-coolant accident. The proposed system is not capable of coping with hydrogen generation rates in accordance with current AEC criteria unless the primary containment is inerted. Therefore, the Committee believes that the containment should be inerted and that the hydrogen control system should be designed to maintain the hydrogen concentration within acceptable limits using the assumptions listed in AEC Safety Guide 7, "Control of Combustible Gas Concentrations in Containment Following a Loss of Coolant Accident."

Other problems related to large water reactors have been identified by the Regulatory Staff and the ACRS and cited in previous ACRS reports. The Committee believes that resolution of these items should apply equally to the Newbold Island Station.

The Committee believes that the items mentioned above can be resolved during construction and that, if due consideration is given to these items, the Newbold Island Nuclear Generating Station Units Nos. 1 and 2 can be constructed with reasonable assurance that they can be operated without undue risk to the health and safety of the public.

Additional comments by Dr. H. O. Monson, Dr. D. Okrent and Dean N. J. Palladino are attached.

Additional Comments by Dr. H. O. Monson, Dr. D. Okrent and Dean N. J. Palladino

Although the large, low pressure, high in-leakage secondary reactor building proposed by the applicant for Newbold Island Units 1 and 2 represents an improvement over reactor buildings currently employed for BWRs at sites with lower surrounding population densities, we believe that further improvement is appropriate. The relatively small volume of the steel pressure-suppression type primary containment introduces some crowding of equipment and some attendant problems in the simultaneous accomplishment of full protection against violation of primary containment by possible missiles, jet forces, and pipe whip, and accomplishment of full access for in-service inspection. Some further protection would be provided against extremely low-probability accidents involving a concurrent loss of primary system integrity and a limited violation of primary containment by the use of a large, relatively high-pressure (of the order of 10 psi, as has been proposed for a BWR at another site having a comparable surrounding population density), low-leakage, secondary containment building. Such a high-pressure, secondary containment, coupled with a pressure-suppression primary containment, provides a combination which can tolerate a fairly substantial violation of primary containment arising from the same event which caused a loss of coolant, as well as further protection against unforeseen events. We believe that this improvement in safety capability is warranted for a more densely populated site like Newbold Island, and recommend that the issuance of a construction permit be contingent on the use of a high-pressure, low-leakage secondary containment.

For postulated loss-of-coolant accidents involving small break sizes, the high-pressure coolant injection system (HPCI) arranged so as to inject into one of the core spray loops is predicted by the applicant to be highly effective in limiting peak clad temperatures to moderate levels. We believe that for a high power, high-power-density reactor at a site as densely populated as Newbold Island, the applicant should give further consideration to the use of an HPCI system on the second core spray loop. The purpose would be to provide redundancy of this means of protection in the event that the single HPCI system became ineffective because of failure of an HPCI component or because the accident arose from rupture of the core spray line into which the HPCI injects. The automatic depressurization system which together with the low-pressure emergency cooling systems constitutes an alternate means for coping with small breaks, albeit by introducing a larger opening, would continue to serve as a backup.

15

Some Aspects
of the Regulatory Process

General Design Criteria

All the LWRs conceived and proposed to the AEC for construction permits from the Shippingport reactor in 1953 to Dresden 2 in 1965 were generated without a set of safety criteria that the design was required to meet. AEC review of these reactors was on an ad hoc, case-by-case basis, and although a general design pattern evolved, there were no criteria against which the design could be compared, and there was essentially no review of the detailed design approach, which is vital to the actual safety level achieved.

Actually, individual design criteria had been evolving over the years prior to 1965. Typically, with each new construction permit review (and with the review of some difficult sites, such as Corral Canyon for the Los Angeles Department of Water and Power) new design criteria were introduced as the result of something specific to a site or reactor. These arose from questions concerning possible low-probability events not previously considered, and from an unusual operating experience which had generic implications.

A simple example of the development of a site-related criterion relates to tornadoes. When the General Electric Company proposed to build the Southwest Experimental Fast Oxide Reactor (SEFOR) in Arkansas, the need to provide suitable protection against tornadoes came into focus. It was necessary to establish design basis tornadic wind speeds and to consider other possible effects such as tornado-induced missiles. As a result of this review and additional reviews of the matter in the ensuing months, it became clear that there was a "tornado belt" in the United States which included Arkansas. But it was soon observed that for almost all regions east of the Rockies tornadoes occurred at a frequency which was smaller, but not radically so. Very destructive tornadoes had occurred in Massachusetts and other places far from the "tornado belt." Thus, tornado requirements were applied to all future power reactors east of the Rockies; and this later spread to the entire United States.

Floods were taken into account somewhat in the reactors constructed in the early 1960s; however, such criteria became more stringent with the evolution of the probable maximum flood approach in the late 1960s and early 1970s.

Hurricanes did not become a focus of site-related phenomena until review of the Turkey Point site in Florida in 1966. Afterwards, hurricanes began to receive increased emphasis for all East Coast and Gulf Coast sites; later they were also considered as a possibly serious cause of flooding at many inland sites.

Seismic criteria evolved in a more complex way. There was little if any seismic requirement beyond the uniform building code for reactors east of the Rockies in the 1950s. Then, the matter received considerable attention for sites in California, beginning with the site reviews for the Los Angeles Department of Water and Power and Southern California Edison and progressing rapidly with the San Onofre, Bodega Bay and Corral Canyon (Malibu) sites. By 1964, seismic design began receiving detailed attention in the eastern United States as well.

Engineered safety features also evolved considerably during the period 1960–1965. Dresden 1, for example, a 1950s vintage reactor, had as many as six off-site power lines feeding the plant, but no on-site emergency alternating current power source. In the 1960s a swath of tornadoes knocked out all lines concurrently; fortunately no accident ensued. Hence, with the passage of time, first one small on-site diesel, then a larger diesel, then redundant diesels to drive containment-related safeguards became the standard. And in 1966, redundant on-site power had to be available to power the ECCS, requiring still larger diesels (or their equivalent).

In the spring of 1965, the regulatory staff began developing general criteria for nuclear power plants under pressure from the commissioners and with the expectation or knowledge that an AEC regulatory review panel planned to recommend considerable emphasis on the development of criteria.

On May 26, 1965, Mr. Price forwarded a draft document entitled "Criteria for Nuclear Power Plant Construction Permits" to the ACRS for comment. The staff stated that design criteria were needed more urgently than criteria for the operating stage;* hence, priority was being given to criteria for construction.

At the first subcommittee meeting on the subject, June 5, 1965, Mr. Price indicated that the reasons for development of guides at that time were as follows:

(a) Continued pressure from the nuclear industry for better definition of the information required at the various stages of review. The Oyster Creek and Nine Mile Point Atomic Safety and Licensing Board decisions had emphasized the need for clarification.

(b) The desire of the commission to clarify its position by a redefinition of the rules rather than a policy statement.

(c) The increased workload anticipated in the future which would require a set of standards against which proposals could be judged. In addition, the guides would help to identify important areas so that submissions by applicants would be more to the point with a reduced amount of extraneous material which must be reviewed.

The general statement preceding the draft criteria sent to the ACRS began:

> Attached hereto are a number of proposed criteria for the design of a nuclear power facility which, upon the supplying by an applicant of adequate information to permit a conclusion that the criteria are fulfilled, would by definition justify a finding of reasonable assurance that the facility can be built and operated without undue risk to the health and safety of the public.

The ACRS had a number of problems with this philosophy. They generally felt that the criteria, after they had been worked over and found acceptable, might represent necessary but not sufficient design requirements, and that there would be a continuing need for engineering judgment.†

Reactors had been receiving construction permits based on a commitment to meet rather general criteria, but the criteria had been rapidly increasing in number and in detail. And, as the passing of time would soon show, the regulatory process was on the verge of shifting from the requirement of general criteria and the plausibility of their being met, to a continually increasing interest in specific design information prior to issuance of the construction permit.

*Unfortunately, operating reactors did not receive appropriate emphasis prior to the Three Mile Island accident in 1979, 14 years later.

†This proved to be a justified concern. The nuclear industry adopted the position that meeting the regulatory criteria (or more generally speaking, the licensing requirements) was adequate, and it usually did not perform its own assessment of possible specific weaknesses.

At the special ACRS meeting, June 18, 1965, the committee was divided as to the need for such criteria. Some believed the criteria would be of use to the Atomic Safety and Licensing Boards but not particularly to applicants. Others believed that the criteria were worthwhile. ACRS member Thompson thought that the time was too early for issuance of such design criteria and separately advised the commissioners to that effect. At the next subcommittee meeting, July 15, 1965, ACRS member Rogers was concerned that the existing list might be insufficient, and that the criteria should not be completed until four or five reactors situated near cities had been reviewed and analyzed.

At the July 15 subcommittee meeting, Mr. Price was asked by ACRS member Okrent how the criteria handled the problem of different requirements for sites in the country, near cities, and in cities. Mr. Price stated that the only item related to this matter was the amount of credit to be allowed for engineered safety features in reducing the calculated off-site doses (in the arbitrarily postulated MCA which assumes an intact containment).

The minutes of the July 21–22, 1965, subcommittee meeting record considerable discussion on the criterion dealing with containment design basis. Dr. Doan of the regulatory staff wanted to define the maximum credible accident as the accident involving complete loss of coolant, followed by meltdown of the entire core, and the occurrence of a metal-water reaction. It was noted that in the regulatory staff safety evaluation of the San Onofre 1 site, the staff had given credit for the ECCS preventing more than 6% of the core from melting. Nevertheless, Dr. Doan believed that reactors should be designed to withstand the worst consequence of any credible accident, should the other engineered safeguards fail.

At the end of the discussion it was decided to reword the particular criterion as follows:

Provisions must be made for the removal of heat from within the containment structure to maintain the structure within design limitations. The containment must be designed on the basis of complete depressurization of the primary system, melt-down uninhibited by core protective systems, and the occurrence of a metal-water or other chemical reaction. If engineered safeguards are needed to prevent containment vessel rupture due to heat release by an accident including these assumptions or by decay heat, at least two independent, differently designed systems which can accomplish the necessary functions must be provided. Each of these systems must be redundant in vital components so that no credible failure of a single component could prevent either of the systems from functioning properly.

This acknowledged that credit would be given for the effect of engineered safeguards in deciding the required containment pressure, but not how much credit.

The staff's early drafts of the general design criteria reflected what appeared to be the then accepted practice of the regulatory staff. As the criteria went through hard review and evaluation in successive drafts, they were made, more and more, to reflect what the regulatory staff and the ACRS thought should be employed as general criteria for upcoming plants. The process of trying to write the criteria led to rethinking of the adequacy of much of what had been accepted on past plants.

At its sixty-seventh meeting, October 7–9, 1965, the ACRS reviewed a comparison of the proposed design criteria with the design bases for the upcoming Dresden 2 BWR, and then gave a favorable committee opinion concerning the publication of the latest draft of the criteria in the Federal Register for public comment.

On November 22, 1965, the Atomic Energy Commission issued a press release announcing the proposed criteria and requesting public comment (AEC 1965b). Following receipt of the public comments on the proposed design criteria of November 22, 1965, considerable effort on redrafting the criteria ensued by the regulatory staff with frequent interaction with the ACRS.

Other events also transpired. On November 24, 1965, two days after publication of the draft criteria, the ACRS sent a letter to Chairman Seaborg of the AEC concerning pressure vessels (see Chapter 7) which initiated major efforts by the AEC and the nuclear industry on improved pressure vessel design, fabrication, and in-service inspection. And, in the summer of 1966, the China Syndrome question arose during the construction permit review of the Dresden 3 and Indian Point 2 reactors, leading to ACRS recommendations for major new efforts on improved primary system integrity and major emphasis on greatly improved ECCS, all as part of the recognition that the containment building was not an independent barrier.

Hence, there were several major changes in the revised set of proposed general design criteria which were issued for comment by the AEC on July 10, 1967 (AEC 1967).

The original 27 criteria had grown to 70, but this of itself was not especially significant, since some separation of the original criteria had taken place. However, there now was a separate cluster of criteria on ECCS, and there were a large number related to primary system integrity. The containment design basis (criterion 49) did not explicitly include coping with full meltdown of the core; it contained a vague phrase— "including a considerable margin for effects from metal-water or other chemical reactions that could occur as a consequence of failure of emergency core cooling systems."

Criterion 44 called for "at least two emergency core cooling systems preferably of different design principles, each with a capability for accomplishing abundant emergency core cooling." And the single-failure criterion, which had appeared in the 1965 draft in slightly different words, was prominent in the July 1967 version of the general design criteria.

The press release stated that the design criteria were intended to provide flexibility, and, with few exceptions, they did. Hence, they still served primarily to define the general safety areas which needed to be addressed. They did not provide a quantitative safety goal to be reached, or some other quantitative basis for establishing the adequacy of any particular specific design. A commitment by an applicant to meet the general design criteria provided no basis for assessing the safety level he would seek to provide. Nor did the criteria, for the most part, establish the safety level that the regulatory staff would require in order to approve construction and operation of a reactor. Not that this was good or bad. Quite consciously, the general design criteria left most matters up to "engineering judgment."

Except for the qualitative reference to a margin for chemical reactions in containment design, the general design criteria made no reference to problems arising from core melt or to methods of coping with or amelioration of the consequences of core melt.

The "proposed" criteria of July 10, 1967, provided "interim guidance" to the regulatory staff and the nuclear industry for several years, even though they were not formally adopted as an AEC regulation.

On February 20, 1971, the AEC published a revised set of general design criteria which became Appendix A to Part 50 of the AEC's regulations after 90 days (AEC 1971a).

The 1971 criteria, now 64 in number, were generally drafted to conform with the design features of LWR plants which had received permits during the previous few years. They still remained general in nature. The word "appropriate" was used very often, e.g., "Appropriate consideration of the most severe of the natural phenomena that have been historically reported" and "Fire detection and fighting systems of appropriate capacity and capability shall be provided," etc.

Criterion 35, no longer required two ECCSs, each capable of providing abundant cooling; rather, the criterion now said "A system to provide abundant emergency core cooling shall be provided", and further on, the single-failure criterion was imposed. This represented a weakening of the equivalent criterion (number 44) in the 1967 version; it represented an acceptance of the actual situation on the reactors which were being approved in 1971.

None of the criteria related to accidents involving large-scale core melt.

An example of the problems encountered in going from the general design criteria to specific implementation is available from criterion 3—fire protection. All of the reactors approved in the few years preceding 1971 and for a few years after 1971, were evaluated by the regulatory staff as having met this criterion. However, when the Browns Ferry fire occurred in 1975, a full-blown technical review of the adequacy of protection against fire ensued, and major specific modifications were required, even on existing plants, without any change in the wording of criterion 3.

The statement of considerations (or introduction) to the 1971 general design criteria pointed out several safety considerations for which specific requirements had not as yet been sufficiently developed and uniformly applied in the licensing process to warrant their inclusion in the criteria at that time. Most of that list, including all those related to redundancy, common mode failures, systematic, non-random failures, and passive failures, continued to remain open issues for the ensuing years.

Despite their lack of specificity and the existence of some gaps, the general design criteria proved to be useful over the years and to have a continuing relevancy.

The development of detailed regulatory positions with regard to the matters covered in the general design criteria (and with regard to other safety matters) began rather actively around 1967–1968. A considerable, continuing growth was occuring in size of the regulatory staff, and with this came increased breadth and depth of knowledge in specific technical areas. The less-than-satisfactory experience with operating license reviews for reactors where only general design information had been available at the construction permit stage, and the difficulty in implementing major design changes in an already constructed plant, automatically led to growing emphasis on getting more design detail at the construction permit stage.

Since a process of promulgating new AEC regulations covering specific design criteria was expected to be slow and cumbersome, a more flexible approach was sought for the documentation of specific approaches acceptable to the regulatory staff (and the ACRS). The regulatory staff began to develop internal documents which specified acceptable detailed design approaches to specific problems. In 1970, the regulatory guide method to public documentation of acceptable design approaches was initiated with the issuance of the first such guide. (This dealt with the requirements for net positive suction head for ECCS pumps* and arose as a way of implementating a philosophic safety concern that the function of the ECCS was not to be compromised because of a loss in containment integrity, which, were it to occur, could change a relatively moderate accident to one having severe effects.)

The regulatory guide approach flourished so that by 1978 more than 100 such guides had been published. The guides did not carry the force of law like the general design criteria; however, for the most part utilities found it

*The sources from which pumps draw their water must be at a high enough pressure to avoid cavitation in the pumps.

simplest to follow an approach prejudged acceptable by the regulatory staff.

Pressure Vessels: Inspectability, Inspection, and Integrity

The ACRS report of November 23, 1965, on pressure vessels produced sharp verbal reactions from the nuclear industry and the AEC, as was discussed in Chapter 7. With the next few cases, Brookwood, Millstone Point, and Indian Point 2, and later with Newbold Island, the ACRS established a pattern of dealing with the question of "coping with vessel failure" which has remained essentially unchanged. More specifically, for the more rural sites, the ACRS accepted the reactors without measures to deal with vessel failure, while for the Newbold Island site, which was somewhat more populated than Indian Point 2, the ACRS asked for limited protection against pressure vessel failure, including a capability for cooling the core.

To some extent, the advent of the China Syndrome matter, with the recognition of the direct correlation between core melt (for any reason) and a loss of containment integrity, lessened earlier emphasis on pressure vessel failure as possibly the most significant source of a large reactor accident. Nevertheless, the ACRS remained very interested in improving the reliability of pressure vessels.

Since neither the nuclear industry nor the American Society of Mechanical Engineers (ASME) code committee responded with great rapidity to oral urging by the ACRS for more stringent requirements during fabrication and for a program of in-service inspection, the ACRS decided to initiate a program within the AEC regulatory groups for the development of detailed requirements. This was an unprecedented action. At its special meeting, December 2–3, 1966, the ACRS voted to ask Mr. Price and Mr. Shaw of the AEC whether members of their groups could and would work with one or more ACRS members to develop additional requirements for ASME Section III nuclear pressure vessels:* these were to be prepared quickly so they could be considered by the ACRS within two or three months.

At this meeting, there was also considerable discussion on the inspectability of pressure vessels, and the lack of access for inspection in BWRs. The ACRS agreed that such access was needed in future reactors. The question was, "How soon could and should new requirements in this regard be imposed?"

*The American Society of Mechanical Engineers (ASME) was the body with the responsibility for producing codes to be used by industry in the design and fabrication of pressure vessels. ASME Section III applied to nuclear pressure vessels.

These matters were discussed further at the regular ACRS meeting on December 8–10, 1966 (the eightieth). The Browns Ferry BWRs were coming up for review soon, and for BWRs, there was also some concern about access to the bottom interior of the reactor vessel for inspection. (This access was very difficult in existing designs.) The following ACRS motion was adopted at the meeting:

With regard to the inspectability of reactor pressure vessels for pressurized and boiling water reactors, the following is the position of the ACRS.

1. The interior of the vessel, including the bottom head, should be accessible for general visual observation on a scheduled periodic basis. Such observation has as its objective detection of mechanical damage or structural failure of reactor internals.

2. Practical means of access should be provided to essentially 100% of a reactor pressure vessel surface, either from the inside or outside or a combination thereof; the purpose of this access is to permit thorough inspection of the vessel at appropriate intervals by visual means and ultrasonic or other suitable methods.

3. The ACRS realizes that it may take time to achieve these aims, but expects to see them fully achieved in plants for which construction permit applications are filed more than one year after announcement of this position.

4. The foregoing should be announced formally to the nuclear industry within the next several months. The target date for the announcement should be selected at the February, 1967 ACRS Meeting.

The ACRS chairman was asked by the committee to discuss this position with Mr. Price, and to explore with him possible means for informally advising the industry in advance of the date to be selected.

Both initiatives, on more stringent fabrication requirements and on provision for inspectability, moved surprisingly rapidly through the regulatory systems*; and with this stimulus the ASME code committee responded quickly, and began to develop similar though not identical requirements in the nuclear pressure vessel code. (Later, a new portion of the pressure vessel code, Section XI on Inservice Inspection, was developed.)†

Another relatively prompt response to the November 24, 1965, ACRS report on pressure vessels lay in the initiation of new research and develop-

---

*In August, 1967, the AEC issued tentative regulatory supplementary criteria for ASME nuclear pressure vessels for comment by code groups and the nuclear industry. In early 1968, the AEC formally requested public comment on such supplementary criteria. This prompt action not only improved pressure vessel integrity, it established the principle that the AEC could and would impose requirements over and above the code, if deemed necessary.

†The AEC–ACRS primary system review group had developed draft requirements on pre-operational and inservice inspection of pressure vessels by October, 1967.

ment programs. The AEC initiated a major safety research program on heavy section steels, which was to continue for more than a decade. And the Pressure Vessel Research Council, a quasi-industry group, took various initiatives to improve knowledge and eventually codes and standards relating to pressure vessel behavior.

One of the ACRS consultants, Dr. Paul Paris, had expressed reservations, in the absence of good experimental information, about the state of knowledge of the ductility of thick-walled pressure vessel sections after extensive irradiation. Considerable theoretical and experimental effort was subsequently devoted to this subject, and, while Paris's worst fears were not confirmed, a considerable effect of wall thickness on the fracture toughness (or nil-ductility transition temperature) was found, leading to a significant change in the previous requirements concerning an acceptable relationship between the pressure and the metal temperature, one which would not impose excessive stress when the metal was not fully ductile.

Generally speaking, most of the results arising from the Heavy Section Steel Test Program, which was conducted largely by Oak Ridge National Laboratory, indicated that at operating tempertaures, thick-walled pressure vessel steel was very tough and not inclined to undergo rapid fracture.

Within the United States, there was only one modest effort reported to the ACRS on efforts to cope with the possibility of gross pressure vessel failure. During the late 1960s, the Consolidated Edison Company informally submitted a report on a design prepared by Combustion Engineering of a pressure vessel with two steel walls. Consolidated Edison thought that it would need to be able to cope with pressure vessel failure in order to obtain approval for constructing a reactor at a site much more populated than Indian Point 2. The design was reviewed only briefly by the regulatory staff. The proposal was not part of a formal application, and the staff did not try to examine the potential merits of the preliminary design in great detail to see if it offered significant promise of enhanced safety. Nothing ever came of this proposal.

Outside the United States, at least two approaches to protection against pressure vessel failure have been considered. In Sweden, a fairly detailed conceptual study was performed on an underground reactor design intended for a possible urban site (the so-called Vartan reactor). However, this proposal was not pursued.

The German ACRS adopted a philosophy of increased safety by design for the BASF (Badische Anilin-und Soda-Fabrik AG) site near Ludwigshaven, which was more heavily populated than previously approved German sites and more densely populated than the Burlington, New Jersey, site which was rejected by the ACRS and the AEC staff. The German ACRS pursued this philosophy systematically and in detail for BASF, with specific provisions to cope with vessel failure. However, about the time for approval of construction, the application for the BASF reactor was withdrawn.

In early 1971, the ACRS initiated a detailed reevaluation of the reactor pressure vessel matter. It was initially proposed by some committee members that the study consider both the status of pressure vessel reliability and the possible measures available to cope with vessel failure. However, the ACRS decided to restrict the study to an evaluation of pressure vessel reliability. A fairly large subcommittee was established with H. Etherington as its chairman, and the subcommittee obtained the services of several consultants who were expert in specific facets of vessel integrity. After very many meetings the subcommittee completed a rather long report in late 1973, which the full ACRS adopted as a committee report and forwarded to AEC Chairman Dixy Lee Ray in a letter dated January 14, 1974. A principal conclusion of the report, which the AEC published soon thereafter as WASH—1285, *Report on Integrity of Reactor Vessels for Light Water Reactors* (ACRS 1974), related to the probability of pressure vessel failure, as follows:

> The Committee has reviewed available statistics of boiler drums and other nonnuclear vessels and concludes that the 99 percent confidence upper bound for the probability of disruptive failure (pertinent to reactor vessels) of such vessels is less than 1×10^{-15} per vessel-year.
>
> The Committee believes that the disruptive failure probability of reactor vessels designed, constructed, and operated in accordance with ASME Code Sections III and XI is at least one order of magnitude lower than that of the non-nuclear vessels evaluated. Accordingly, the Committee concludes that there is reasonable assurance that the disruptive failure probability of such reactor vessels is less than 1×10^{-6} per vessel-year.
>
> As defined, disruptive failures of reactor vessels include failures of various magnitudes and descriptions, not all of which would exceed the capability of the engineered safety features. Accordingly, the probability of reactor vessel disruptive failure beyond the capability of Engineered Safety Features is lower than the probability of 10^{-6} per vessel year for disruptive failures of all kinds.

Not too long after issuance of the WASH–1285 report (ACRS 1974) the regulatory staff issued its own report on pressure vessel integrity, in which it basically agreed with the ACRS conclusions (AEC 1974b). And in the WASH–1400 report (NRC 1975a), the median failure probability of pressure vessels was taken to be 10^{-7} per reactor-year, which led to the conclusion that pressure vessel failure was not a principle contributor to risk from LWRs.

However, a number of events and situations have kept the matter of pressure vessel integrity from disappearing from sight. The ACRS report did not conclude what the effects of transients might be on pressure vessel integrity. For example, the matter of anticipated transients without scram* (ATWS) remained to be resolved, and ATWS provided a possible mechanism for vessel failure. Also by 1973, when the ACRS report was already completed, there had been a few incidents of reactor overpressurization while the primary system was cold. By 1977, this had grown to be a large number of incidents (about 20), and the regulatory staff took generic action with all PWR operators. The staff requested that short-term measures be taken to correct the breakdown in administrative controls (and other human errors) which were leading to overpressurization events, and that longer-term measures be considered to reduce the reliance on administrative controls.

A concern raised in the ACRS report related to older vessels for which no provision for in-service inspection had been made and for which many parts of the vessel might be inaccessible. Also, there was poorer knowledge of the original status of the vessel quality. The regulatory staff reviewed this matter and a few years later issued a report which basically stated that the existing situation was acceptable (with the exception possibly of one reactor). This conclusion was largely a matter of judgment, since a quantitative estimate of the vessel reliability is difficult to obtain.

Among the other matters which have kept the matter of pressure vessels in sight was the controversy in the United Kingdom. Professor Alan Cottrell, who has served as Science Advisor to the British Government, questioned that pressure vessels for LWRs could be made with acceptable integrity. After some considerable argument and study, a report was issued in October 1, 1976, by a committee officially constituted by the British government and headed by Dr. W. Marshall, which generally endorsed the integrity of LWR pressure vessels, although it made several recommendations for improved quality assurance. However, worries still persist concerning the ability of existing methods of inspection to consistently find fairly large flaws.

Other questions which have arisen include the finding of multiple small cracks in nozzle weld regions by one of the German research groups, and the potential for multiple concurrent failure in the large number of small instrument lines entering the bottom of a PWR. Also, the recent Three Mile Island accident suggested that overpressurization of a "cold" vessel with reduced ductility could occur during a lengthy transient in which the reactor vessel was subject to substantial cooling.

It is to be expected that with time, as more and more vessels undergo in-service inspection, relatively large flaws will be found. And extremely

---

*Scram is nuclear reactor jargon for the act of inserting all the reactor control rods at once as fast as the mechanisms will permit in order to terminate the fission reaction chain and reduce the reactor power to that generated by the radioactive decay of fission products.

difficult decisions will have to be made concerning continued operation, since the costs of repair may be very large. The Pilgrim 1, Massachusetts, reactor has already provided a very small example of the problem which arises when sizeable flaws are found in the pressure vessel of an operating reactor.

Oyster Creek: Quality Control and Backfitting

Review of the Oyster Creek construction permit application was completed by the regulatory staff well before the existence of the general design criteria and before the development of more detailed design criteria. At that time the AEC had developed no criteria with regard to quality assurance in the fabrication and construction of nuclear power plants.

The Jersey Central Power and Light Company had submitted on March 26, 1964, an application for authorization to construct a 1,600 MWt BWR at the Oyster Creek, New Jersey, site. Within five months, the regulatory staff and the ACRS had completed the construction permit review, and on August 25, 1964, the committee issued a brief letter favorable to construction of the reactor.

The subcommittee minutes for the construction permit review note that the preliminary safety analysis report for Oyster Creek was about "the skimpiest" the ACRS had been asked to review, and the ACRS letter noted that many details of the design had not been completed.

On January 25, 1967, Jersey Central submitted a final safety analysis report in support of its request for an operating license. In response to a letter sent to the utility by the regulatory staff on October 22, 1966, which outlined the problem of "backfitting"* for improved emergency core-cooling capability, it also requested a priority review of the emergency core-cooling system (ECCS). There had been relatively little attention paid to the ECCS during the construction permit review for the Oyster Creek reactor, but with the advent of the China Syndrome and the emphasis on improved emergency core-cooling systems to prevent core melt in the event of a loss-of-coolant accident (LOCA), the situation had changed.

At the eighty-first ACRS meeting, January 12–14, 1967, Mr. Price told the committee that the regulatory staff had sent letters to all companies with operating reactors, or reactors under construction, with power levels in excess of 50 MWt. The letters had asked for a review of the core-cooling question. Because of the strong response from industry, it had become evident to Price that the commissioners should have been informed prior to

*The term "backfitting" is used to describe the retrofitting of new or changed safety requirements to reactors whose construction has been essentially completed or whose operation has already begun.

sending the letters. The industry reaction was generally one of concern over the costs which might be involved. Mr. Price suggested that the standards currently employed by the staff and ACRS be articulated so that the industry would have an idea of what backfitting would involve.

Dr. P. A. Morris of the regulatory staff observed that in the case of ECCS the staff had established an industry-advisory group to provide guidance. In other areas where improvements were desired in existing plants, the staff and their consultants should determine what constituted the present criteria. The operating plants and those under construction could then be reviewed to see what should be done by way of improvement.

Commissioner Ramey advised the committee at this meeting that during the past few months, utilities and equipment manufacturers had held discussions with the commission and had expressed their concern over snowballing safeguards requirements leading to increased costs. They had also indicated their feeling that increasing requirements for redundancy in safety systems was leading to the point where overall safety was degraded. They felt some balance was necessary. Commissioner Ramey told the ACRS that the concern of the commission was with the sense of urgency in the letters sent out by the staff concerning ECCS. These letters indicated that it was a matter of the highest priority that the information be gathered, and this indicated to some people in industry that major changes were to be instituted with some urgency.

ACRS member Hanauer observed to Commissioner Ramey that in the case of plants in the early stages of construction, there might well be some cause for urgency so that they could make changes while flexibility still existed.

In its report number 1 to the ACRS on the Oyster Creek reactor, dated June 16, 1967, the regulatory staff reviewed the background of the LOCA–ECCS problem, provided several conclusions on required changes in ECCS, and identified areas needing further study. The Oyster Creek plant did not utilize the jet pump reactor design typical of the current class of General Electric BWR plants starting with Dresden 2. In the jet pump design, the large recirculation pipes penetrated the vessel near the top of the core, and following a LOCA, the emergency core-cooling water would not automatically leak out the bottom of the vessel. This allowed the reactor vessel to refill and the core to be recovered, which was advantageous. In the Oyster Creek plant, there were five large recirculation pipes which entered the vessel below the core, and rupture of a pipe would make core reflooding impossible. Hence, while for Dresden 2 and subsequent BWRs, a core-flooding system had been added in addition to core sprays which added water on top of the core, the ECCS for Oyster Creek included only core sprays.

The staff noted that at the time the Oyster Creek reactor had been reviewed for a construction permit, emphasis had been placed on the primary containment building with the assumption that it represented the "true final boundary." Since that time, consideration of the core melt problem had caused a shift in emphasis to improved core-cooling capability. This has created a sizable gap in the equipment provided for Oyster Creek as compared to present standards for other reactors.

The ECCS proposed for the Oyster Creek plant at the construction permit stage had consisted of two single core-spray loops without redundancy in active components like pumps. Since then, 100% redundancy as well as booster pumps for higher-pressure operation in each core-spray loop had been added. At the construction permit stage, it had been assumed that for smaller, more probable leaks, in which the reactor system pressure stayed high for a long time even though the primary system was losing vital water, off-site power would remain available and the existing feedwater pumps could supply emergency core cooling. The core-spray systems were low-pressure systems and would not be effective for such an accident.

General Electric had now added a semi-automatic depressurization system, so that if there were a small LOCA and off-site power were lost, a series of valves on the primary system piping could be opened, changing a small LOCA to a medium-size LOCA which would cause depressurization of the primary system fast enough that the core-spray system could function and be effective in time to cool the core before it became excessively hot.

The regulatory staff felt that these changes, made "voluntarily" by General Electric after the China Syndrome issue was raised, did not bring the Oyster Creek ECCS capability up to the level being proposed for Dresden 2 and subsequent plants. This raised the question of backfitting.

The regulatory staff decided that the Oyster Creek plant required the following additional changes in its ECCS: provisions necessary for redundant on-site alternating-current and direct-current power sources; multiple signals for actuation of the ECCS; addition of a high-pressure injection system; that the ECCS be immune from passive element failures; and programming of all four pressure relief valves to open on the initiation signal.

The regulatory staff also thought that further study should be carried out on the following topics: containment flooding; diversification in ECCS initiation signals; and low-pressure flooding or an alternate emergency cooling system.

The regulatory staff pursued the ECCS matter with high priority, and in their second report to the ACRS, dated April 18, 1967, emphasized that "additional emergency core cooling protection was required for the small break region to prevent reactor vessel blowdown via the auto relief system for very small breaks in the primary systems," that is, a high-pressure injection system.

The staff noted that they were attempting "by means of extraordinary effort to complete our safety evaluation of the entire facility by the December 1967 ACRS meeting." This was in response to pressure from the applicant not to be delayed in starting up the reactor.

The ACRS supported the staff position on the ECCS, and after considerable prodding the reactor vendor finally proposed additional high-pressure emergency cooling adequate to meet the staff requirements. It was impractical to provide an additional ECCS system to reflood the core, as was done for Dresden 2 and the succeeding BWRs, because of the different reactor design of Oyster Creek in which "bottom" piping breaks could empty the reactor vessel. Containment flooding also posed problems. However, redundant on-site direct-current and alternating-current power sources were possible and were backfitted.

Many other safety aspects differed in the Oyster Creek plant from those the regulatory staff were requesting of new plants under consideration for construction in 1967. The following is a partial list of items which were reviewed for the Oyster Creek reactor and would have been necessary for it to meet the new criteria. However, they were not backfitted.

(1) The two isolation valves in each steam line were both outside the containment building, instead of one in and one out, creating the possibility of a non-isolable steam-line rupture outside the containment building.

(2) Tornado protection was much less than the 1967 requirement.

(3) A sidewise whipping of large pipes from the dynamic reaction forces, following rupture conditions, could damage containment integrity.

(4) Access for periodic inspection of the pressure vessel and primary system was limited.

Thus, while many backfitting questions arose during the review, it was found to be impractical to make major hardware changes. Even changes in electrical circuitry were required by the regulatory staff only when a very strong safety connotation was involved.

The experience with backfitting questions during the Oyster Creek review was similar, though probably more difficult, to that encountered with the other operating license reviews undertaken after the summer of 1966. Because of the new emphasis on ECCS and the development of draft general design criteria, there were many differences apparent between the plants that were already built, and the requirements for new plants. The problem remained a very thorny one, with decisions made on each reactor on a judgmental basis which involved both the safety significance and the ease of making modifications. This experience led the regulatory groups to require more detailed design information at the construction permit stage. It also brought large protests from the nuclear industry, to which Congressman Hosmer referred in the speeches described in Chapter 12. The controversy eventually resulted in the adoption by the AEC in 1970 of the so-called "Backfitting Rule" which required that, in order to require a backfit, the regulatory staff had to make a finding that "such action will provide substantial, additional protection which is required for public health and safety or the common defense and security." The onus for backfitting was thus placed on the regulatory staff, and reactor licensees exercised rather little initiative in this regard, taking the approach that their plants were already acceptably safe. The "Backfitting Rule" unquestionably exercised an inhibiting effect on the regulatory staff in this regard, and perhaps led to a tendency for several years for the staff not to "look" for backfit questions on older plants.

The first ACRS subcommittee meeting related to the operating license review for the Oyster Creek reactor was held on July 28, 1967, and included a tour of the facility, as usual. During that meeting various site-related safety issues, as well as matters pertaining to ECCS, were discussed. But the most significant thing to arise from the subcommittee visit was a considerable concern by ACRS members about the quality of the field construction and fabrication.

This ACRS concern with the quality control of welding, with the sloppiness of the construction site, and with the lack of an adequate quality assurance program, in general, is recorded in the minutes of the ninety-first ACRS meeting, November 2–3, 1976. In fact, between the July subcommittee visit and the November ACRS meeting, a specific major quality control problem had turned up. During a hydrostatic test of the reactor pressure vessel (prior to reactor operation) on September 29, 1967, a small leak was noted near one of the control rod drive housings. Subsequent examination revealed that the vast majority of all the control rod housings, over one hundred, had suffered cracks in certain welds, and that there was a generic deficiency in both the design and welding. A major program was required to define an acceptable repair procedure and to reweld all the deficient joints.

This very major set of flaws turned out to be only the beginning of a long story. The regulatory staff, on being made aware of the ACRS concern about the overall quality assurance, decided to give more detailed attention to inspection of the Oyster Creek reactor than was then customary. The problem of inadequate quality assurance turned out to be so severe that start-up of the reactor was delayed for at least a year beyond the date which the utility had been insisting was vital.

In a report to the ACRS dated November 29, 1968 (a year later), the regulatory staff documented a large number of deficiencies in the quality assurance program, some of which still remained to be resolved at that time. For example, the report listed many deficiencies in the installation of circuits for instrumentation and power, involving violation of the criterion requiring separation of redundant circuits. Most of these could be remedied in a practical fashion, fortunately.

Further deficiencies in quality assurance turned up in 1969, when some cracked valves were investigated, and it was found that at one point the quality assurance program had lapsed sufficiently to permit the purchase and installation of second-hand valves of unknown condition.

The findings on the Oyster Creek reactor provided a very great impetus for the generation of a generally applicable, structured, quality assurance program and for an expanded staff role in auditing what was done. The AEC gave high priority to the development and adoption of a rule on quality assurance. Also, pressure was exerted by the AEC on the utilities to see that improved quality assurance programs were instituted. The commissioners decided that the regulatory staff (and not the ACRS) would have the responsibility for determining that adequate quality assurance was actually employed on each reactor.

The question of quality assurance has remained troublesome. The Browns Ferry fire, the fabrication problems arising at the North Anna plant, and the Three Mile Island accident, are three examples from a rather long list.

The Browns Ferry fire illustrated a failure of the utility to institute adequate quality assurance practices, the staff or the industry to develop adequate fire protection codes or regulations, and the regulatory staff to act in timely fashion on an identified problem. It also provided a good example of the thorny decision-making process that occurred whenever backfitting was involved.

Actually, the evolution of concern about quality assurance dates back at least to 1963. The minutes of the forty-sixth ACRS meeting, January 31, February 1, 2, 1963, record a meeting between the ACRS and AEC Chairman G. T. Seaborg and AEC Commissioners J. G. Palfrey, J. T. Ramey, L. J. Haworth, and R. E. Wilson. ACRS Chairman David Hall presented a draft of a letter to the commission recommending improvement in the quality and reliability of materials and equipment in order to ensure better engineered safeguards for reactors. Mr. Price, who had not seen the draft, said he preferred not to comment. Chairman Hall noted that surveillance by the AEC and the utility to insure satisfactory quality assurance at the plant during construction and operation was difficult. Although some of the commissioners believed such a formal letter might be useful, it was concluded, following Commissioner Wilson's comments, that bringing the matter to the attention of the commission orally was sufficient; Commissioner Wilson feared that the letter might be misunderstood by the public and reflect unfavorably on the reactor program.

Six years later, following the Oyster Creek episode, quality assurance became a major aspect of the AEC's effort to improve both the reliability and safety of reactors, an effort which has had mixed success.

The backfitting issue has remained a thorny problem. There do not exist quantitative criteria for safety that all reactors must meet. In other aspects of society, safety improvements are frequently not retrofitted unless a clear deficiency which represents an unacceptable risk is found in existing factories, airplanes, etc. A generally accepted approach to backfitting for nuclear power reactors remains to be defined.

Prairie Island and the Accident
Involving a Break in the Steam Line

The ACRS completed its review of the construction permit for the two 1,650 MWt PWRs at Prairie Island, Minnesota, at its ninety-fifth meeting, March 7–9, 1968. That review was known most for the so-called "Prairie Island position" on separation of protection and control instrumentation which is discussed in Chapter 16 in the section entitled "Separation of Protection and Control."

The regulatory staff report to the ACRS of February 27, 1968, on the Prairie Island site included the usual analysis of the off-site consequences of a postulated steam-line break accident. On September 28, 1972, the regulatory staff forwarded to the ACRS its safety evaluation report in favor of the proposed operation of the Prairie Island plant.

The staff report pointed out no major problems or controversies between it and the applicant, and it appeared that the Prairie Island case would be a relatively routine ACRS review. The staff report stated that the Prairie Island reactors were designed and constructed to meet the AEC's general design criteria, as proposed in July 1967. The applicant had not been asked to reanalyze the plant against the general design criteria adopted in 1971; however, the staff said "our technical review did assess the plant against the General Design Criteria now in effect and we are satisfied that the plant design generally conforms to the intent of these criteria."

The ACRS subcommittee on the Prairie Island site met on October 24, 1972, and the case was scheduled for full committee review at a special ACRS meeting on October 26–28, 1972.

On the evening of October 25, 1972, the ACRS office provided to ACRS members the following two-page letter just received from sources unknown, in an envelope postmarked Philadelphia, Pennsylvania:

There are several safety related unresolved items between DRL (Division of Reactor Licensing) and the applicant on both Kewaunee and Prairie Island Projects. Some of these are:

1. Non-fulfillment of old criterion 20, 21, 40, 41 and 42.

2. A rupture anywhere (everywhere) outside the containment in the steam line including the rupture of a relief header produces intolerable consequences in that the walls, floors and ceilings will collapse in less than 60 milliseconds with loss of all electrical and mechanical equipment. A massive destruction of the Aux Building or the control room will result, and, therefore, safe shutdown of the reactor is jeopardized.

3. Containment over-power protection systems are very different from other two loop plants and indeed reduce the safety on the plant in the event of a LOCA or steam line break inside the containment.

4. DRL has not done any independent containment pressure transient analysis nor have they performed any compartment pressure analysis.

5. Several Electrical and Instrument & Control Changes affecting safety systems have not been received by DRL e.g., Accumulator & RHR (Residual Heat Removal) Valve interlocks Boric Acid System which is full of holes.

6. Loss of a situation Battery (Single Failure) past LOCA will prevent both diesels from startup and also prevent several 2 out of 3 logic from performing properly.

7. Turbine overspeed missile* will go right through the containment dome, the Aux Bldg. Roof, Spent Fuel Pool Roof, etc...

8. Changes in the Westinghouse ECCS Model especially Reflood and Blowdown have not been reviewed extensively by the Staff.

9. Seismic Analysis of Critical components have not been satisfactorily completed.

10. Primary Containment leakage of 0.1% day is very easily attainable from present day technology and, therefore, there is no need to allow higher leak rate and thereby expose the public to a higher off-site dose.

11. Several potential leak paths have no Iodine Filtration and go directly to the atmosphere. These paths should be modified to fall within the Secondary Containment boundary. This path alone contributes to more than 50% of the total off-site dose.

12. Containment pressures of 60 psig (75 psia) are very real in an accident situation and, therefore, over pressure test at 80 or 85 psia should be performed to assure containment integrity.

13. The Containment Free Volume has never been verified by DRL and is definitely well below the assumed values.

With just these many items alone it is not in the public interest to allow further actions until complete review by DRL.

The anonymous memorandum was discussed briefly by the ACRS in executive session on the morning of October 26, and it was decided that

---

*If the turbine controls fail, the turbine can be driven to destructive speeds, rupturing the turbine rotor and generating large, high-speed missiles.

members would individually try to interpret the significance of the listed items and explore them during the upcoming discussions with the regulatory staff and the applicant.

Some of the items listed, such as the effect of a turbine overspeed missile, represented known effects or phenomena which were already being studied on a generic basis to determine their potential for causing serious accidents and whether changes in current or future plants might be required to cope with them. Some appeared to have been met acceptably, or represented areas where it was reasonable to expect the regulatory staff to handle the matter prior to their actual issuance of an operating license; and this was confirmed by questions to the staff or the applicant.

ACRS member Etherington made some crude calculations concerning the possible pressure buildup in the auxiliary building from a gross steam-line rupture, and judged that the point raised in the anonymous memorandum might have some validity. When this issue was raised with the regulatory staff and applicant, it quickly became clear that the applicant had not evaluated such a rupture in his safety design, nor had the regulatory staff reported on the issue in their review. It also confirmed that a potentially serious safety problem could be associated with the rupture of steam lines and other high-energy lines outside containment.

The ACRS decided not to complete action on its operating license review for Prairie Island, and forwarded the anonymous memorandum to the regulatory staff for review and comment. The staff met with the applicant on November 3, 1972, and on November 8, 1972, the staff forwarded an information report to the ACRS in which they concluded that the overall matter of steam- and feedwater-line breaks required further evaluation and advised that they were requesting considerable additional information from the applicant.

At the one hundred fifty-first meeting, November 9–11, 1972, the regulatory staff reported to the ACRS on the issue of a steam-line break for the Prairie Island site, as follows:

> Many modifications to the existing Auxiliary Building will be required to provide pressure relief in the event of steam line failure, and the Applicant, Northern States Power Company, has not yet completed the evaluation of the consequences of a double ended break of the main steam line in the building as requested by the Regulatory Staff. His current analysis identifies a single-ended rupture of the largest (6 inch) main steam line branch connection as the design basis accident. The Committee was informed that engineered safety features and other critical equipment in compartments exposed to the steam atmosphere following a break would have to be qualified for service in such an environment.

When other PWRs besides Prairie Island were examined with regard to this safety issue, the problem was found to apply to most of the reactors under construction or in operation to a greater or lesser degree. Each reactor was, therefore, reviewed and evaluated specifically and in detail for adverse effects from pressure, forces, and the high-temperature steam environment which would accompany large ruptures of such lines.

The steam-line break itself should be an event of fairly low probability; however, the probability was not thought to be so low that the potential effects could be neglected. Some of the reactors required large changes at considerable expense in money and time in order to reach an acceptable level of protection against such breaks.

Why the matter had slipped through the regulatory process is obscure. The ACRS assumed that the general design criteria required consideration of such postulated failures. Apparently, the architect-engineers, who had designed this portion of the nuclear power plant, had been using an old design basis for such postulated ruptures, one which had been accepted prior to 1966. And, while the regulatory staff was reviewing in increasing depth the effects of postulated ruptures of the largest pipe in the primary system, they had not extrapolated this approach to the secondary system of a PWR.

By March, 1973, the Prairie Island applicant had established what design changes and other measures were needed to cope with steam-line and other high-pressure process-line breaks outside the containment vessel, and at its one hundred fifty-sixth meeting, April 12–14, 1973, the ACRS wrote a report in favor of the operation of the Prairie Island reactors.

All in all, the main steam-line break issue was confronted and resolved in a relatively expeditious manner, not typical of all safety issues arising in the regulatory process. The next chapter will discuss an issue in which the opposite extreme applies, namely, an issue which remained unresolved ten years after it was first raised.

16

Anticipated Transients Without Scram (ATWS)

Separation of Protection and Control

S. H. Hanauer, who became an ACRS member in 1965, was the first member to have a strong personal background of experience in reactor instrumentation and control. He had worked in the subject for many years at Oak Ridge National Laboratory (ORNL), prior to becoming a professor at the University of Tennessee. He brought to the ACRS not only much practical knowledge about the difficulty of getting reliability from components like motors and valves which are actuated by instrumentation, but also a strong personal conviction that in good reactor design the safety systems provided to shut the reactor down should not be interconnected to the systems used to control the reactor. This was a philosophical approach, like "separation of church and state," which Hanauer and others at Oak Ridge National Laboratory had developed as the sound way to avoid unexpected, hidden interactions between the control instrumentation and safety instrumentation which might negate safe reactor shutdown. There existed a history of failures to support this point of view, and Hanauer began raising the matter in connection with specific reactor projects under review by the ACRS in 1966 and 1967. The ACRS specifically identified the issue in its report on the Westinghouse reactor at Point Beach, Wisconsin, in May 1967. The four light water reactor vendors each had adopted their own approach to instrumentation design, and it was especially the Westinghouse practice of using the same sensors and subsystems for both control and protection (safety) purposes that troubled Hanauer.

It had been customary on many of the earlier reactors for the instrumentation design to be unavailable for review at the construction permit stage, and the regulatory staff and the ACRS had accepted this practice.

In connection with review of the PWR proposed for Diablo Canyon, California, in the fall of 1967, Hanauer noted that Westinghouse had still not submitted an instrumentation design, although they proposed to submit reports giving some details by the end of 1967. He felt it would not be wise to approve still another reactor on this basis. Westinghouse was therefore prompted to provide partial information on the design during the ACRS review of the Diablo Canyon site. At its ninety-second meeting, December 7–9, 1967, the ACRS tried to complete its construction permit review for Diablo Canyon. The site was relatively remote, and, at that time there were no known large faults nearby, so that the seismic design basis was not a matter of controversy. The Diablo Canyon reactor was to be one of the first of the high-power PWRs built, however, which made it a logical reactor on which not only to look for new, previously unanticipated issues, but to resolve some that had been ongoing.

At the ninety-second meeting, Hanauer noted that the Westinghouse design did not meet the general design criterion which asked for two emergency cooling systems, preferably diverse. More importantly, the control and protection instrumentation were intermingled. Westinghouse claimed their design met the proposed new IEEE–279 standard for Nuclear Power Plant Protection Systems (IEEE 1967); but Hanauer had major reservations concerning the adequacy of the proposed Westinghouse protection system design, and in the committee caucus did not think a letter favorable to construction of the Diablo Canyon reactor could be written.

The matter received considerable discussion at the ninety-second meeting, and the ACRS finally decided to write a report dated December 20, 1967, which included a strong reservation on this topic, as follows:

> The Committee believes that control and protection instrumentation should be separated to the fullest extent practicable. The Committee believes that the present design is unsatisfactory in this regard, but that a satisfactory protection systems can be designed during the construction of the reactor. The Committee wishes to review an improved design prior to installation of the protection system.

The Prairie Island, Minnesota, case represented the next Westinghouse PWRs to be reviewed for a construction permit. The regulatory staff, who had previously accepted the proposed Westinghouse design, asked the

Prairie Island applicant to respond to the above paragraph from the ACRS report on Diablo Canyon. Only one minor design change was proposed by Westinghouse.

In their safety evaluation report to the ACRS on the Prairie Island site, the regulatory staff proposed the following criterion:

As an absolute minimum, each variable monitored for protection should be instrumented by sufficient channels independent of control to meet the single failure criterion (generally this means three channels connected in two of three coincidence). Furthermore, the applicant may elect to provide additional channels of protection which are not independent of control. If he elects to follow this route, the applicant should provide a rigorous failure mode analysis to show that there can be no interaction between the control system and the independent protection channels through the shared channel(s).

The ACRS Opinion on Prairie Island

The ACRS, the staff and Westinghouse discussed this matter at the ninety-fifth meeting, and then the ACRS included the following comment in its report of March 12, 1968, on the Prairie Island site and reactors:

The applicant has proposed using signals from protection instruments for control purposes. The Committee continues to believe that control and protection instrumentation should be separated to the fullest extent practicable. The Committee believes that the proposed protection system can and should be modified to eliminate or reduce to a minimum the interconnection of control and protection instrumentation. The modified system should be reviewed by the Regulatory Staff.

In the next few months, the ACRS wrote construction permit reports on other Westinghouse reactors, including Surry in Virginia and Kewaunee in Wisconsin, with similar recommendations on the separation of protection and control.

In the meantime, Westinghouse continued to argue that their design met the proposed IEEE–279 criteria (IEEE 1967) and was adequate. However, they prepared a somewhat modified design which they proposed as a possible alternative at a meeting with the regulatory staff on May 3, 1968. They said they would offer their customers a choice of the original design and the modified design, but they would recommend the original as providing better control. In a letter dated May 30, 1968, to the ACRS, committee consultant E. P. Epler found that the modified Westinghouse design still posed problems of interconnection of safety and control, and he recommended further steps to reduce such interconnection. He appended a draft article entitled "Identical Systems for Protection and Control" in which he reviewed the bad previous experience with such systems.

On May 8, 1968, the ACRS held a briefing on control and safety instrumentation, at which experts from the Argonne, Brookhaven, Los Alamos, and Oak Ridge National Laboratories gave their opinions. The ORNL representatives explained their reasons for feeling that separation of control and protection was important, particularly in the instrumentation systems used to measure neutron flux (which is proportional to the power produced by fission). They acknowledged, nevertheless, that some mixing was tolerated under certain conditions even in the ORNL designs.

All the experts agreed that systems designs which purposely intermixed control and protection must be carefully designed and analyzed. However, there was not a unanimous position in favor of full separation; nor was there agreement that adequate analysis of a mixed design could be accomplished.

One problem facing the ACRS was the implied threat by Westinghouse that if forced to go to complete separation, they would go to a single channel for control. This would lead to less reliable control and more plant transients, which would decrease safety since each transient represented a challenge that the safety systems had to counter successfully.

Hanauer Explains His Position

In a memorandum to ACRS members dated May 22, 1968, S. H. Hanauer tried to summarize the problem as he saw it. The memo is reproduced below.

Interaction of Protection Instrumentation With Control

In this paper, I have tried to set down concisely my reasoning in the current discussions on Westinghouse designs.

1. The basic problem is to avoid having an event requiring protection action somehow inhibit the needed protection. In the classic example (HTRE-3),* design defects in the neutron-flux instruments prevented an increase in current when the reactor power increased; the too-small current both told the servo to withdraw the rods and blinded the protection system to the resulting power increase. The core was destroyed.

It is for this reason that common use of a detector signal for both control and protection is automatically suspect.

2. The situation is complicated by the fact that three or four channels of instrumentation are provided for each protection variable, rather that just one. Random single failures do not fail such a system, because of the redundancy. However, such duplication does not prevent system failure due to design deficiencies, since all can be expected to fail for such a reason.

*HTRE-3 (high temperature reactor experiment) was an experimental reactor built at the National Reactor Testing Station in Idaho.

The use for control of signals from one, or two, or all the redundant protection channels is the point at issue.

3. If the only postulated faults are random component failures, it is easy to show, using statistical analysis, that control use of protection signals won't hurt you. A single failed channel causes an excursion; the other channels, assumed unaffected by the fault, provide the needed protection. The channels must be truly independent for this assumption to be valid.

This reasoning is the origin of the IEEE criterion (Section 4.7) that for such a failure the unfailed channels must meet the single-failure criterion. (The present Westinghouse designs do not meet Section 4.7 in every respect.)

4. In my opinion, random component failures are not the only postulated faults against which a protection system should have a defense. We learn from the history of various accidents, incidents, and near-misses that mistakes are made in design, execution, and operation of instrumentation systems. Such non-random (systematic) failures are not amenable to statistical analysis. They are the reason for employing diversity in protection-system design; one hopes that if a mistake is made, it will not involve dissimilar types of equipment.

The use of redundancy doesn't reduce the probability of occurrence of systematic failures, so the redundant nature of the protection signals does not in this instance justify their use for control.

5. As a minimum position, the regulatory staff has proposed requiring enough protection channels completely independent of control to satisfy the single-failure criterion; usually, this number is three. Additional channels in which control and protection could be mixed would also be allowed, provided they could be demonstrated independent of the sacred three.

This is a strict interpretation of the meaning of IEEE Section 4.7. Westinghouse does not agree that the staff position is the correct interpretation.

My own feeling is that mixing control and protection by using the same signal for both functions is wrong in principle, but that the proposed staff position is probably as far as one can go in requiring separation of a recalcitrant applicant.

6. "Protective overrides" are *control* functions, not protection, put in for the laudable purpose of forestalling the protection action by initiating a milder action first. Examples are (a) blocking rod withdrawal on high flux to inhibit a potential increase which, if continued, would require a scram; (b) shutting off feedwater on high boiler water level to avoid carrying a slug of water to the turbine, which could cause a turbine trip and scram. Overrides fall into two classes: blocking rod withdrawal can't cause an excursion requiring protection-system function, but shutting off the feedwater surely can. Thus it should be allowed to use redundant signals from the protection system to block rod withdrawal, but not to shut off feedwater.

7. Let there be no illusion; the protection system cannot be made "completely independent" of control. The two systems related to the same reactor plant, are located in the same control room, are watched and manipulated by the same operator, are serviced by the same maintenance men, have cables in the same containment. What I am objecting to is the deliberate use of signals from protection instruments for control purposes, leading to the potential that faults could initiate

an excursion and consequentially cripple the needed protection for that excursion, to the detriment to health and safety of the public.

8. As a postscript, it seems worth pointing out that concern should also exist where identical devices, albeit independent, are used for both protection and control. A design, maintenance, or operating error in such identical components, or their exposure to a deleterious, unusual environment, would have the same effect as their interconnection. Thus, diversity as well as independence should be required of protection and control instrumentation. This matter is in urgent need of further study.

Mangelsdorf Disagrees

At its May 1968 meeting, on the urging of member Mangelsdorf, who questioned the Hanauer position, the ACRS requested the regulatory staff to provide the committee with a background paper on the whole subject, including the staff point of view. Such a paper was prepared and the subject was discussed at the ninety-eighth meeting, June 5–8, 1968.* The staff report, which provided an excellent review of the matter, concluded as follows:

In summary, we believe the determination of which approach is preferable depends on a judgment as to whether protection channels independent of control are required to protect against subtle control and protection system interactions.

We have been asked what the DRL position on control and protection interaction would now be if the ACRS comment on Diablo Canyon had not been made. Since we recommended approval of a number of designs based on the IEEE criterion during the year before the Diablo Canyon letter, we would probably still be using the IEEE criterion. Our deliberations since the Diablo Canyon letter have, however, made us question one of our recommendations on Diablo Canyon. We recommended acceptance of certain scram channels which did not meet IEEE–279 based on their not being required for safety. This gives rise to what have been called first class and second class scrams. This could cause confusion as to which protection channels should be relied on for safety and which should not be relied on. The present DRL criterion takes care of this situation. If however, we returned to using IEEE–279, we would attempt to develop criteria for channels used for equipment protection rather than safety. We also will have to develop criteria for protective over-rides whether we continue with our present approach or return to the IEEE criterion.

Subsequent to the meeting with the regulatory staff, the ACRS considered three alternative positions.

*In the discussion it was noted that the British and Canadians required separation of protection and control; also, that General Electric employed it.

(1) The IEEE single-failure position plus over-rides.

(2) The regulatory staff position on the Prairie Island reactor where some channels were permitted to be interconnected while others were independent.

(3) Separation to the maximum extent practical.

Interestingly, ACRS members Hanauer, Hendrie, and Zabel, who had the most experience with such systems, all indicated they would prefer position 3, that is separation to the maximum extent practical, if starting anew. However, because of other considerations, such as the existence of tested designs by each reactor vendor, they did not recommend that choice for the committee position.

The ACRS finally agreed to a modified position, as follows:

The Committee believes that systematic, non-random, concurrent failure of redundant elements should be considered in the design and review of reactor instrumentation systems. Systematic failure of the protection system and, particularly, common failure modes of control and protection systems, are of concern. The Committee suggests that the Staff consider supplementation of the IEEE criteria for use in review of instrumentation, and propose criteria which take the possibility of systematic failure into account.

On July 5, 1968, the regulatory staff provided a short discussion paper to the ACRS. They agreed that there was a need to go beyond IEEE–279 (IEEE 1967) in the consideration of potential sources of systematic or common-mode failures.

At the one hundredth ACRS meeting, August 8–10, 1968, the regulatory staff provided another status report which included several alternative recommendations. ACRS members Hanauer and Mangelsdorf took strongly opposing views on the staff proposals.

An ad-hoc subcommittee consisting of members Hanauer, Mangelsdorf, and Hendrie (serving as chairman) met on September 4, 1968, to consider the matter in connection with review of the Russelville, Arkansas, plant which was on the ACRS agenda for September.

The three members arrived at three different positions: Mangelsdorf favored acceptance of the Westinghouse proposal; Hanauer favored a position like that adopted for Prairie Island; and Hendrie favored separate flux instruments for protection and control.

The Emphasis Shifts to Common-Mode Failures

The summary minutes of the one hundred first ACRS meeting, September 5–7, 1968, record the adoption of the following compromise committee recommendation to the regulatory staff.

For reactors under construction, and approved prior to Diablo Canyon, the ACRS believes it reasonable to accept designs in accordance with IEEE-279, modified as necessary for protective overrides and channels used only for equipment protection. In view of the desirability of having similar instrumentation systems in the units of a multi-reactor plant, Point Beach No. 2 may be considered on the above basis.

For Diablo Canyon, and reactors approved subsequently, the applicant should be required to show that any interconnection of control and safety instrumentation will not adversely affect plant safety in a significant manner when considering the possibility of systematic component failure.

The Committee reiterates its belief that systematic, non-random, concurrent failure of redundant elements should be considered in the design and review of reactor instrumentation systems. Systematic failure of a protection system and, particularly, common failure modes of control and protection systems are of concern. The Committee suggests that the Staff continue its program to supplement the IEEE criteria for use in review of instrumentation, and to develop criteria which take the possibility of systematic failure into account.

This remained the position of the ACRS for the next several meetings. The emphasis had shifted from a request for as much separation of protection and control as practical to a conscious effort to account for systematic (common mode or common cause) failures by appropriate design approaches.

The Birth of ATWS

An anticipated transient is an occurrence which is expected to occur once or more during the life of the reactor and to interrupt its normal operation. Some simple examples include a trip of the turbine (a sudden closure of the steam valves, stopping rotation of the turbine), or a loss of off-site power which leads to tripping (sudden shutdown) of the primary system pumps. For some such transients it is important that the control and safety rods be rapidly inserted into the core (a reactor scram) to shut down the chain reaction. The transients requiring scram are generally not identical for a PWR and a BWR. In a BWR, for example, several transients result in isolation of the core from the turbine with a resulting increase in reactor pressure which collapses the steam bubbles in the core and would lead to a larger than desirable power increase unless scram occurred.

In a letter dated January 21, 1969, to R. F. Fraley, executive secretary of the ACRS, committee consultant E. P. Epler suggested that the reliance being placed on reactor safety systems in BWRs provided inadequate protection for the health and safety of the public. More specifically, Epler argued that reactor scram was needed to prevent core meltdown and a loss of containment integrity following a routine operating event such as loss of electric load, which might occur about once a year. Epler argued that a scram system unreliability smaller than 10^{-4} per demand could not be expected because of systematic failures, and that, therefore, core melt and a major release of radioactivity might occur with a probability larger than 10^{-4} per reactor-year.

In a memorandum enclosed with his letter to Fraley, Epler mentioned that public figures like Alvin Weinberg, the Director of ORNL, and Chauncey Starr, then Dean of Engineering at University of California, Los Angeles, and formerly President of Atomics International, had publicly indicated that the probability of a serious reactor accident was similar to that of a jet airliner plunging into Yankee Stadium during a World Series game, which Epler estimated as roughly 10^{-7} per year. However, because of the lack of measures to cope with the China Syndrome, and because of his own estimate of a relatively high scram failure rate, Epler felt that the actual probability of a serious accident may be a factor of 1,000 higher.

The matter raised by Epler was placed on the agenda of the next ACRS meeting, the one hundred sixth, February 6–8, 1969, and Epler was present to discuss his concern. The committee decided to refer the matter to a subcommittee for review and provided copies of Epler's letter to the regulatory staff.

At a special meeting, May 2, 1969, the ACRS discussed ATWS briefly during its first review of the proposal for construction of the Hatch BWR in Georgia. As a result, the committee decided to pursue the matter and arranged for a subcommittee meeting with General Electric about ATWS to be held on May 7, 1969, immediately prior to the regular May full committee meeting. During discussions between the ACRS and the staff at the subcommittee meeting, Dr. Peter Morris of the staff agreed that the matter of ATWS needed study and that the designer ought to look at the consequences and potential remedial measures for this type of accident. He said that General Electric felt that they should concentrate on improving the reliability of the scram system; however, he said that General Electric had also started to look at the consequences of a failure to scram. Dr. Morris stated that Mr. Price had advised General Electric that protection against failure to scram would not be required on the Hatch and Brunswick plants, that is, the plants currently being considered.

ACRS member Okrent asked the staff if this would not be a bypassing of the intent of the Atomic Energy Act, if the regulatory staff knew of a condition that could present a hazard to the health and safety of the public and did not require the applicant and designer to correct it. Dr. Beck of the staff said that a rigorous interpretation of the act would mean shutting down a number of operating reactors. ACRS member Hanauer asked what regulatory alternatives the staff had. If the ATWS problem was identified in the Hatch safety evaluation, would it preclude the issuance of a construction permit? Could the staff issue a construction permit only if it did not identify ATWS as an unresolved safety concern? Dr. Morris said that there should be some middle ground in this area. Dr. Morris reported that General Electric was currently estimating the cost of backfit items at approximately $50 million per plant.

General Electric then presented some preliminary information to the ACRS on their evaluation of the behavior of present BWR designs and alternative designs, assuming a failure of the reactor to scram after a transient which called for scram. General Electric had not yet reached a conclusion on these matters and was going to continue their study and present the results in a topical report. General Electric would not leave copies of the material presented to the subcommittee.

The ACRS Calls Out the ATWS Problem

At its one hundred ninth meeting, May 8–10, 1969, the ACRS completed action and wrote letter reports on the Hatch Unit 1 BWR and on the application to build two BWRs at the Brunswick Station in North Carolina. In each report the ACRS included the following paragraph:

A study should be made by the applicant of further means of preventing common failure modes from negating scram action and of design features to make tolerable the consequences of failure to scram during anticipated transients.

Thus, the ACRS had decided to identify the issue as one requiring study but not to recommend yet that changes in design were needed.

At the May 1969 meeting the ACRS also began asking questions concerning the effects of an ATWS on the PWR that happened to be in for review (the Ginna reactor near Rochester, New York). Thus, Epler's original question about failure to scram a BWR following a turbine trip, which had already been broadened into other transients for BWRs, was now examined for PWRs. And the committee recommended in July 1969 that ATWS be studied for the PWR proposed for construction at Three Mile Island, Pennsylvania.

Because of the issue of separation of protection and control that was first brought into full view with the ACRS letter of December 1967 on Diablo Canyon (which later evolved into the concern with systematic failure as enunciated in a letter on the Russelville, Arkansas, plant in September 1968), the regulatory staff had been engaged for almost a year in studies of systematic failures in reactor protection systems. In a letter to ACRS Chairman Hanauer, dated August 4, 1969, Dr. Morris provided a status report on these studies. He indicated a preference for separating the study of systematic failures from the study of ATWS in order not to delay the former, but noted that at a subcommittee meeting some ACRS members and consultants indicated that they saw the two matters as strongly connected. Morris requested comment on his proposed plan of action, which involved having each vendor do separate (but related) studies on the two topics.

In a memorandum from Fraley to Mr. Price, dated December 15, 1969, the ACRS accepted Morris's proposal but expressed concern with the "time required to resolve the safety questions."

We shall see that this memorandum was prophetic, and that time stretched out to more than a decade without resolution of the matter.

While the regulatory staff began its study of ATWS, the ACRS continued to conduct relatively short reviews of the matter with each applicant and to insert a paragraph concerning the matter in each ACRS case letter.

At the one hundred twenty-first ACRS meeting, May 7–9, 1970, Babcock and Wilcox, the vendor for the Midland reactor in Michigan, stated that they had recently discussed with the regulatory staff the results of the their studies of various systematic failures. Babcock and Wilcox added that the staff had additional requirements regarding the studies, and, therefore, Babcock and Wilcox would have to wait for a clarification of these requirements before continuing the studies. Babcock and Wilcox said they had not analyzed a number of anticipated transients, e.g., loss of all feedwater or primary system pumps or off-site power. The committee mentioned the possible benefit of additional safety valves and of rapid injection of boron into the reactor coolant.

The unwillingness by Babcock and Wilcox to exercise initiative in deciding what transients needed study and to rapidly develop the relevant information was generally characteristic of all the vendors, and contributed greatly to delay in the first few years of consideration of ATWS.

The ACRS recommended in its report on the Midland reactor that "the applicant accelerate his study of means of preventing common failure modes from negating scram action and of design features to make tolerable the consequences of failure to scram during anticipated transients." The committee also noted that the applicant stated that the engineering design would maintain flexibility with regard to design measures to provide a greater pressure relief capacity for the primary system and a diverse means of scramming the reactor.

At its one hundred twenty-second meeting, June 11–13, 1970, the ACRS held a general discussion of ATWS. Member Hanauer summarized the information presented by the four reactor vendors to the instrumentation and control subcommittee on June 3, 1970. He stated that the consequences of a failure to scram after an anticipated transient appeared to be as follows for each of the reactor vendors. (a) For PWRs a loss of all feedwater or a loss of off-site alternating-current power appeared to have the potential for causing excessively high pressures in the primary system if there was a failure to scram. Westinghouse already had more capacity to relieve pressure than was required by the pressure vessel code. Hence, it appeared that they could resolve the problem by slightly increasing this capacity. Another possibility was to provide a boron injection system. (b) Babcock and Wilcox might have to provide more significant measures than Westinghouse because of the reduced water inventory in the secondary side of their steam generator and the smaller relief valve capacity in their primary system. (c) Combustion Engineering had not provided any information on possible design changes. (d) In the General Electric BWRs there were only a few seconds in which to mitigate a failure to scram. General Electric believed that tripping out the recirculation pumps, when the anticipated transient occurred, would solve their problem.

The minutes of the one hundred twenty-third meeting, July 9–11, 1970, note that General Electric stated that they "need AEC criteria if more is required of them regarding ATWS." They found it "a problem to have to document their studies in the public record."

Dr. Hanauer, who had been an ACRS member for about five years while on the faculty of the University of Tennessee, was asked by Mr. Price, the Director of Regulation, to join the AEC regulatory staff on a full-time basis in mid-1970. Thus, the minutes of the July 1970 ACRS meeting show Dr. Hanauer reporting on behalf of the regulatory staff to the ACRS of which he was no longer a member. Dr. Hanauer commented that "some information which is needed to resolve the ATWS matter is still lacking, but it is questionable how much more will be forthcoming."

The Staff Develops a Tentative Position

At its one hundred twenty-fifth meeting, September 17–19, 1970, the ACRS discussed ATWS with the regulatory staff. The staff had submitted a report in September on ATWS, which included the following recommendations:

The reliability of scram systems of current design demonstrated to date and the occurrence rate of anticipated transients leads to the conclusion that anticipated transients without scram having serious consequences will occur at an unacceptably high rate when there are a large number of reactors in operation.

The consequences of anticipated transients without scram should be shown to be acceptable or design changes should be made.

Where changes are needed, they can be provided by improving the reactivity reduction systems or by modifying the plant so the consequences of not reducing reactivity are acceptable.

Applicants should be required either (1) to demonstrate that with their present designs the consequences of anticipated transients without scram are acceptable, or (2) to make design changes to improve significantly the reliability of the scram system.

The BWR and PWR manufacturers should be informed of this decision.

The analysis of ATWS in accordance with the guide (Appendix D) originally prepared for use by the reactor designers should be pursued with applicants. This should be done with applicants on current and future construction permit cases and with holders of construction permits for which the AEC safety evaluation and the ACRS letter identified this problem. Backfitting of other cases should be considered on a case by case basis.

However, at the September 1970 meeting, the committee was not ready to endorse the regulatory staff recommendations and decided it needed more discussion on the subject. ACRS member Mangelsdorf was particularly disinclined to act.

At the one hundred twenty-sixth ACRS meeting, October 27, 1970, there was a long discussion with representatives of Westinghouse and of the North Anna reactors in Virginia and the Trojan reactor in Oregon. These groups had previously said they would maintain flexibility in the reactor designs to accommodate any changes to resolve the ATWS issue. Now, they "clarified" what they had meant by maintaining flexibility. Specifically, they said they were not maintaining flexibility for hardware changes; were these to be required, the AEC would have to make a finding that such changes would provide substantial additional protection which was required for the public health and safety, in accordance with the AEC Regulation 10 CRF 50.109—Backfitting. So that, except for possible changes which would affect the reliability of the instrumentation and control system, the reactors were being proposed (and implicitly or explicitly accepted) as is, despite the lack of resolution of ATWS.

The ATWS Issue Moves Slowly

By this time, at least some ACRS members were getting concerned about the delay in resolving ATWS and the apparent loss of real flexibility to make future changes on plants currently receiving favorable construction permit reports. The minutes of the one hundred twenty-seventh meeting, November 12–14, 1970, record a committee discussion on how to accelerate the pace. The regulatory staff had still not sent the four reactor vendors a list of questions the staff had prepared concerning ATWS. The vendors said they had to have this list before they would supply more information. At the request of the committee, ACRS member Okrent prepared for the ACRS members a short discussion paper on ATWS, dated November 30, 1970, which is reproduced below.

A Discussion of ATWS*, per Request of the
Committee at the November, 1970 Meeting

1. Safety Objective—The objective of a total probability of less than 10^{-7} per reactor year for a very bad accident (an accident much worse than Part 100) is postulated. For 1,000 reactors, this yields a probability of 1 in a 100 each century, and the U.S. now anticipates about 1,000 reactors by the year 2000. If this objective is missed by a factor of 100, there will probably be several very bad accidents in the world during the next century.

There are a variety of ways (more than 10) whereby a very bad accident may occur. Therefore each of these must have a probability much less than 10^{-7} (say 10^{-8}) per reactor year, if the combined probability is not to exceed the obective of 10^{-7}.

Thus, an objective of 10^{-8} per reactor year for a very bad accident from ATWS is suggested, with a need to meet this objective within a factor of ten.

2. The Staff report references various experts who have estimated an unreliability of scram between 10^{-3} and 4×10^{-4} per demand. At the ACRS Subcommittee meeting on August 26, 1970, General Electric stated that experience with GE reactors led to a failure probability of 8×10^{-4} with a 95% probability. It was stated that to demonstrate empirically an unreliability of 10^{-7}, approximately 300,000 reactor years with a zero failure history would be required.

Frequent testing can improve the failure probability somewhat, say a factor of ten, for some common mode failures, but not all can be or have been detected by testing. The Regulatory Staff concludes that the failure probability is much larger than 10^{-7}.

All of the ACRS consultants offering opinions agreed with the Staff and with the previously published opinions that the unreliability of current systems probably falls in the range of 10^{-3} to 10^{-4}.

3. Historically, this is now an old issue which represents, in part, the evolution of the issue of separation of safety and control which was first raised in an ACRS letter in December, 1967. The question of common mode failures followed, and ATWS, per se, was called directly to ACRS attention about two years ago in a letter from Epler, and soon thereafter in a publication in the Nuclear Safety Journal. The first Subcommittee meeting with GE on ATWS was held 18 months ago.

On or around June, 1970 the ACRS took the position that ATWS was a high priority item requiring resolution and asked the Staff for a report with recommendations. A draft Staff report was provided in early July. At that time questions were raised concerning how the Staff was handling probabilities—so a Committee decision was put off, and a Subcommittee meeting held in August.

In September a formal Staff analysis was received, reaffirming the July draft. It

---

*The Staff report of September, 1970 on ATWS merits re-reading. This memo's major function is to call it (and the subject) again to your attention.

concludes that scram reliability is far from adequate and that positive steps should be taken.

The Committee took no action in September. In October the Committee approved the Staff's sending out a list of questions to vendors, but took no decision. The matter was referred to the vendor subcommittee to continue work on the ATWS question.

4. What information does the ACRS seek? Is each vendor subcommittee supposed to get new information which will provide a different basis on which to judge scram unreliability? If so, how? For example, what empirical information is there on which to judge B&W or CE power reactors, neither of which have yet operated? It appears that no vital source of new information which can contradict the Staff position on reliability has been identified. In fact, during the recent past, another failure has actually been experienced at Hanford and a partial failure at SEFOR thereby reinforcing the Staff position.

5. I would like to propose the following course of action to achieve what can be accomplished in a practical manner on a reasonable time scale:

A)   The ACRS vendor subcommittees should not spend any significant amount of time in discussion of the reliability or unreliability of scram systems. Instead their efforts should be directed toward:

1)   Establishing requirements for the possible improvements in the methods for initial and periodic testing of safety systems.

2)   Establishing procedures for maintenance and periodic adjustment of safety systems that will minimize the introduction of common mode failures (e.g., adjust only a limited number of channels during any single shutdown, replace components in a limited fraction of installed channels until performance of the new components is demonstrated).*

3)   Consideration of improvements in systems design that will further reduce the possibility of common mode failures such as:

a) Use of diverse components, functional signals, physical separation, etc. (Note: Westinghouse, B&W and GE reports regarding signal diversity have already been received. A report by CE has been promised.)

b) Separation of control and safety systems.

B)   In addition to A above, the vendors should be required to (1) demonstrate that with their present designs the consequences of anticipated transients without scram are acceptable or (2) describe design changes which render the consequences of ATWS acceptable.

C)   The ACRS vendor subcommittees should plan to report to the full Committee within 6 months with respect to A & B above and the Committee should then recommend implementation of appropriate procedures and design changes as are considered necessary.

The minutes of the one hundred twenty-seventh meeting in November 1970 also note that member Okrent attached additional remarks to an

*This matter should also be reviewed with representative utilities.

ACRS report concerning a power level increase in the Oyster Creek BWR; one of the remarks pertained to ATWS. (The committee had discussed this and some other "backfit-type" matters for Oyster Creek and had chosen to omit any reference to ATWS in its report; Okrent was unhappy with the slow pace taken on the ATWS problem.)

At the one hundred twenty-eighth meeting, December 10–12, 1970, the ACRS discussed the ATWS matter further in executive session. The committee decided to send a memo to Mr. Price which indicated the current committee position regarding ATWS:

1) Reliability of present protection (scram) systems cannot be accurately established at this time.

2) The Staff is urged to send out the ATWS questions to vendors and get answers expeditiously.

3) The Committee and Regulatory Staff should maintain a detailed and continuing review of present plants, including exploration of what fixes are possible if they are needed.

Newbold Island Is Required to Deal with ATWS

The ACRS was continuing to move with all deliberate speed. At its one hundred twenty-ninth meeting, January 7–9, 1971, the ACRS decided that an ad hoc group should be formed to pursue the ATWS matter. The committee also agreed that the ATWS matter should be considered generically, that is, in terms of each reactor type, rather than case by case by the regulatory staff. However, the ACRS agreed that the staff should attempt to get reasonable assurance that the critical areas of ATWS would be answered for the high-population-density Newbold Island BWRs prior to completion of that construction permit review.

At the next ACRS meeting General Electric proposed implementation of a recirculating pump trip as a backup to reactor scram for the Newbold Island reactor. However, a month later at the March 1971 ACRS meeting, Commonwealth Edison and General Electric maintained that no modification such as pump trip was needed for the Quad Cities BWRs which were located on the Mississippi River in Illinois and were undergoing an operating license review.

On July 30, 1971, the regulatory staff forwarded a report to the ACRS in which they concluded that, specifically with the Newbold Island reactors in mind, automatic tripping of the recirculation pumps provided a substantial increase in the probability that the facility could withstand an ATWS event. The staff noted that further analyses were required concerning the diversity of components in the pump trip, and criteria were required with regard to provision of a capability for injection of neutron absorbing liquid into the core to reduce the reactor power fast enough to maintain the suppression pool* temperature at an acceptably low temperature. At the one hundred thirty-sixth ACRS meeting, August 5–7, 1971, both the Newbold Island and Limerick stations were committed to the use of the pump trip.

Beginning in early 1971, the ATWS subcommittee, chaired by ACRS member H. O. Monson, met frequently. First a series of meetings were scheduled with each of the four LWR vendors. All of the vendors argued that their scram systems were much more reliable than Epler or the regulatory staff suggested. Unreliabilities less than 10^{-6} or 10^{-7} per demand were obtained by the vendors when they performed reliability analyses of their respective systems. General Electric reported an unreliability as low as 10^{-15} per demand. How to include systematic or common-mode failures in such analyses was admitted to be troublesome, however.

It turned out that each of the PWR vendors had to develop new computer code models and systems in order to analyze ATWS events with the high degree of sophistication required. Simpler calculations which employed assumptions intended to provide an upper bound on the calculated pressures, had yielded unacceptably high pressures for some transients. Also, the different vendors exhibited varying degrees of cooperation and speed in analyzing the set of anticipated transients agreed upon by the regulatory staff and the ACRS. One or more of the PWR vendors insisted that a loss of all main feedwater was not an anticipated transient and for many months would not analyze it. (It was expected and turned out that this transient would be one of the most difficult ATWS events for a PWR to ride through without unacceptable overpressurization.)

The subcommittee meeting minutes show dissatisfaction by Chairman Monson with the slow pace of the vendors and the staff. They also show that the ACRS was still divided, with member Mangelsdorf strongly questioning the need for any action on ATWS. The minutes also indicate division among the regulatory staff.

The Staff Recommends a Position on ATWS

Finally, on April 28, 1972, L. Manning Muntzing, the new Director of Regulation, transmitted to the ACRS the regulatory staff position on ATWS which incorporated the following recommended position:

---

*The suppression pool is a large pool of water which is capable of absorbing large amounts of heat and of condensing steam which is discharged into it as the result of a loss-of-coolant accident or the opening of pressure relief valves during an ATWS.

Applicants should be required to (1) demonstrate that with their present designs the consequences of anticipated transients without scram (ATWS) are acceptable, or (2) make design changes which render the consequences of anticipated transients without scram acceptable; or (3) make design changes to improve significantly the reliability of the scram system.

This was basically the same as the staff position of September, 1970. At its one hundred forty-fifth meeting, May 4–6, 1972, the ACRS agreed with the staff position and a letter giving general approval was dispatched to Mr. Muntzing on May 10, 1972.

After more than three years, the matter appeared to have been resolved. Reactor designers would have to demonstrate an ability of their reactors to tolerate ATWS, unless by changes in design they could provide convincing arguments that the scram systems were 100–1,000 times more reliable than currently estimated. The AEC expected that the alternative that would be used was to make the reactor design such that the consequences of ATWS would be tolerable.

A Change of Mind

However, the regulatory staff continued to hold internal discussions with regard to their position on ATWS. In a draft report dated November 30, 1972, they reversed their stance on the recommended solution for ATWS; namely, for reactors whose construction permit applications were to be filed a few years in the future, improved reliability in the shutdown system (i.e., two independent systems) would be required, rather than an ability to tolerate the consequences of ATWS.

The ATWS subcommittee met with the regulatory staff to discuss this revised position on January 10, 1973. The problem of how and whether to backfit safety features to avoid or mitigate ATWS at plants in operation or under construction was a particularly difficult one, in view of the considerable problems involved in making changes in valves, piping, etc., in an existing plant. This problem had been aggravated by the increased number of plants now in this situation, as compared to 1969. The subcommittee questioned the proposed approval by the staff of the higher, and hence, less conservative "faulted stress conditions" rather than "emergency stress conditions" as an acceptable limit for backfitting ATWS.

On January 22, 1973, Mr. Muntzing formally forwarded to the ACRS the revised staff approach of November 30, 1972, as the recommended licensing positions on ATWS. And at its one hundred fifty-sixth meeting, April 12–14, 1973, the ACRS wrote a letter to Mr. Muntzing giving general approval to the new staff position, more to get action than because the new position was preferable to the old one. The ACRS letter of April 16, 1973, to Mr. Muntzing was signed by ACRS Chairman Mangelsdorf and concluded "The Committee also continues to believe that it is timely to begin implementation of the proposed ATWS position."

The AEC Announces Its Position on ATWS

In September 1973 the regulatory staff issued a report, WASH–1270, called "Technical Report on Anticipated Transients without Scram for Water-Cooled Power Reactors," in which they publicly adopted the position they had taken in January 1973 (AEC 1973a). One very important aspect of the WASH–1270 report was that it defined an overall safety goal, as well as a quantitative goal for ATWS. Specifically, it stated the following:

In establishing the boundary between accident sequences that are to be within the design basis envelope, and hence for which engineered safety features are provided, and accidents that reasonably may be assigned to that small residuum for which no further protective features are considered necessary, the regulatory staff uses the safety objective that the risk to the public from all reactor accidents should be very small compared to other risks of life such as disease or natural catastrophes. The Staff believes this safety objective is met by requiring a design basis accident envelope that extends to very unlikely postulated accidents, and by establishing the further objective that accidents not included in the design basis envelope should have an average recurrence interval of at least a thousand years for all nuclear plants combined.

For an anticipated population of about one thousand nuclear plants in the United States by the end of the century, the safety objective will require that there be no greater than one chance in one million per year for an individual plant of an accident with potential consequences greater than the Part 100 guidelines. Since plants now being designed and constructed are expected to have service lives approaching 40 years, and may thus be part of the century-end population, the Staff believes it appropriate to consider their designs in the light of this future requirement. In view of the difficulty of determining such a low probability, the Staff regards this number as an "aiming point", or design objective, rather than as a fixed number that must be demonstrated for a given plant design.

The staff went on to say that since there are many potential paths to a serious accident, they propose to allocate only one-tenth of their objective to any one source or path; hence, the safety objective for ATWS was that it not lead to an accident with serious off-site consequences more frequently than 10^{-7} per reactor-year.

This same safety objective was subsequently applied to other accident sources subject to some degree of quantification.

With the issuance of the WASH-1270 report in September 1973 (AEC 1973a), the regulatory staff had taken a position on ATWS and it was seemingly resolved except for implementation. The ACRS letter of February 13, 1974, on the status of generic items moved the ATWS matter into the resolved column on exactly this basis. In the period 1974–1975 all the reactor vendors submitted analyses on ATWS in general response to the requirements set forth in the WASH-1270 report.

On July 3, 1974, an ACRS subcommittee meeting was held with representatives of the regulatory staff to discuss the status of ATWS and the ATWS criteria that the staff were developing to deal with plants already in operation, under construction or to be constructed. The staff advised that no firm proposals had been submitted by the reactor vendors for Class A plants (those for which a construction permit application would be filed after October 1, 1976, and for which diverse shutdown systems of high reliability would be required). Dr. Hanauer of the regulatory staff stated that several vendors had already submitted analyses which showed protection system unreliability values smaller than 10^{-7}, but that these were not considered acceptable by the staff because they did not satisfy the intent of the WASH-1270 report (AEC 1973a) or the staff criteria. The staff also stated that the vendors did not consider the staff criteria to provide adequate guidance for the initiation of new designs.

The Staff Proposes a Change in Position

In September of 1975, the regulatory staff asked to meet with the ACRS concerning a possible major change in the approach to ATWS adopted in the WASH-1270 report for Class A plants. At a subcommittee meeting held October 8, 1975, the staff listed four alternatives, as follows:

(1) Implement the original WASH-1270 position (i.e., no change in philosophy).

(2) Make a reevaluation of the positions of WASH-1270.

(3) Accept partial conformance with the WASH-1270 position.

(4) Change the position for Class A plants, namely, require that consequences of an ATWS be made tolerable by other design changes if necessary.

According to the staff, alternative 1 had as an objective the elimination of ATWS as a design basis accident by reducing the probability of an ATWS to an acceptably low value. The staff noted that if this were accomplished for Class A plants, older (Class B) plants would become a controlling factor on the overall risk from ATWS in all LWRs since there would be a large number of Class B plants,* roughly 200 or 300, for which the approach was to make ATWS tolerable (with some residual probability of exceeding safety limits). Warren Minners of the staff, in discussing alternative 2, said the objective here would be to determine if any new information had changed the conclusions of the WASH-1270 report (AEC 1973a); the staff did not expect that a re-evaluation would reveal that the probability of an ATWS in plants of current design was significantly different from that in the report.

With regard to alternative 3, Mr. Minners said that the PWR vendors had proposed additional shutdown systems which conformed to the independence and diversity requested in the WASH-1270 report. Such systems could reduce the probability that the system would not de-energize the control rods but could not assure scram, and the safety objective might not be satisfied.

The staff recommended alternative 4, namely, making the consequences of ATWS acceptable (also working on improving the reliability of the existing protection system). According to Minners, this alternative was based partly on the argument that if the safety of Class B plants could be made adequate, additional requirements for Class A plants were unwarranted; and, in fact, the provisions for Class B plants might be better able to cope with situations which were currently unrecognized. Implicitly, there appeared to be doubt among the staff that diverse shutdown systems could or would be proposed and developed to the point where the staff could agree that the probability of ATWS was acceptably low.

The ACRS sent a letter to Mr. Lee Gossick (NRC Director of Operations) dated October 17, 1965, in which they concurred with the staff proposal to revise the criteria for Class A plants, along the lines of alternative 4. Effectively, the staff had reverted to their original approach, that proposed prior to the WASH-1270 report (and the one favored by the ACRS).

In documents dated December 9, 1975, the regulatory staff prepared status reports on ATWS for each of the four LWR reactor vendors. In these documents the staff took positions with regard to "acceptable fixes" for Class B plants and identified outstanding issues. The ACRS held three days of subcommittee meetings and had an extensive discussion with the staff and the reactor vendors (who disagreed strongly with many staff positions) at the one hundred eighty-ninth ACRS meeting, January 8-10, 1976.

*Class B plants were those for which a construction permit application was filed before October 1, 1976, and for which operation began after ATWS had first been identified as a safety concern.

After this meeting, the ACRS issued a report to NRC Chairman William A. Anders endorsing the general approach and safety objectives adopted by the staff, which included aiming for a maximum probability of 10^{-7} per reactor-year for an ATWS with unacceptable consequences.* However, the ACRS indicated that the arguments outstanding between vendors and staff concerning what constituted an acceptable way of meeting the criterion must still be resolved.

So, ATWS was "almost resolved," but there remained many complicating factors. Many representatives of the nuclear utilities and the reactor vendors turned to the results to be found in the draft WASH–1400 report (AEC 1974a), and the final version (NRC, 1975a), as a strong demonstration that ATWS was not a major contributor to risk for LWRs. They concluded that, hence, the existing situation was satisfactory and no design modifications were needed to improve either the reliability of scram systems or the ability of the reactors to accept an ATWS without intolerable consequences.

General Electric concluded that implementation of the mitigating requirements defined in the staff status report of December 9, 1975, would be very expensive,† and in a document dated September 30, 1976, proposed a different recommended solution to ATWS, namely, incorporating a proposed alternative reactor scram system as backup to the existing system so that the NRC's ATWS safety objective of 10^{-7} per reactor-year could be met.

In a memorandum to the committee dated November 24, 1976, ACRS consultant Epler discussed the General Electric proposal and remained unconvinced that the claimed reliability could be achieved thereby. And apparently, the regulatory staff also remained unconvinced.

---

*In the minutes of the one hundred ninety-second meeting, April 8–10, 1976, an ACRS position on acceptable risk, which was prepared for a letter responding to ten questions posed by G. Murphy, Executive Director of the staff of the Joint Committee on Atomic Energy (JCAE), was recorded as follows:

"The probability of an accident having serious consequences to public health and safety should be less than 10^{-6} per reactor-year. A serious accident is one having consequences similar to that of the crash of a mid-sized jet airliner (approximately 150 passengers). It was generally agreed that simply exceeding the limits of Part 100 would not necessarily constitute 'serious consequences.'" In a letter dated April 12, 1976, to the JCAE, the ACRS stated its belief that for reactors to be constructed in the next several years, a probability of less than one in a million per reactor-year for a serious accident was suitable as an interim objective.

† In a letter to the NRC dated September 28, 1976, the Long Island Lighting Company stated that "an expenditure of approximately $50 million would be required for the total installed cost of additional equipment and logic, but excludes costs of financing, outages or delays."

ATWS remained a very controversial issue between the NRC and the industry.

Beginning in the fall of 1976, a series of reports entitled "ATWS: A Reappraisal" was published by the Electric Power Research Institute (EPRI 1976a, 1976b, 1977, 1978). In summary, the EPRI reports reevaluated the probability of failure to scram and estimated the risk to the public from ATWS. Using their assumptions and choice of data, the authors of the reports concluded that the probability of failure to scram was much lower than 10^{-4} per demand (by a few factors of 10) and that ATWS posed insignificant risk to the health and safety of the public.

In March 1977 the NRC formed a task force on ATWS in an effort to finally resolve the matter. In July 1977 the staff reported once again to the ACRS, reiterating their general position of December 1975 that scram unreliability could not be shown to be acceptably low and that measures were required to mitigate the consequences of ATWS.

It is in a sense curious that with this relatively unchanged position by the staff over a period of many years, little was done to implement any actual changes with regard to ATWS. The question first arose for boiling water reactors, and at least one mitigating design change, the recirculating pump trip, was included as a partial backup to scram in all BWRs receiving construction permits after the Newbold Island review in 1971. However, many, if not most, of the operating BWRs had not incorporated this feature by 1976. When the ACRS learned of this in early 1976, it sent a memorandum to the regulatory staff recommending that steps be taken promptly to include this change in operating BWRs, unless it was not needed to make the consequences of ATWS tolerable. The regulatory staff shortly thereafter issued letters to all operators of BWRs asking for their specific plans to incorporate this feature.

However, the year 1977 passed without issuance of a new regulatory staff position on ATWS, and this generic item remained unresolved eight years after its inception. Much had been learned about the subject. There obviously remained wide differences of opinion concerning the safety significance of the matter. Reactors continued to be designed and receive construction permits without incorporation of mitigating features. And few of the operating BWRs had implemented the recirculation pump trip by the end of 1978.

A New Staff Position

In April 1978 the regulatory staff issued a new report, NUREG-0460, entitled "Anticipated Transients Without Scram for Light Water Reactors" (NRC 1978c), which provided a changed position. An excerpt from this report follows:

The Staff now concludes that transients that would result in serious consequences if accompanied by scram failure could be expected to occur in the future population of plants at a rate of five to eight per reactor-year. We also estimate that the probability of scram failure, based on nearly 700 reactor years of operating experience in foreign and domestic commercial power reactors with one observed potential scram failure, is in the range of 10^{-4} to 10^{-5} per demand. Thus, the expected frequency of ATWS events that could result in serious consequences is approximately 2×10^{-4} per reactor-year. We recommend that a safety objective of 10^{-6} unacceptable ATWS events per reactor-year is more appropriate, and therefore, that some corrective measures to reduce the probability of consequences of ATWS are required.

The staff report proposed a change in safety objective from that in the WASH-1270 report (AEC 1973a), namely, from 10^{-7} per reactor-year for an unacceptable ATWS event to 10^{-6} per reactor-year. This was apparently based on the relatively large frequency of core melt predicted in the WASH-1400 report (NRC 1975a) (5×10^{-5} per reactor-year)—which did not seem to pose an unacceptable risk. The staff employed a mixture of deterministic and probabilistic methodology to prescribe the design approaches that would be needed to meet the new safety objective for each LWR vendor.

The new staff proposals were again opposed very strongly by the industry, and after many meetings between the ACRS, the staff, and representatives of the nuclear industry, strong differences of opinion still existed. The ACRS was not prepared to endorse the proposed staff position.

In early 1979, after issuance of the NUREG/CR-1400 report of the Risk Assessment Review Group or Lewis panel (NRC 1978a), which criticized the WASH-1400 report (NRC 1975a), and after endorsement of the Lewis Panel report (NRC 1978a) by the NRC commissioners, the staff proposed a greatly revised position on ATWS, one which strongly reflected the difficulties in backfitting an operating plant or even a plant under construction. For such plants, emphasis was placed on those changes in circuitry that were relatively easy to accomplish and that might provide increased scram reliability. For plants that were to be constructed, the emphasis was maintained on hardware changes to mitigate the consequences of an ATWS should it occur, that is, to keep pressure and temperatures below acceptable limits. In arriving at their new position the regulatory staff stated they were now using engineering judgment since the commissioners had stated that probabilistic methods could not be used to provide a quantitative basis in licensing.

In the spring of 1979, the Three Mile Island accident introduced additional questions on the behavior of PWRs which caused the staff to reevaluate their ATWS position for PWRs. In early 1980 the staff proposed a more stringent position with the stated intention of trying to resolve ATWS once and for all. The industry once again disagreed with the staff, and the ACRS, in a letter to NRC Chairman John Ahearne dated April 16, 1980, took a position which would require less backfitting than was being proposed by the staff. More than eleven years after the letter by Epler, the ATWS issue remained unresolved.

On June 28, 1980, at the Browns Ferry Unit 3 BWR, 76 of 185 control rods failed to insert fully into the core when a scram was called for by the reactor operator. Fortunately, this occurred during a routine shutdown from power, rather than during the kind of reactor transient in which complete and rapid scram of all the rods might have been very important. Occurrence of this event provides strong support to those, including me, who have been arguing that measures should be taken to mitigate ATWS.

17

The Seismic Safety
of Nuclear Reactors

In this chapter we shall see that seismic safety considerations were largely overlooked for the first several power reactor built east of the Rockies. Then, in the period 1963–1965, in connection with review of each of the proposed reactors at Bodega Bay, San Onofre, and Malibu, California, the originally proposed requirements for seismic design were made two or three times more stringent; and an awareness of the need for seismic design criteria for all reactors emerged.

The ACRS and the regulatory staff differed in their final positions on the Bodega Bay reactors, with the regulatory staff recommending against the site. (The reactor proposal was then withdrawn by the utility.) Soon after, the regulatory staff lost their case in favor of construction of the proposed Malibu reactor first with the ASLB and then with the AEC commissioners, who upheld the intervenors and ruled that design for surface faulting was required, which effectively ruled out the site.

Beginning about 1965, the ACRS consistently tended to be somewhat more conservative than the regulatory staff in their approach to seismic design criteria. Also, the ACRS began to make strong recommendations for more seismic safety research.

Finally, we shall see that although in 1975 the reactor safety study, WASH–1400 (NRC 1975a), concluded that seismic events represented a very minor contributor to accident risk from a nuclear reactor, ensuing

261

developments have led to a fairly strong case that the seismic contribution
to risk from LWRs may be appreciable.

Early Developments in Seismic Safety

One of the first mentions of seismic matters in the minutes of the
statutory ACRS occurred at its twelfth meeting, December 11–13, 1958. In
the discussion of the site for the Carolinas-Virginia Tube Reactor (CVTR;
25 miles northwest of Columbia, South Carolina) mention was made of the
severe Charleston, South Carolina, earthquake of 1886. No seismic design
basis for the proposed reactor was mentioned.

The minutes of the nineteenth meeting, September 10–12, 1959, reported
that the proposed site for the Humboldt Bay, California, reactor was in a
moderate to heavy earthquake region and the seismic safety factors were
slightly more conservative than those used for conventional power plants.*

The absence of special seismic design requirements for light water reactors constructed in the eastern United States during the 1950s is noted in a
report by Professor H. Shibata of the University of Tokyo (Shibata 1970).
He summarized the results of a report made by a Japanese study group on
nuclear reactor safety on behalf of the Japan Atomic Industrial Forum.
They observed that studies and investigations on "nuclear reactors and
earthquakes" were underway in the United States in 1959; however, aseismic design was hardly studied in the construction of four nuclear power
plants in the eastern part of the United States, because earthquakes in the
region were considered to be small.

At a special meeting, March 5, 1960, the ACRS prepared a hurried
response to a letter from AEC General Manager A. R. Luedecke concerning the suitability of sites in southern California. The meeting minutes
indicate the relatively crude approach then proposed for seismic design.

Mr. Booth of the Staff reviewed briefly the earthquake features of the California
area as to location, frequency, and the scale for expressing the intensity. The larger
shocks have given an acceleration in the horizontal direction of about one-third of
gravity, "g," (measurements are such that this might be in error by a factor of two);
the vertical component has always been less than horizontal. The problem of
vertical shock has not been considered in structures to date, because they are, of
course, designed to resist one "g." The probability of quakes is about constant over

---

*The seismic design accepted for the Humboldt Bay reactor in 1960 was later questioned in
1968. Then, with the discovery of active faults in the vicinity of the site, the adequacy of the
seismic design was questioned once again in 1977, leading to shutdown of the reactor.

California, except in the extreme northeast portion and the upper part of the San Joaquin Valley.

Mr. Booth concluded that simple construction precautions should eliminate the possibility of release of radioactivity by earthquakes. However, although the frequency is less, earthquakes have been known to occur in nearly all parts of the United States, and since special seismic design features are not usually included in reactors, earthquake damage to these other reactors might lead to more serious consequences than would a shock in California.

In its letter of March 6, 1960, to the AEC the ACRS said the following:

> With respect to seismic considerations, we understand that it is present utility industry practice in California to locate generating stations at least one mile from known surface faults, and to design and construct these stations using local codes supplemented by special analyses and increased seismic design factors for those critical plant components necessary to maintain the station on the line. In addition, in the case of a nuclear reactor facility, special analyses and increased seismic design factors are needed for those reactor plant systems whose failure could result in a release of radioactive material. With these precautions, the Committee believes the reactor facility would be adequately protected against seismic disturbance.

The ACRS did not at that time initiate steps to make seismic design a serious consideration for reactors east of the Rockies.

The Bodega Bay Review

In the early spring of 1963, the ACRS completed (for the first time) its review of the application by Pacific Gas and Electric Company to construct a 1,008 MWt BWR at a remote site on Bodega Bay, Sonoma County, California. The unique aspect of the review lay in the fact that the site was only about 1,000 feet west of the San Andreas fault zone.

The minutes of the ACRS subcommittee meeting held concerning Bodega Bay on March 20, 1963, report briefly on seismic matters, as one of seventeen items discussed at the meeting. It was reported that:

> The plant is designed for motion in all directions. The same design criteria will be used at Bodega Bay as have been used at other (non-nuclear) plants which have satisfactorily withstood fairly severe earthquakes in this region.

In its report to the ACRS on the Bodega Bay site dated April 2, 1963, the regulatory staff approved the proposed design basis for acceleration caused by an eathquake of 0.3 g,* subject to the qualification that no faults existed under the plant. Their total discussion of seismic matters was rather brief.

The ACRS reviewed the construction permit application for the Bodega Bay reactor at its forty-seventh meeting, April 11–13, 1963. While there was discussion about seismic aspects of the proposed plant, this was not the principal focus of the ACRS review. The ACRS membership did not include any professional seismologists, geologists, or seismic engineers, and at that time the committee did not have any such consultants. The ACRS reported favorably on construction of the Bodega Bay reactor in a letter dated April 18, 1963, to AEC Chairman Seaborg. Much of the letter related to engineered safety features in a general fashion. The paragraph on seismic-related matters is reproduced below.

The requirements that are imposed on plant design because of location in an active seismic area have been considered by the applicant, and the referenced documents contain the recommendations of seismologists who have been consulted on this question. Tentative exploration indicates that the reactor and turbine buildings will not be located on an active fault line. The Committee believes that if this point is established, the design criteria for the plant are adequate from the standpoint of hazards associated with earthquakes. Careful examination of the quartz-diorite rock below should be made during building excavation, to confirm this point. Furthermore, the Committee suggests that, during design, careful attention should be given to the ability of emergency shutdown systems to operate properly during and subsequent to violent earth shocks, and to the stress effects that might be introduced because the reactor building and the turbine building are to be anchored in different geological formations. The need for earthquake-induced shutdown and isolation of the primary system can be considered at a later time.

With hindsight, both the ACRS review and that of the regulatory staff appear to have been relatively limited, in view of the nearness of the proposed site to the San Andreas fault. In effect, this site represented one where there was a high probability that the ability of the reactor to accomplish safe shutdown in the face of a major earthquake would be challenged during the lifetime of the plant.

Although the schedule projected by Pacific Gas and Electric for approval of construction of the Bodega Bay reactor showed issuance of a construction permit by July 1963, this schedule slipped. The foundation excavation for the Bodega Bay reactor containment proceeded, and in the process of examining the exposed surfaces, several small faults were found;

---

*A method of seismic design and analysis was developed whereby specification of the peak value of the horizontal acceleration caused by the earthquake (such as 0.3 g or 30% g) determined all the vibratory parameters.

the one which received the most attention was labeled the Shaft fault (named for the shaft which was dug for the reactor containment building). Since the original review was carried out on the assumption that the reactor would not be located on an active fault, the entire matter was opened for re-review.

The regulatory staff asked the United States Geological Survey to review the Shaft fault, attempt to evaluate its past history as a function of geologic time, and provide guidance regarding its potential for future movement. They also asked the United States Coast and Geodetic Survey to advise them on seismological matters and on possible tsunami effects from off-shore or distant marine earthquakes. Additionally, they obtained Professor Nathan M. Newmark of the University of Illinois and Dr. Robert A. Williamson of the Holmes and Narver Company to advise them on seismic engineering design. The ACRS obtained the consulting services of a seismic engineer, Mr. Karl V. Steinbrugge.

J. Schlocker and M. G. Bonilla of the United States Geological Survey studied the matter and concluded that the Shaft fault had displaced sediments that were more than 42,000 but less than 400,000 years old, and that the last movement on the fault took place during the last 400,000 years. Their report went on to say that, because the possibility existed that the faulting had occurred during the past few hundred years, it was prudent to predict that faulting was a possibility at the site during the next 50 to 200 years (which meant that differential permanent displacement of the rock under the plant might occur and possibly rupture the containment building or cause other safety-related failures). Bonilla and Schlocker also discussed the past history of "sympathetic" faulting, or the creation of new faults, close to the San Andreas fault and concluded that this possibility had to be considered for any site close to the San Andreas fault, even if no fault were already present. Their report concluded that the possibility of a two to three-foot offset at the site should not be ruled out. This would represent a very difficult design basis, one not previously attempted, if it were required that the reactor could withstand such a shift.

Other interpretations of the available geological information also existed, however. Professor Hugo Benioff, a consultant to Pacific Gas and Electric, stated that auxiliary faults would slip only a matter of inches when the main San Andreas fault ruptured. Dr. Don Tocher, another Pacific Gas and Electric consultant, estimated that differential motion on the Shaft fault would be less than an inch, given a major earthquake on the San Andreas fault. Benioff, Tocher, and Mr. E. Marliave, another Pacific Gas and Electric consultant, all concluded that the Bodega Bay site was acceptable. The ACRS appeared to think that the arguments of the Pacific Gas and Electric consultants were plausible.

On March 27, 1964, the large Alaskan earthquake occurred. In addition to very considerable damage in Alaska, particularly due to landslides, this earthquake produced an appreciable tsunami wave at Crescent City, California, providing further empirical evidence of this phenomenon. How much impact this earthquake had on the Bodega Bay review is difficult to assess. However, it seems to have led to consideration of somewhat larger design basis accelerations at Bodega Bay.

On April 30, 1964, the regulatory staff submitted a report on the Bodega Bay site to the ACRS for consideration at its fifty-fifth meeting, May 6–8, 1964. The staff had concluded that the site was not suitable.

The staff noted that Pacific Gas and Electric had doubled its proposed design basis for seismic vibration from $0.33g$ to $0.66g$ for structures and components important to safe shutdown of the facility. The staff's consultants thought that some peak accelerations as high as $1.0g$ might occur; however, the staff believed that the potential effects of ground vibration could be adequately handled. The staff also noted that Pacific Gas and Electric (PG&E) had not discussed seismically induced tidal waves (tsunamis).

Most importantly, the staff stated that the United States Geological Survey believed that the possibility of two to three feet of permanent displacement on Bodega Head in the event of a large earthquake on the San Andreas fault was low, but not low enough to be ignored in the design of the facility. The staff went on to say:

> The question then becomes, can a reactor facility of this power level be designed and built which would safely accommodate the differential ground movements of the magnitude that could be expected. The company has proposed a design concept which would accommodate some relative displacement. However, such an engineering safeguard has not been used in practice, nor has it been proven for structures approaching the size and complexity of the Bodega reactor facility with its complicated and sensitive components.
>
> Although the design criteria proposed by PG&E are not in agreement with criteria suggested by our consultants to protect against vibrational effects and possibly the effects of tsunamis resulting from a large earthquake occurring on the San Andreas fault at or near Bodega Head, it appears that these problems might be adequately resolved by changes in design criteria and more complete analyses. However, the risk could be effectively eliminated by moving the plant to a location a couple of miles distant from the main fault zone. On this basis, we have concluded that the site proposed by PG&E is not suitable for a reactor of the general type and power level proposed.

The consultants to PG&E and to the regulatory staff continued to exhibit large differences in opinion at the fifty-fifth ACRS meeting in May,

1964. The ACRS apparently was satisfied either that a large displacement would not occur under the reactor, or that such displacement could be designed for. It prepared a letter favorable to construction of the reactor at the proposed site, but the letter was held up and not dispatched to the AEC at the request of the regulatory staff. Some of this letter, written in May 1964 but never issued, follows:

The Committee has been advised that during the life of the proposed reactor there is a high probability that it will experience at least one major earth shock. There is associated with such an earthquake a remote probability that the plant will be subjected to the effect of a shearing motion in the rock on which it is built. The Committee is of the opinion that designs of the nature proposed can be made to withstand the effects of the anticipated earthquakes such that at the worst the reactor can be shutdown and cooled without undue release of fission products.

The Committee recognizes the presence of earthquake hazards at this site and believes that special measures will be necessary. The Committee would like to be kept informed regularly as to the developments in areas closely related to the safety considerations arising from these hazards. Among others these should include:

(1) Provisions to accommodate possible earth movements and effects of displacement along the fault. The applicant has proposed orally to design the building to withstand up to two feet displacement along the discovered fault. The Committee believes that the engineering principles are sound and, if extended to take into account the possibility of the same motion in any direction, will afford the degree of assurance required for protection of the reactor.

(2) Consideration of model testing or other experimental verification of novel design features associated with earthquake protection.

(3) Redundant provisions to assure emergency cooling water in case of damage to normal and emergency supply systems by earthquakes or tsunamis.

(4) Design and tests of critical plant components such as instrumentation and control rod operating mechanisms to withstand earthquake damage.

The ACRS letter was thus favorable toward construction of the Bodega Bay plant, with provision in the design for surface displacement.

At a meeting arranged by the regulatory staff on June 17, 1964, Mr. Whelchel of PG&E agreed to design the plant for a movement of three feet, including the necessary engineered safeguards.

On August 17, 1964, the ACRS received a written report from its consultant Karl Steinbrugge. His conclusions include the following:

The proximity of the San Andreas fault is not, in itself, an adequate reason for prohibiting the construction of the proposed Bodega Head facility. The provisions for 3-foot radial clearance around the walls of the containment structure is satisfactory, in the writer's opinion, for any credible fault displacement beneath the structure. It would be reasonable to request that the detailed earthquake design be left in the hands of a firm specializing in earthquake engineering.

The foregoing conclusions and recommendations have been directed towards specific problems at Bodega Head. The findings are believed to be conservative, and are consistent with the present knowledge regarding the state of the art of earthquake engineering.

On October 6, 1964, the regulatory staff issued a report to the ACRS concerning seismic considerations for the Bodega Bay site. The staff report summarized the differences of opinion among the experts and discussed the novel aspects of the design concept proposed to cope with up to three feet of displacement under the reactor. The staff agreed with their consultant, Professor Newmark, that it was technically feasible to meet the proposed seismic design basis, but stated that the existing difficulties in confirming seismic design adequacy at other plants would be aggravated for the Bodega Bay site.

The staff report concluded with a statement that the Bodega Bay reactor raised policy issues as to whether the public benefits of operating the reactor were high enough to justify constructing the reactor close to the San Andreas fault zone. But it did not draw a conclusion.

The staff report was accompanied or followed by opinions from its various advisors. The United States Coast and Geodetic Survey recommended that the plant be designed for a maximum ground acceleration on rock of $2/3g$ and that ground accelerations as high as $1 g$ should be taken into account. They also recommended that the design take into account 50 feet tsunamis from nearby severe marine earthquakes and 30 feet tsunamis from distant generating areas. Consultant Newmark wrote a fairly lengthy report in which he concluded that:

The structural integrity and leak tightness of the containment building can be maintained under the conditions postulated. However, certain precautions must be considered, especially in the design of umbilicals and of penetrations to the containment building.

The ACRS completed its re-review of the construction permit application for the Bodega Bay reactor at the fifty-eighth meeting, October 7–10, 1964. Professor George Housner, a consultant to Pacific Gas and Electric, stated that the plant could be designed to withstand two to three feet of faulting. Steinbrugge agreed and Newmark also stated that he conceived no technical problem in designing the reactor to resist large displacements.

Dr. Doan of the regulatory staff took the position that although novel designs to resist three feet of displacement could be realized, they were untestable. He believed that the public benefits resulting from such a reactor had to be balanced against the risks. ACRS member Osborn pointed to the inability to test other reactor arrangements. ACRS member Rogers said that other structures, e.g., bridges and dams, had been built to resist earthquakes without the benefit of testing; hence, the present plans for the Bodega Bay reactor did not seem to him to be a very great extension in existing engineering practice. ACRS member Thompson stated that since cooling water would be found primarily at the ocean in California, rejection of this site could lead to a decision against reactors in the entire state. ACRS member Silverman said the site was very good from the population point of view; if it could be designed to safely withstand three feet of earth movement, it should be accepted.

The ACRS completed and issued a letter report favorable to construction of the Bodega Bay reactor at its fifty-eighth meeting. The letter, dated October 20, 1964, was sent to AEC Chairman Glenn T. Seaborg and signed by ACRS Chairman H. Kouts. It is reproduced below.

ACRS Report to the AEC on Bodega Bay Atomic Park—Unit no. 1

At its fifty-fifth meeting on May 7–9, 1964, at Argonne, Illinois, and at its fifty-eighth meeting on October 7–10, 1964, the Advisory Committee on Reactor Safeguards again considered the proposal of Pacific Gas & Electric Company to construct and operate a 1008 MW(t) boiling water reactor on Bodega Head north of San Francisco, California. The Committee had the benefit of oral discussion with representatives of the applicant and its consultants, with the AEC Regulatory Staff and its consultants, including staff members of the U.S. Geological Survey (USGS) and the U.S. Coast and Geodetic Survey (USC&GS) and of the reports cited below. Subcommittee meetings were held July 31, 1962 and March 20, 1963 and members of the Committee again visited the excavated site on June 3, 1964. Numerous information meetings were held with the applicant, the AEC Regulatory Staff, and with consultants.

This proposal had been considered at the Committee's forty-seventh meeting and reported on in its letter of April 18, 1963 which stated:

"Tentative exploration indicates that the reactor and turbine buildings will not be located on an active fault line. The Committee believes that if this point is established, the design criteria for the plant are adequate from the standpoint of hazards associated with earthquakes. Careful examination of the quartz-diorite rock below should be made during building excavation, to confirm this point. Furthermore, the Committee suggests that, during design, careful attention should be given to the ability of emergency shutdown systems to operate properly during and subsequent to violent earth shocks, and to the stress effects that might be introduced because the reactor building and the turbine building are to be anchored in different geological formations. The need for earthquake-induced shutdown and isolation of the primary system can be considered at a later time."

The exploration suggested in the above comment has been completed, and the geologic features discovered have led to further structural considerations in the design. These geologic features include fractures in the underlying rock. One has been identified as the so-called "shaft fault". The character, extent, and age of the most recent activity of this fracture are controversial. Nevertheless, the applicant has considered its significance in the proposed structural protection.

Proximity of the site to the San Andreas fault system has been given careful consideration. The Committee has been advised by several consultants that, during the life of the proposed reactor, there is a high probability that the reactor site will experience at least one major earth shock. There is associated with such an earthquake a remote possibiity that the plant will be subjected to the effect of a shearing motion in the rock on which it would be built. The USGS and USC&GS have proposed values for the intensity and accompanying earth motions, including shear, which could be anticipated during the worst earthquake. Determination of these values has been hampered by lack of authoritative historical records and reliable measurements. The applicant and his consultants believe that lower values are more realistic. The Committee considers that the USC&GS and USGS values are conservative.

The applicant has proposed methods for mechanical and structural design to meet the predicted seismic occurrences. The applicant also has proposed to design the building to withstand up to three feet of shear aisplacement along any plane at the site. The Committee believes that the engineering principles and general design proposed to incorporate them are sound. These considerations afford that degree of assurance required for protection of the reactor in the unlikely event of the predicted maximum earthquake.

The USC&GS has recommended a design height for tsunami run-up at Bodega Head. The applicant stated that the facility design and safeguard procedures will be such the plant would withstand such a tsunami safely.

The Committee is of the opinion that the applicant's design objectives may be accomplished within the scope of present engineering knowledge.

Many details of the proposed design have not yet been completed. It is understood that the applicant will continue to give careful attention to the following items during design and construction: limitations on the maximum reactivity of individual control rods; provisions to accommodate possible seismic earth movements and shear displacement; consideration of testing or other experimental verification of structural design features associated with earthquake protection; provisions to assure adequate cooling water in case of damage to normal and emergency supply systems; core behavior during earthquakes; design and tests of critical plant components such as instrumentation, isolation valve, and control rod operating mechanisms to withstand earthquake damage; additional considerations which may be needed if zirconium clad fuel is to be used.

The Committee recognizes that the applicant has accepted very conservative values for earth shear movement, earthquake magnitudes, and tsunami heights as design criteria. These criteria should not be construed as precedents for use elsewhere.

With due consideration being given to the items discussed above, the Advisory Committee on Reactor Safeguards is of the opinion that the power reactor facility as proposed may be constructed at this site with reasonable assurance that it may be operated without undue hazard to the health and safety of the public.

The AEC issued a public announcement in Washington, D.C., on October 26, 1964, about the Bodega Bay reactor, which follows:

The Atomic Energy Commission today is making public two reports concerning the safety aspects of a nuclear power plant proposed by Pacific Gas and Electric Company at Bodega Head, approximately 50 miles north of San Francisco.

One report is from the Commission's Advisory Committee on Reactor Safeguards, a group established by law to advise the AEC on safety matters involved in reactor construction and operation. The ACRS has concluded that there is reasonable assurance that the proposed reactor can be constructed and operated at the Bodega Head site without undue hazard to the health and safety of the public.

The other report, by the Division of Reactor Licensing, has been issued by the AEC Director of Regulation, whose staff makes safety reviews of reactor licensing applications. The Regulatory Staff has concluded that "Bodega Head is not a suitable location for the proposed nuclear power plant at the present state of our knowledge."

Under AEC regulatory procedures, a decision by the Commission on PG&E's application will not be made until after the holding of a public hearing and issuance of an initial decision by a three-member atomic safety and licensing board.

The conclusions from the summary analyses of the regulatory staff follow:

The containment and all of the emergency equipment for shutting down the Bodega reactor and maintaining it indefinitely in a safe condition in the absence of seismic disturbances are designed on the basis of well-established engineering principles. They can also be tested to ascertain that the design objectives have been achieved. Consequently, there is a high degree of assurance that the reactor can be built and operated without undue risk to the health and safety of the public in the absence of seismic disturbances.

The seismic design of the reactor structure to withstand purely vibrational effects is also based on well-established engineering principles which in some cases at least have been verified in the presence of earthquakes. Thus, while it is not possible to carry out any measurements on the finished structure to assure that the seismic design objectives have been accomplished, there is sufficient experience background to justify a conclusion that the specified seismic vibrational criteria can be achieved and that the plant can therefore be safeguarded against any credible earthquake that does not rupture the foundation rock.

We believe there is room for reasonable doubt, however, that a comparable situation exists with respect to that particular aspect of the proposed seismic design of the Bodega reactor structure intended to assure that the containment and reactor shutdown functions will remain intact in the event of a shear displacement of its foundation bedrock as great as three feet in any direction. While the proposed engineering principles appear reasonable, experimental verification and experience background on the proposed novel construction method are lacking. If approved, this would, to the best of our knowledge, be the first attempt on record to design a building structure and its associated vital equipment to withstand the effects of substantial movement in its foundation with the vibration accompanying a severe earthquake. Because of the magnitude of the possible consequences of a major rupture in the reactor containment accompanied by a failure of emergency equipment, we do not believe that a large nuclear power reactor should be the subject of a pioneering construction effort based on unverified engineering principles, however sound they may appear to be.

The Advisory Committee on Reactor Safeguards has reached the conclusion that the reactor can be constructed and operated at the proposed location without undue risk to the health and safety of the public. We have carefully considered the views of the ACRS. We have the highest respect for those views and we do not lightly reach an opposite conclusion. This is a kind of case, however, on which reasonable men may differ. In our view, the proposal to rely on unproven and perhaps unprovable design measures to cope with forces as great as would be produced by several feet of shear ground movement under a large reactor building in a severe earthquake raises a substantial safety question.

In all respects except one the proposed design of the Bodega Nuclear Power Plant provides reasonable assurance that the plant can be built and operated without undue risk to the health and safety of the public. However, the single exception is quite important if one accepts the credibility of an earthquake of sufficient magnitude to cause a major displacement of foundation rock underneath the plant. Although there is a wide difference of expert opinion on the credibility of such an earthquake, prudent judgment favors accepting the conservative recommendations of the USC&GS and the USGS. On this basis and for reasons given above, it is our conclusion that Bodega Head is not a suitable location for the proposed nuclear power plant at the present stage of our knowledge.

Thus, the final positions of the ACRS and the regulatory staff disagreed with regard to the acceptability of the Bodega Bay site. Such a disagreement had not occurred previously and it created a considerable stir when it occurred.

Although the regulatory staff position of April 30, 1964, had been to reject the Bodega Bay site, the regulatory staff position in their report to the ACRS for the October 1964 ACRS meeting had been less definite, stating that there were policy considerations involved. No final conclusion was drawn in the report, and in their discussions with the ACRS at the October meeting, the regulatory staff did not state that their final position was to reject the site.

Undoubtedly, the ACRS thought that its recommendation was going to become that of the regulatory staff. The ACRS appeared to be rather skeptical that displacements as large as two to three feet were an appropriate design basis; and when Housner, Newmark, and Steinbrugge all said that a design could be accomplished, the committee accepted this as adequate.

Pacific Gas and Electric withdrew its application in the face of the regulatory staff decision. Looking back with roughly 15 years of hindsight, it appears likely that the proposed design bases for vibratory motion might not now be acceptable. As more strong motion accelerograms have been obtained from locations near the source of major earthquakes, such as that at San Fernando in 1971, and as increased knowledge of earthquake generation has developed, accelerations much larger than $2/3g$ for sites so close to a major fault have been suggested. Additionally, since the probability of a large earthquake at the site would be close to unity over the reactor lifetime, a high degree of assurance would be needed concerning the reactor's ability to accomplish safe shutdown after such an earthquake.

The difference in final opinion between the ACRS and the regulatory staff came as a surprise to the ACRS, and there was considerable discussion concerning the procedural and technical aspects of the matter. It was agreed that, in the future, steps would be taken so that the final positions of each group were known to both groups prior to issuance of final reports.

The Malibu Nuclear Plant

The Malibu (or Corral Canyon) reactor site and reactor had received preliminary consideration by the ACRS and regulatory staff as part of the review of potential reactor sites and reactor concepts which was conducted for the Los Angeles Department of Water and Power (LADWP) in 1962. By mid-1964, when a construction permit review was reaching its culmination, additional seismic questions had arisen, partly from matters directly related to the site, such as landslides, and partly from the increased consideration of seismic matters in California arising from the Bodega Bay review and the Alaskan earthquake of 1964.

The ACRS subcommittee meeting of June 18, 1964, on the Malibu site and reactor considered several seismic design questions, but no very difficult obstacles seemed to arise. The regulatory staff had brought the United States Geological Survey and the United States Coast and Geodetic Survey into the case as advisors, and the staff report to the ACRS, which was received on July 1, 1964, concluded that a seismic design acceleration of 0.3g was acceptable and that the probability of potential hazard to the public from differential ground movement due to an earthquake at the site was low enough to be disregarded. At the fifty-sixth ACRS meeting, July 9–11, 1964, there was considerable discussion of seismic matters, particularly the potential height of tsunami waves at the site. The ACRS concluded it could write a letter favorable to construction of the Malibu reactor and a report dated July 15, 1964, was sent to AEC Chairman Seaborg.

In the months following July 1964 there was a very considerable discussion between the applicant and the regulatory staff and its consultants concerning the actual seismic engineering criteria, stress limits, and analytical methods to be used. This was probably the first reactor to receive such detailed evaluation of seismic engineering considerations, and out of this review evolved much of the approach which was generally adopted for upcoming reactors.

By December 1964 the regulatory staff had satisfied itself with the seismic engineering approach which had been developed. In that month the United States Geological Survey issued a report which accepted the proposed seismic design bases and concluded that "the probability of permanent ground displacement by faulting in the Corral Canyon site in the next 50 years is negligible," although faulted deposits, probably less than 100,000 years old, had been exposed in a recently opened test trench at the site near where the reactor was to be located.

At the sixtieth meeting, December 10–12, 1964, and the sixty-first meeting, January 14–16, 1965, the ACRS again reviewed the Malibu reactor, and in a letter dated January 25, 1965, concluded that the seismic engineering approach was adequate and that, subject to previous reservations, the reactor could be constructed with reasonable assurance that it could be operated without undue risk to the health and safety of the public.

Construction of the Malibu reactor was contested at the hearing of the Atomic Safety and Licensing Board (ASLB) where the adequacy of the seismic design was one of the major points of contention. The intervenors in the case had the benefit of several well-qualified consultants in the field, and a considerable technical discussion ensued.

The testimony of Professor H. Bennioff, this time a consultant to the intervenors rather than the utility, as at Bodega Bay, raised a serious question about the previous earthquake magnitude considered at the site. He predicted a magnitude of up to 7.25 might occur, in contrast to the 5.5 magnitude chosen by the United States Coast and Geodetic Survey (USC&GS). The United States Geological Survey (USGS) later concluded that, if a magnitude 7 to 7.5 had to be considered, permanent ground displacement of some unpredictable amount would occur.

Excerpts from a memorandum, dated September 28, 1965, from R. Fraley, the ACRS Executive Secretary, to ACRS members, summarized the situation as follows:

It should be noted that the draft USC&GS report considered at the 56th meeting included the following statement:

"All of the known surface ground displacement on the Malibu Coast fault zone is prehistoric—that is, more than 200 years old. If the band of deformed rocks just south of the Malibu Coast fault trail is considered to be part of this fault zone, the most recent found displacement occurred sometime between about 200 and 400,000 years ago.

"The likelihood of ground displacement at the site due to earthquakes depends on the frequency and severity of earthquakes along the Malibu Coast and related faults.

"The Malibu Coast fault is considered to be part of an active system that includes the Newport-Inglewood zone. Only 3 to 5 magnitude shocks have been associated with the Malibu Coast fault; none of these has resulted in known displacement at the ground surface in historic time. However, in prehistoric time faulting at eight known localities along the general trend of the Malibu Coast fault has displaced rocks no older than 400,000 years. It can be inferred from these data that similar faulting may have occurred within the site but the displacements have not been detected because of generally poor exposures. On the basis of this record the probability of ground displacement at Corral Canyon in the next 50 years is very low."

The Committee in its letter of July 15, 1964 interpreted this report and discussions at the meeting as follows:

"The Committee was informed that the geology of the site was suitable for the proposed construction. It was reported that no active geological faults are present at the site."

The USC&GS report considered by the Committee at the 60th meeting (January 1965) included the following summary:

"Based on available geological evidence the probability of permanent displacement on the ground surface by faulting in the Corral Canyon site during the next 50 years is negligible. Seismic shocks can be expected at the Corral Canyon site; more than 54 seismic events of magnitude 4 to 6.3 have been recorded within 62 miles of the site in the past 112 years."

Exploratory trenches were dug at the site during April, 1965 at the suggestion of the ASLB, and a subsequent USC&GS report was issued in July 1965. The following comments were included in this report:

"Faults of several magnitudes are present in bedrock of the Corral Canyon site. The Malibu Coast fault, about 800 feet north of the reactor location is of regional significance and large magnitude of displacement; where well exposed, its trace is marked by a zone of brecciated and sheared rock as much as 75 feet wide. Intraformational faults, such as fault F, exposed in Corral Creek and trench 3 (the reactor-location trench) can be traced tens to hundreds of feet; their displacements are probably on the order of tens of feet. All demonstrable fault movement in the Corral Canyon site is pre-Recent (more than about 10,000 years) in age."

"Comparisons of degree and time of deformation in different parts of the fault system indicate that future faulting is at least as likely to occur in the Malibu Coast zone as in any other part. The available seismic record is not sufficient to establish the recurrence interval for large-magnitude faults in the system; this interval is greater than the approximately 200 years of historic time and it may exceed the approximately 10,000 years of Recent time. As this recurrence interval is large compared to 50 years, the probability that a large-magnitude shock with center near Corral Canyon will occur during the next 50 years is very low. This very low probability, coupled with the lack of evidence for surface faulting in the Malibu Coast zone during Recent time, indicates that the probability of permanent displacement of the ground surface by faulting at Corral Canyon during the next 50 years is very low (this same very low probability was described in the U.S. Geological Survey report of 1964 as negligible, which was used there in the sense of very low). This assessment implies no judgment of public risk; it is not intended as a judgment of the consequences of surface faulting in any particular utilization of the Corral Canyon site."

"Fault F is about 35 feet northeast of the center of the reactor building which means that is passes under the reactor containment structure."

Summary of Matters Identified by the Malibu ASLB

1. "Since the containment building is not specifically designed to withstand ground displacement, has it been established and on what basis: (a) what is the ground displacement that it can withstand; (b) what would be the amount of fission products released if a displacement greater than that identified in (a) occurred; and (c) what would have to be provided in the design to give such resistance?

"The Board suggests that experience with relative ground movement be used as the basis in the Southern California area for the selection of useful values in these parameters.

2. "Is it acceptable to grant approval on the basis that the structural requirements are 'within the range of accepted practice and established knowledge' even though the detailed design has not been presented?

3. "What is the meaning of the phrase 'without undue risk to the health and safety of the public' as understood by all participants, especially the word 'undue'? References to appropriate authority are requested.

4. "The board interprets the ACRS Report of July 15, 1964 as carrying 'the admonition that this reactor should not be located over an active fault.' The Board requests standards or suggestive standards to measure an active fault."

On July 14, 1966 the ASLB issued a ruling that it would be necessary for the plant to be designed to withstand differential ground displacement if it were to be constructed, but no quantitative figure for the displacement was specified. The ASLB thereby sided with the intervenors and against LADWP and the regulatory staff. The regulatory staff appealed the decision of the ASLB to the AEC commissioners themselves.

On November 21, 1966, the possibility of a congressional investigation was raised by Senator George Murphy of California in a letter to the AEC. He questioned the AEC's site-selection criteria and the behavior of the AEC regulatory staff in connection with the reactor proposed by the Los Angeles Department of Water and Power at Malibu, California. Senator Murphy stated that going any further with the project would undermine the public's confidence in the nuclear power industry. He recommended to the commission that the AEC "Staff should be candid with the applicant," that its site-selection program be revised, and that the AEC take a "very close look" at its staff procedures. The senator questioned whether the AEC staff had made a "hasty" and "ill conceived" judgment in favor of construction of the Malibu reactor and now desired to "save face" by taking exception to the decision of the Atomic Safety and Licensing Board, which recommended plant construction only if the design criteria provided for permanent ground displacement.

Responding, AEC Chairman Seaborg told Senator Murphy, in a letter dated December 6, that:

The Commission's regulatory staff has the same right to appeal a preliminary decision of a licensing board as any other party.

Under the licensing procedures established by the Atomic Energy Act, the Commission is obliged to consider the issuance of a construction permit for a proposed facility at the site selected by the applicant.

On March 28, 1967, the AEC commissioners issued a final decision in which they upheld the decision of the ASLB. This was a landmark decision, in which an intervenor against a nuclear plant won his case. Furthermore, the AEC commissioners were establishing a benchmark on "How safe is safe enough?" by not adopting as a basis for an acceptable risk the data indicating that movement had not occurred in the last 10,000 years and possibly for 180,000 years. In other words, if one could use past geologic history to assess the likelihood of such faulting during the near future, the commissioners were saying that a risk of one in 10,000 per reactor-year of a large radioactivity release was unacceptable.

Seismic Safety—Late 1963 to 1979

In the second half of 1963, the San Onofre Unit 1 reactor on the Camp Pendleton Marine Corps Base near San Clemente, California, was reviewed for a construction permit. At the forty-eighth ACRS meeting, July 11–13, 1963, representatives of Southern California Edison described their planned approach to seismic design in terms of an acceleration of about 0.2 g. Southern California Edison stated that they did not feel that a nuclear plant needed to be designed with any more attention toward earthquakes than a conventional steam plant, with the exception of the critical sections, e.g., safety injection system. Mr. W. R. Gould of Southern California Edison said the present plan was to use a seismic design basis of 0.2 g, but that a higher figure might be necessary after the results of a study to be completed in a few weeks were available.

By the forty-ninth ACRS meeting, September 5–6, 1963, the seismic design approach for San Onofre was more specific and more conservative. Southern California Edison stated that the plant would be designed for a safe shutdown if an earthquake caused an acceleration of 0.5 g. (During this same time-period, the proposed seismic design basis for the Bodega Bay reactor was also being increased from 0.3 g to 0.66 g.)

In late 1963 and early 1964, the question of an appropriate seismic design basis for an eastern reactor (Connecticut Yankee) introduced some relatively modest controversy. At the ACRS subcommittee meeting held on December 13, 1963, Mr. L. Minnick, speaking for the applicant, pointed out that earthquakes are generally not considered in New England. There had been minor earthquakes in Connecticut in the last 150 years, and the site was stated to lie on the boundary between seismic zones 1 and 2.* He, therefore, suggested that a 0.1 g design basis for safe shutdown of the reactor would be appropriate.

The regulatory staff obtained the advice of the United States Coast and Geodetic Survey who recommended that a larger peak acceleration by used as a design basis for safe shutdown, namely, 0.17 g. This contrasted with the recommendation in a letter sent by Father John J. Lynch, a consultant to the applicant, who stated that 0.03 g was a suitable safety factor for the design.

Mr. Roger Coe agreed to the design basis recommended by the regulatory staff on behalf of the applicant, but complained about the extensive and costly engineering necessary to study seismic safety for the staff's recommended value. The committee did not disagree with the staff in its report of February 19, 1964, on the Connecticut Yankee reactor. However, at least some of the ACRS members appeared to question the need for as high an acceleration as proposed by the staff. At the special ACRS meeting held on February 24, 1964, the committee wrote AEC Chairman Seaborg suggesting "further consideration of the question of earthquake probabil-

---

*The United States has been roughly divided into four seismic zones, using a scale which ranges from 0 for no seismic activity to 3 for large earthquakes.

ity and magnitude for regions where seismic activity is low and of the question of adequate engineering for small movements."

The ACRS also requested advice from an independent consultant as to what constituted a suitable seismic design basis for the Haddam Neck (Connecticut Yankee) site. The consultant, Dr. Perry Byerly, in a letter dated August 12, 1964, stated "the design factor of 0.17 g is reasonable and just. I would not accept a lower factor." Thus, Dr. Byerly supported the recommendation of the United States Coast and Geodetic Survey, and disagreed strongly with Father Lynch's view. The Connecticut Yankee review showed a wide variation in expert opinion on seismic events for the eastern United States; and it brought the need for increased attention to seismic design for eastern sites into focus.

In May 1964 Dr. R. A. Williamson prepared a report for the regulatory staff in response to ACRS interest in two questions.

(1) To what extent do seismic considerations influence reactor costs?

(2) How much seismic resistance is available in facilities not designed to withstand seismic forces?

Dr. Williamson estimated that seismic engineering might add 20% to 50% to the engineering costs, but only 2% to 4% to the total construction costs. He also estimated that the effects of seismic design on actual construction costs would be very small, except for a reactor like Bodega Bay, where there might be an increase of 5% to 15%. He advised that containment vessels would have a considerable inherent degree of seismic resistance, but that equipment items and related systems tended to be vulnerable and were not normally constructed to withstand seismic forces. Dr. Williamson concluded:

It is strongly recommended that no projected power reactor facilities for location in areas currently regarded as seismically inactive be designed without complying with certain minimum standards for earthquake resistance.

On September 28, 1964, the ACRS held a meeting of a newly formed seismic design criteria subcommittee. Dr. C. R. Williams, the subcommittee chairman who had undergraduate training in geology, opened the meeting by expressing concern over the policy of the regulatory staff to accept standards recommended by other government agencies without any substantive technical backup. He was also unhappy about the fact that many reactors in the past had been approved with little review of seismic effects and that the points which got major emphasis by the staff appeared to be determined by the intervenors.

ACRS consultant Steinbrugge recommended that a set of seismic design standards be developed for reactors as soon as possible. This would provide guidance badly needed by applicants and would establish a basis for consistent and comprehensive evaluation of plant design. Steinbrugge expressed concern over the fact that many of the studies and designs of the day considered only major systems and components. "Smaller" systems were being constructed by conventional practice (e.g., piping runs were laid out by pipefitters rather than designers) with no appreciation of their importance to plant safety.

ACRS consultant Byerly noted that the AEC must decide whether it was willing to gamble (probability approach) or take into account the worst disturbance which would occur. Byerly recommended use of the maximum recorded event at a specific site as the basis for design. Dr. Beck of the staff noted that this presented a problem at locations where there was no recorded history. Since records concentrated on loss of life and money, bad shocks in remote areas often were unrecorded. Dr. Beck also noted that there had occasionally been "bad" shocks in areas of previously recorded low intensity.

It was suggested by the ACRS that the regulatory staff should also consider other natural phenomena such as tsunamis, forest fires, wind waves, land slides, and floods in the development of criteria.

In retrospect, this was a very important meeting. Not only was there a general consensus on the need for seismic design of reactors and the establishment of seismic criteria or guidelines; it was a beginning of the development of criteria for all natural phenomena.

The Development of Seismic Criteria

The regulatory staff initiated work with its consultants to develop more specific seismic engineering criteria in 1965. In a letter dated May 22, 1967, Edson Case, then Director of Reactor Standards in the AEC regulatory staff, forwarded to the ACRS a draft document entitled "Seismic and Geologic Siting Criteria for Nuclear Power Plants," and requested ACRS review and comment, initially via an appropriate subcommittee. To assist in accomplishing this task, as well as to enable an independent review of sites having difficult seismic questions (such as the proposed Bolsa Island site in California), the ACRS obtained the services of several well-qualified consultants in seismology and geology.

The draft criteria (dated May 11, 1967) represented a considerable effort by the regulatory staff and their consultants, and provided a very good starting basis for the development of criteria. The criteria included a minimum design basis (or floor) of 0.1 g for the acceleration caused by an earthquake. The seismic engineering criteria were not included in the draft and were stated to be under development.

One of the most controversial aspects of the first draft related to a curve, based on a report under preparation by Bonilla of the United States Geological Survey, which related surface displacement to magnitude of earthquake on a surface fault. Also controversial was the choice of the distance from the center line of a fault within which a reactor should not be located. Another item which received much scrutiny and comment was the definition of a "capable fault," meaning a fault with surface expression, which was deemed capable of exhibiting permanent relative displacement on the two sides of the fault as the result of an earthquake.

A very large number of meetings were held between the ACRS and the regulatory staff, and many revised drafts were prepared. In the latter half of the 1960s, the regulatory staff seemed to require that it obtain comment and preferably concurrence from the AEC Division of Reactor Development and Technology (DRD&T) on such criteria, although the latter represented the AEC promotional side.* During the period when the Bolsa Island project was active, the seismic criteria were held in a state of abeyance by Mr. Price while their potential impact on Bolsa Island (a project important to the AEC) was assessed.

One important potential difference of opinion between the ACRS and the regulatory staff was discussed by the committee at its ninety-fourth meeting, February 8–10, 1968. It was reported in executive session that all the consultants to the seismic subcommittee had expressed concern over whether the approach being taken to the safe shutdown earthquake (SSE)† east of the Rocky Mountains was sufficiently conservative. It was reported that Mr. Case of the regulatory staff had agreed to examine the possibility of defining a higher base-level (or minimum) acceleration east of the Rockies.

The matter of what "floor" on seismic design should be used remained an open question between the staff and the ACRS for a long time; it was

---

*For example, at the ninety-eighth meeting, June 5–8, 1968, ACRS member Okrent asked the regulatory staff the status of the seismic design criteria. Mr. Price reported that a draft of the proposed criteria had been sent six weeks previously to Mr. Shaw, the Director of DRD&T, and that he (Price) was still awaiting comments. Okrent asked for a copy of the present draft and noted that even if Shaw didn't have a conflict of interest, DRD&T should not control the regulation of criteria. He suggested that if there was a serious disagreement between the regulatory staff and Shaw, the disagreement should be taken to the commissioners. Mr. Price said that there seemed to be some differences between the regulatory staff and Shaw, and also that he was not sure how these criteria would effect the Bolsa Island project.

†The reactor was to be designed so that if an earthquake as large as the safe shutdown earthquake occurred, all portions of the plant important for safety would still perform any needed function, the chain reaction could be terminated, and fission product decay heat could be removed without any abnormal plant behavior or release of radioactivity.

eventually resolved (temporarily) by ACRS acceptance of the original staff position of 0.1 g as the minimum SSE, without benefit of a comprehensive study of the matter.

It was November 1971, after very many major re-drafts, before the Atomic Energy Commission finally issued a Notice of Proposed Rule-Making to amend the commission's regulations, 10 CFR Part 100, by adding a new appendix; Appendix A: "Seismic and Geologic Siting Criteria for Nuclear Power Plants." The proposed criteria reflected the practice which had been followed in actual construction permit reviews and were reasonably specific in their definition of a "capable" fault. They also gave guidance as to the general extent of the geologic and seismic investigation required. However, they did not provide a quantitative criterion for establishment of the design basis (or safe shutdown) earthquake, and as time went on, the wording used proved to provide great flexibility, i.e., required the exercise of judgment wherein seismic experts could and would continue to differ greatly.

In 1972, seismic design basis again became a matter of some controversy in connection with review of the Virgil Summer plant in South Carolina. The ACRS seismic consultants generally favored a larger seismic design basis than the 0.15 g proposed by the regulatory staff. However, except for one dissenting member (Okrent) the ACRS supported the position of the regulatory staff.

Appendix A to 10 CFR Part 100 was adopted by the AEC in 1973. In the same way as the 1971 proposed criteria, the 1973 Appendix gives general, not specific, advice on the measures to be followed in determining the seismic design basis for a nuclear power plant. It is subject to much interpretation.

The ACRS Again Wants to Raise the Floor

In agreeing to adoption of the seismic criteria in 1973, the ACRS again raised the question of using a higher minimum design basis accleration. The ACRS memo, dated June 11, 1973, was sent to L. Manning Muntzing, Director of Regulation. It is reproduced below.

Proposed Seismic and Geologic Siting Criteria
Appendix A to 10 CFR Part 100

The Committee believes it acceptable to publish the draft Seismic and Geologic Siting Criteria, dated June 7, 1973. However, the Committee notes that the minimum acceptable acceleration currently in use (0.1 g SSE) frequently leads to considerable debate among seismic experts concerning an acceptable seismic design basis, even in regions of low seismicity. The Committee recommends that a study be initiated to determine the practicality and cost of raising the minimum design basis acceleration to 0.15 or 0.20 *g* for future reactors, except where foundation conditions and geological and seismic information are recognized by a clear consensus of expert opinion to warrant a lesser value. The Committee expects that an increased minimum design basis acceleration would assist in standardization, would provide increased margin to cover uncertainty, and would facilitate seismic aspects of regulatory review for many sites, without involving an important increase in cost, if factored into the original design.

This ACRS request was acknowledged in a response by Lester Rogers of the staff to ACRS Chairman Mangelsdorf, dated July 3, 1973. A study as requested was promised; however, no results of any such study by the staff surfaced in the next several years.

Seismic engineering criteria had been developed and were being applied on construction permit reviews during this period. However, rather than being incorporated into the AEC rules and regulations via a rule-making procedure, they were made public and implemented via a simpler procedure, namely, regulatory guides.

While the need for seismic criteria was recognized relatively early in the regulatory process, and with time such criteria were developed, only a very modest seismic safety research program was funded by the AEC during this time-period, despite ACRS recommendations beginning as early as 1965. In a report dated January 17, 1968, the ACRS noted the formative stage of the AEC seismic safety research program and recommended work in several areas, including soil-liquefaction and soil-structure interaction. The ACRS also recommended development of a detailed program of earthquake engineering research; however, there apparently did not exist strong support for such a program within the AEC.

In a letter to Mr. Price dated May 17, 1971, the ACRS noted that its geological and seismological consultants had expressed concern that seismic conditions in the eastern United States were poorly understood and had recommended that emphasis be placed on the early development of information that would aid the AEC regulatory groups in the determination of conservative parameters for the SSE. The letter identified the Charleston earthquake as requiring accelerated study, and made some other specific recommendations.

The thoughts in this memorandum were repeated in an ACRS report to AEC Chairman Dixy Lee Ray, dated May 16, 1973, which discussed the entire eastern United States, and pointed out areas warranting special attention, including South Carolina, the St. Lawrence Valley, southeastern Missouri, western Ohio, and the Cape Ann region of Massachusetts. During the review of the Greenwood Energy Center in Michigan at the one hundred seventy-second meeting, August 8–10, 1974, Dr. Carl Stepp of the staff estimated that the return frequency (or probability per year) of the proposed safe shutdown earthquake was 10^{-5} per year for this site, and said that the same return frequency was reasonable for other sites recently reviewed. The ACRS wrote to Mr. Muntzing, asking how an SSE return frequency of 10^{-5} per year was compatible with the 10^{-7} per year identified in the staff report on ATWS, WASH–1270 (AEC 1973a), as the objective for the probability per year of a serious accident from any specific source like earthquakes.

The regulatory staff response to the ACRS was two-fold. First, the staff argued that the use of Appendix A did not introduce any requirement for a quantitative criterion on return frequency for the SSE. Secondly, they argued (qualitatively) that there were many large safety factors* inherent in the seismic engineering design so that, overall, serious reactor accidents due to earthquakes should have a very low probability per year of occurrence. The staff also stated that there were very large uncertainties in any estimates of the return frequency of the safe shutdown earthquake and concluded that "the margins of safety inherent in use of our seismic design requirements are not currently quantified, nor do we believe that they can be, at this time, with sufficient rigor to be used in making specific licensing decisions."

Since Appendix A did not provide any real guidance on how to allow for the fact that only a very limited amount of empirical data was available (i.e., the recorded history of earthquakes in the United States is very short, measured in geologic time), the matter was left to the judgment of the individual reviewer, and differences of opinion existed in the return frequency which corresponds to the SSE. A graphic illustration of how large such differences could be was provided by a survey of expert seismic opinion (Okrent 1975) in which seven experts varied by factors of 1,000 to 10,000 in their estimates of the return frequency of the same large earthquake at 11 different reactor sites.

As more return frequency estimates were provided to the ACRS by the staff or the applicants for construction permits following the Greenwood review, the estimated frequency tended to become larger, and values larger than 10^{-4} per year for the probability of exceeding the SSE were not infrequent. This further compounded the problem of judging what constituted an adequate seismic design basis.

Some seismic engineers began to provide their estimates of a rough quantification of the risk to nuclear reactors from earthquakes. For exam-

*The staff noted that, among other things, the real damping of the vibratory motion was underestimated, that the combination of seismic loads was done in a way which overestimated the total stress, and that the analytic representation of the earthquake exceeded the vibratory effects from an earthquake having the intensity of motion assumed in establishing the design basis value of g.

ple, Professor R. Whitman of Massachusetts Institute of Technology, speaking for the Atomic Industrial Forum Ad Hoc Committee on Seismic Design Bases at a meeting held with the regulatory staff on September 19, 1974, in Bethesda, Maryland, discussed conservatism in seismic design and listed a chain of steps to a seismically induced reactor accident and their probabilities. In his opinion these factors combined to give him an overall risk of 10^{-7} to 10^{-8} per reactor-year of a serious accident from an earthquake (which met the criterion defined in WASH–1270 (AEC 1973a) on ATWS).

A Dissent on Seabrook

In 1974, the review of the Seabrook, New Hampshire, reactors again provided a focal point for questions of seismic design adequacy. Additional comments by member Okrent to the ACRS letter on Seabrook of December 10, 1974, to AEC Chairman Dixy Lee Ray follow below:

The Seabrook Station site is near what is generally recognized as the Cape Ann-to-Ottawa Trend. Mechanisms for earthquake generation in the New England area are not well understood, and expert opinion differs concerning the potential for and probability of relatively large earthquakes at or near the site.

The Regulatory Staff have ultimately based their judgment as to an acceptable safe shutdown earthquake on the application of 10 CFR Part 100, Appendix A, rather than a probabilistic estimate of earthquake size versus recurrence interval. It is of interest to note that Appendix A provides only general guidance; furthermore, it specifically refers to the possible choice of a safe shutdown earthquake larger than that found in the historical record for a tectonic structure or province.

During the ACRS review the Regulatory Staff did state that the seismicity of the tectonic region applicable to the Seabrook site could be interpreted to be about an order of magnitude larger than other tectonic provinces having a similar maximum historical seismic event. Furthermore, a member of the Regulatory Staff stated that his estimate of the probability per year of occurrence of an earthquake of intensity MM VIII* at the Seabrook site is about 10^{-4}, and the Staff did not rule out the possibility of a larger earthquake occurring within the region under consideration. They stated that conservatism in analysis, stress limits, and other factors decreases the overall probability of failure of seismic Class 1 structures and piping by a few orders of magnitude and hence the overall probability of a seismically induced accident exceeding 10 CFR Part 100 would be acceptably low. However, earthquakes are almost unique in their ability to fail each and every structure, system, component, or instrument important or vital to safety, and, in my opinion, the Staff evaluation of additional margin available from stress limits, methods of analysis, etc., did not consider all such systems, e.g., D.C. power or emergency A.C. power.

---

*Modified Mercalli Intensity VIII; a measure of the shaking produced at a specific site; one based on qualitative estimates of the damage produced in past earthquakes. The MM scale goes from I to XII, with minor damage beginning at VII.

It is clear that the capability of a reactor to achieve safe shutdown, assuming its SSE occurs, cannot be fully demonstrated by test. Those limited, detailed and independent audits of seismic design of actual plants that have been published indicate that some inadequacies in design and construction exist. Equally or more important, it appears to be unlikely that the plant could survive safely, with a high degree of assurance, a larger earthquake having one or two orders of magnitude lower probability than the proposed SSE.

Given this background, and recognizing the substantial surrounding year-round population density and the very high nearby population during the summer months at Seabrook, I am left uneasy and believe it would be prudent to augment the proposed SSE acceleration of 0.25 g.

I also wish to reiterate my conclusion previously stated in connection with the review of Grand Gulf Units 1 and 2, namely that it would be prudent to provide some additional margin in the seismic design bases for most future nuclear plants sited east of the Rockies.

The WASH–1400 Report Dismisses Seismic Risk

The reactor safety study, WASH–1400 (NRC 1975a), provided support to those who estimated a low risk of a serious reactor accident from seismic causes. The estimate given there was that seismically induced core melt had a frequency of 5×10^{-7} per reactor-year for a reactor on an average foundation, and less than 10^{-7} per reactor-year for a reactor on a rock foundation.

In a preliminary evaluation of the WASH–1400 report issued on November 15, 1975, by a review group from the NRC regulatory staff, the conclusions in WASH–1400 on seismic risk were generally endorsed, as follows:

Comments on the WASH–1400 draft report identified large earthquakes as a potential mechanism for causing multiple system failures and questioned the probabilities given for seismic failure modes. The Study has revised its treatment of this issue to include explicit recognition of the observational data regarding the distribution of earthquakes with respect to intensity. Also examined were the safety margins in a facility when subjected to earthquakes larger than used for design purposes. The calculations reflect the view of acknowledged authorities and current information.

The Study provides a significant perspective on the relationship between basic ground motions, the energy absorption capacity of a structure, and the probability of exceeding design limits. The study conclusions relating to the probability of failure may be open to question in view of the differing seismic margins in the design of electrical, mechanical, and structural components and the potential for common mode failures. While a more detailed application of the Study's methodology would be needed to definitively establish the meltdown probabilities, it does not appear that seismic events are strong contributors to risk.

However, since issuance of the reactor safety study (NRC 1975a), a considerable body of information has been released which suggests that its estimate of seismic risk, and that of Whitman, may be much too low.

At an ACRS subcommittee meeting on seismic activity in the eastern United States, held March 22–23, 1976, a large body of new information appeared, and there was a continuing wide disparity of expert opinion on the adequacy of 200 years of history to predict future, low-probability seismic events. At a next meeting of the seismic subcommittee on February 8–9, 1977, which dealt primarily with matters related to the mechanical interactions between the underlying soil and a large structure, such as a containment building, during an earthquake, many differences of opinion concerning the adequacy of previously accepted practice were expressed.

WASH–1400 Is Questioned

More importantly, papers were published in 1977 and 1978 which disagreed with the methods and the results concerning seismic risk in the WASH–1400 report (NRC 1975a). Risks two or three orders of magnitude larger were estimated. In particular, the doctoral thesis research by Hsieh (Hsieh and Okrent 1977) pointed out errors in logic and arithmetic in the WASH–1400 report (NRC 1975a), and illustrated the potential importance of design errors and degradation of equipment to the seismically induced reactor accidents. Seismic risk estimates performed for the Diablo Canyon reactors and for the proposed Clinch River Breeder Reactor (Clinch River 1977) both disagreed with the conclusion in the WASH–1400 report that the seismic contribution to nuclear plant risk was small, and supported the work by Hsieh, as did a new study by Professor C. A. Cornell of the Massachusetts Institute of Technology and Professor Newmark (Cornell and Newmark 1978).

The controversy over what constitutes an adequate seismic design basis next appeared within the framework of the regulatory process. In connection with the operating license review for North Anna Units 1 and 2, the ACRS noted that its consultants favored a higher minimum seismic design basis ($0.2\,g$) for future reactors in the eastern United States, and recommended that the regulatory staff assure itself that there was a considerable margin in the existing seismic design (SSE = $0.12\,g$ on rock) with regard to equipment needed for safe shutdown of the reactor. The ACRS also made a similar recommendation in its report of January 14, 1977, concerning an operating license for the Davis-Besse nuclear station.

In connection with the ACRS review in 1977 of the construction permit application for the Cherokee-Perkins reactors in South and North Carolina, ACRS member Okrent disagreed with the adequacy of the proposed SSE, and questioned the correctness of the treatment of seismic risk in the WASH-1400 report (NRC 1975a). Okrent's additional remarks to the ACRS letter of April 14, 1977, to NRC Chairman M. A. Rowden on the Cherokee-Perkins reactors follow:

I believe that the philosophy and criteria of Appendix A of 10 CFR Part 100, and their application by the NRC Staff in setting SSE values, should be re-evaluated as part of an early overall re-assessment of the current approach to seismic safety design. I believe that the estimates of the contribution of earthquakes to overall nuclear reactor safety risk, as given in the Reactor Safety Study (WASH-1400) are not without fault, and that seismic contribution to risk is underestimated in that study.

I find the applicant's estimate of the return frequency of the SSE at the Cherokee and Perkins sites of greater than 10^{-4} per year to be unsatisfactorily large, particularly in view of his arbitrary cutoff at MM VII of the earthquakes permitted to contribute to this probabilistic assessment. For Cherokee/Perkins, I find the proposed SSE of 0.15 g marginally acceptable and would prefer that a value of 0.2 g be employed at the foundation level of rock.

In 1977, the Nuclear Regulatory Commission initiated a major new research program in seismic safety including the possible application of probabilistic techniques. In 1978 and 1979 the NRC regulatory staff required re-evaluation of the seismic design bases for several reactors constructed by the Tennessee Valley Authority. This was because the staff re-examined the seismic information using new approaches and increased their estimates of the appropriate SSE. In early 1979, five operating reactors were shut down for an extended period by the NRC in order to permit re-analysis and possible corrections in design because errors had been made in the seismic design of important piping systems. A large number of other reactors have since reported errors in their seismic design.

Prior to 1977, for reasons which are obscure, the regulatory staff had performed few seismic design audits of LWRs, and these were minimal in their scope, at best. The staff had not imposed any requirement for an independent, third-party review of seismic design of each reactor, even though the limited seismic audit performed as part of the reactor safety study (NRC 1975a) on two reactors showed several errors and deficiencies, and there were good reasons to anticipate errors in seismic design, just as in other design aspects.

With the shutdown of five plants in 1979 and other bad experiences, both with regard to design errors and to degradation of equipment, such as snubbers which were intended to damp vibratory pipe motion during an earthquake, the adequacy of detailed seismic design began to receive much more careful attention by the NRC staff.

18

Generic Items,
LOCA—ECCS, and LWR
Safety Research

Generic Items

The first light water power reactors (Shippingport, Dresden 1, and Indian Point 1) were one-of-a-kind reactors which were built in the 1950s with only minimal design information provided to the AEC at the time construction was authorized. These three reactors were not prototypic of future LWRs. The next group of higher-power LWRs (San Onofre 1, Connecticut Yankee, Oyster Creek, and Nine Mile Point), whose construction was authorized in the period 1963–1965, while also not prototypic of the reactors to be built in the following decade, nevertheless, introduced safety issues which would prove relevant (or generic) to all LWRs, to all PWRs or BWRs, or at least to all LWRs having certain components or systems. Again only limited design information was available, and many safety-related issues were raised by the regulatory staff and the ACRS at the construction permit stage but left to be resolved during construction of the plant. The theory was that the applicant bore the risk of having to modify his plant at the operating license review stage if resolution of such issues had not proven to be satisfactory. As the discussion of the Oyster Creek review in Chapter 15 illustrates, backfitting of the reactor design was frequently required, but this was not a straightforward process.

A considerable number of safety issues that were to be resolved during reactor construction were raised for the Dresden 2 reactor in 1965; many of these were generic, that is, applicable to all LWRs or to all BWRs of the

Dresden class. Of course, the recommendations for improved primary system integrity and improved emergency core-cooling systems made in connection with the Indian Point 2 and Dresden 3 reactors were clearly generic and not specific to the individual plants.

However, there did not exist a formal list of generic, unresolved safety issues at that time. And the regulatory process assumed that an adequate degree of resolution would be possible prior to the completion of construction.

The ACRS review of a construction permit application for Browns Ferry Units 1 and 2 early in 1967 can probably be considered to be the genesis of the generic items list. This was a very difficult and controversial ACRS review. While the Browns Ferry site was quite remote, these BWRs represented a large increment in power level over those BWRs previously approved for construction and a very large increase in power above any operational BWRs. They came after the ACRS pressure vessel letter in late 1965 and the China Syndrome matter of 1966, wherein major safety issues had been raised and, at best, had been resolved in only a preliminary or incomplete manner. Furthermore, a higher power density was being proposed for the Browns Ferry reactors, bringing fuel performance limits and potential safety concerns related to fuel melting more into focus. Finally, Browns Ferry was expected to be the prototype of similar reactors soon to be proposed for much more populated sites.

By the time of the eighty-third meeting, March 9–11, 1967, the ACRS had prepared a draft which included a very considerable number of specific matters on which more information was required. The committee members had mixed opinions about the proper course to follow, several expressing doubt that the matters would be resolved by the time construction was complete.

The ACRS decided it could issue a letter with many reservations, and, at the suggestion of the General Electric Company, it asterisked those items that were properly identified as industry-wide problems and not specific to the Browns Ferry reactors. The asterisked items from the ACRS letter of March 14, 1967, on the Browns Ferry site included the following: acceptable peak clad temperatures and the effects of clad failure in a loss-of-coolant accident (LOCA); the possibility and safety significance of fuel melting and fuel failure propagation at normal power; acceptable fuel element damage limits during normal operation and transients; and quality assurance and in-service inspection of the primary system to reduce the likelihood of a loss-of-coolant accident.

When the ACRS completed its review of the Vermont Yankee BWR in June 1967, it repeated several of the items from the Browns Ferry letter, once again labeling them with asterisks. And in July 1967, when the ACRS wrote a report on the Oconee PWRs (supplied by Babcock and Wilcox), it tagged several reservations with asterisks, including two new issues, one concerning the possible effect of thermal shock from ECCS operation on pressure vessel integrity and a second relating to the effects of blowdown forces from a large LOCA on the core and other primary system components.

At its eighty-ninth meeting, September 7–9, 1967, the ACRS discussed the matter of trying to obtain resolution of asterisked items with the AEC commissioners and their senior staff members. The commissioners offered no positive program of action; Mr. Shaw stated that his program was already fixed; and Mr. G. Kavanaugh, the assistant general manager for reactors, said it would not be possible to get additional safety research funds.

The meeting with the commissioners left the ACRS with a general unease about how to get the asterisked items dealt with on a timely basis. It was decided to send letters to Mr. Price and to Mr. R. Hollingsworth, the AEC general manager, asking how each was addressing matters raised by the ACRS as significant to all large LWRs. The letter to Mr. Price stated, in part:

> It will be helpful if the AEC Regulatory Staff addresses itself to these matters, as you proposed, in the reports prepared for the ACRS as part of each project review. Items of particular interest are the adequacy of related research and development efforts, the timing of such programs and the degree of assurance that such programs are likely to resolve these questions satisfactorily. If it appears that the proposed R&D effort may not provide a satisfactory resolution on an appropriate time scale, alternate backup designs should be considered which will provide reasonable assurance that operation of the proposed facility will not present an undue hazard to the health and safety of the public.

By the time that ACRS action was completed in June 1968 on the construction permit application for Zion station Units 1 and 2, it was clear that progress was slow and was going to remain slow on most of the asterisked items. And the list of items continued to grow (for example, the question of anticipated transients without scram, ATWS, was raised in 1969).

During the next few years, by joint effort of the regulatory staff and the ACRS, acceptable regulatory positions were developed on many of the open or unresolved topics, and some of the asterisked, or generic, items began to be resolved with the issuance of safety (or regulatory) guides. Some were resolved by safety research, some by changes in design.

On December 18, 1972, the ACRS issued its first formal report on the status of its generic items, in which it identified 22 matters under the heading "resolution pending" and 25 as resolved, in the sense that a regulatory position had been reached, although it might not have been implemented. The problem of how to backfit, if changes were desirable or necessary, was frequently a thorny one, and usually ended up as a matter of engineering judgment, primarily by the regulatory staff.

Clearly, in 1972, many of the asterisked items identified during the review of the Browns Ferry and succeeding plants were still not resolved, even though the applicants were requesting operating licenses for these reactors. The regulatory staff and the ACRS had to arrive at a difficult judgment as to whether operation was acceptable while a large number of generic items remained unresolved. The judgment made was that such operation would not pose undue risk to the public health and safety, and that when these items were resolved, a backfit decision would have to be made.

The ACRS has reported periodically on the status of generic items in letters to the chairman of the commission. For example, in its sixth such report, dated November 15, 1977, the ACRS identified a total of 48 matters as resolved and about 25 matters in the category "Resolution Pending."

An examination of the generic items covered in the ACRS report of November 15, 1977, both resolved and unresolved, shows that they cover a wide range of topics, and that they vary considerably in their specificity and their potential or probable impact on reactor safety. They vary as well in the way they arose.

One of the resolved generic items related to net positive suction head for ECCS pumps, that is the pressure needed in the water supplied to the pumps to avoid cavitation (the rapid formation and collapse of vapor pockets) within the pumps with consequent damage to the pumps. This matter arose from philosophic considerations. During ACRS review of one or two specific BWRs during the late 1960s, it was ascertained that the designers assumed containment integrity in assessing the containment pressure head available to prevent cavitation of ECCS pumps. The containment is very much a safety-related structure and its ability to withstand LOCA forces is carefully analyzed. Hence, this assumption by designers was not unreasonable. On the other hand, it could be argued that it was desirable that the ECCS function did not depend on containment integrity, so that some low-probability event involving a major loss of containment integrity during a LOCA, e.g., gross failure of a large containment penetration, would not lead automatically to core melt. Furthermore, it was clear that reactor design could equally well proceed without relying on containment integrity for adequate pump suction pressure, if the approach was adopted from the beginning. The ACRS flagged the matter as a desirable change, and it was adopted after a fairly short time.

Another resolved item related to the integrity of flywheels on primary system pumps in PWRs. The potential for missile generation from various sources had received review from time to time. During the ACRS construction permit review of the Indian Point 3 reactor, questions were raised concerning the possible failure of pump flywheels with the generation of large missiles and potentially adverse affects on the integrity of the steam generator and other safety-related components. Experience with flywheels of this type had been good, and it was judged that with proper care in design, fabrication and inspection, the probability of unacceptable affects from gross flywheel failure should be acceptably low. While the existing quality assurance in the industry seemed to be adequate for the most part, there had not been developed an acceptable code of practice. Hence, the matter was identified as a generic item which was later resolved by issuance of a regulatory guide.

Fire protection was made a generic item after occurrence of the fire at Browns Ferry Units 1 and 2. Actually, the matter of how to deal with protection against fires had been a thorny issue for many years. A significant fire had occurred earlier at the San Onofre 1 reactor, in a tray which carried many power cables, due to thermal overload (a design deficiency which was corrected at San Onofre and watched for routinely at all reactors thereafter). The difficulties in postulating the causes of realistic, serious fires and in assessing their consequences left the subject in an ill-defined state. And while progress had been made in improving protection against fires, via criteria on separation of redundant systems, progress had been slow. Opinions varied as to the adequacy of existing approaches, as the minutes of an ACRS subcommittee meeting of fire protection, held on January 19, 1973, show. Ironically, representatives of the Tennessee Valley Authority came to that meeting and discussed Browns Ferry Units 1 and 2 to provide background as to the adequacy of existing practice.

After the Browns Ferry fire, the regulatory staff mobilized a major effort on the matter of fires and fire protection. Identification of the matter as a generic item was pro forma. And its resolution came about with agreement on the pertinent regulatory guide. Actual implementation of the guide preceded its adoption, in part. However, in this matter as in many, there are various possible approaches, each having its pros and cons, and differences of opinion have continued to arise with regard to acceptable detailed practice.

The matter of turbine missiles as a safety concern was raised many times, beginning in 1965 or 1966, before it was identified as a generic issue. Missiles arising from high-speed failure of a turbine rotor might penetrate the reactor containment and could cause damage which would prevent continued safe shutdown of the reactor. In some of the initial discussions on the subject, it was argued by reactor vendors that the existing design provided protection against missiles. Continued consideration indicated that this was not necessarily so for missiles generated at, say, 125% of normal speed, which might be reached during anticipated transients which led to overspeed, and less so if higher-energy missiles were generated by rupture of the turbine rotors at much higher speeds (say, 180%) which the turbine could reach it if were permitted to accelerate until it failed. A statistical analysis of existing turbine experience by ACRS member Bush indicated that historically the failure rate of turbines had been about 10^{-4} per turbine-year. Reactor vendors argued that new turbines were of much higher quality and predicted very much lower failure rates.

Analytical studies indicated that the probability of serious damage to the reactor from turbine failure was reduced markedly if a "peninsular" arrangement (turbine axis pointing at the containment building) rather than a "tangential" geometric relationship (turbine axis parallel to a line tangent to the circular containment building), existed between the turbine axis and the containment. With continued pressure by the ACRS, most new construction permit applications adopted the "peninsular" approach. Hence, while the matter had not been completely resolved, this design change, as well as some improvements in turbine overspeed control, had taken place.

The generic items on ATWS and steam-line breaks outside the containment vessel are discussed in detail in Chapters 16 and 15 respectively. They represent items which arose by other paths, and whose resolution (or lack thereof for ATWS) provide still other examples of the regulatory process. Some of the original Browns Ferry "asterisked" items are also not fully resolved.

The regulatory staff, as a consequence of their more detailed review, developed a much longer list of safety-related issues to be settled. In fact, it is a continuing list. As some matters are resolved, new ones arise, partly as a result of operating experience, partly from changes in design approach, and partly from the surfacing or resurfacing of questions about existing criteria, designs, etc. The Three Mile Island accident also introduced a large number of new or previously non-emphasized generic safety issues.

Some Sidelights on the Loss-of-Coolant Accident and Emergency Core-Cooling Systems (LOCA–ECCS)

Since the coolant in a light water power reactor is normally at a high temperature and pressure (550° F and 1,000 psi for a BWR, 2,200 psi for a PWR), if a rupture occurs in the piping which carries the water from the reactor vessel to the steam generator to the pump and back to the reactor vessel in a PWR (or its equivalent in a BWR), the high-pressure, high-temperature water will literally push itself out the pipe break so that the primary system loses most of its original inventory of water. Unless emergency cooling water were added and then recirculated through a special heat-removal system, the core would be uncovered and would overheat unacceptably. This would be because of the heat generated from radioactive decay of the fission products in the fuel, even though the chain reaction would be terminated by the loss of water, as well as by the insertion of control rods.

The matter of loss-of-coolant accidents (LOCA) and emergency core-cooling systems (ECCS) has probably been the major topic of public discussion in regard to light water reactor safety. It is perhaps the matter which has received the greatest attention in licensing reviews, and it has certainly obtained the bulk of the resources expended in nuclear reactor safety research. It was also the subject of a very long public hearing conducted by the Atomic Energy Commission in connection with the establishment of acceptance criteria for ECCS in 1972–1973.

We shall not try to examine in detail this long and extensive history of LOCA–ECCS or even to cover all major developments up to the present time. Rather, we shall emphasize some of the earlier developments, provide a few selected sidelights, and look at some of the trends.

The first commercial LWRs, such as the Indian Point 1 PWR and the Dresden 1 BWR, had very limited ECCS capability by current standards. The Connecticut Yankee and San Onofre PWRs, approved for construction in 1963–1964, included substantially augmented ECCS, intended to cope with rupture of the largest pipe connected to the primary cooling system, but not rupture of the primary system piping itself. Oyster Creek, Nine Mile Point, and Dresden 2, large BWRs approved for construction about 1965, each had an ECCS comprised of two core-spray systems, with on-site alternating current power capable of operating at least one core spray. The Brookwood PWR and Millstone Point Unit 1 BWR, approved in early 1966, were similar to their immediate predecessors with regard to ECCS design.

During the period through early 1966, the applicants for LWR construction permits presented rather limited information pertaining to the performance capability of the proposed ECCS or the methods employed for performance analysis. The regulatory staff itself did not possess the capability to evaluate the claimed performance. While the regulatory staff and ACRS were beginning to request that some experimental confirmation of ECCS be obtained, ECCS was not treated as a vital safeguard. The ACRS letter dated November 18, 1964 to AEC Chairman Seaborg on engineered safeguards summarizes the regulatory attitude of that era. This letter, which was prepared in response to a commission request for a review of how engineered safety features were being substituted for distance in the siting of reactors, placed primary emphasis on containment and on features to clean up radioactivity from the containment atmosphere. While not crediting ECCS with preventing core melt, the ACRS report of November 18 did not dismiss ECCS as unnecessary.

Another example of the relatively modest emphasis on ECCS in this period is available from the draft "General Design Criteria for Reactors" first published for comment by the AEC in November 1965 (AEC 1965b). Relatively little is stated concerning ECCS design in this draft.

Then a revolution in LWR safety occurred in 1966. As has been discussed in detail in Chapter 8, "China Syndrome—Part 1," the direct correlation between core melt and a loss of containment integrity was recognized, and this strongly influenced the ACRS review of the construction permit applications for the Dresden 3 and Indian Point 2 reactors in 1966. Emphasis shifted from containment buildings to the prevention of core melt; and the LOCA received primary attention as the most probable source of an accident which might lead to core melt. The ACRS reports of August 16, 1966, on the Dresden 3 and Indian Point 2 reactors recommended much greater emphasis on the prevention of a LOCA, and greatly improved capability of the ECCS. Actually, when the core melt issue came to a head, General Electric modified its previously proposed ECCS for the Dresden 3 reactor and added a separate core-flooding system to the two core-spray systems. General Electric argued that it also had redundant and diverse systems for small breaks, in that its single high-pressure core-injection system was backed up by an automatic depressurization system, which could open a set of valves that would reduce the primary system pressure sufficiently to permit the low-pressure core-spray or core-flooding systems to function.

Westinghouse had not proposed an improved ECCS system at the time of the August 16, 1966, ACRS report on the Indian Point 2 and Dresden 3 reactors, and only preliminary examination of the newly proposed General Electric system had been possible. Hence, the ACRS had recommended that both the ACRS and the regulatory staff review the ECCS designs for these two plants before irrevocable commitments were made in fabrication and construction.

Actually, Turkey Point Nuclear Units 3 and 4 in Florida were the Westinghouse PWRs next in line in the regulatory review process, and in the fall of 1966, Westinghouse proposed a modified ECCS, employing accumulators containing large amounts of water under gas pressure, plus a system of pumps. This new ECCS system, while not employing diverse principles (flooding and core spray as in the BWR), was supposed to be redundant and to meet a criterion that the cladding would not reach its melting point for the rupture of any primary system pipe.

This approach to ECCS design was judged to be generally acceptable and was implemented in similar, although not identical, fashion on all succeeding PWRs, including those sold by Babcock and Wilcox and Combustion Engineering.

One interesting aspect of the minutes of the August 1966 ACRS meeting is that the regulatory staff stated, with not much emphasis, "It is conceivable that, in the case of a cold leg break, significant flow from the two remaining accumulators could be led out of the vessel through the inlet plenum."* This slight concern reappeared with a vengeance some years later following a series of simulated LOCA–ECCS experiments on the Semi-Scale Facility, at the National Reactor Testing Station in Idaho in 1971.

Late in 1966, the Quad Cities BWRs were reviewed for construction permits, and the ECCS approach previously outlined for the Dresden 3 reactor (and backfitted to Dresden 2 and Millstone Point 1) was judged to be generally acceptable.

Some of the side effects potentially associated with the postulated sudden, complete break (and possible offset) of a large pipe included pressure waves leading to blowdown forces inside the pressure vessel and dynamic forces in those sub-compartments within the containment building which house major components such as the steam generator.

The matter of whether large primary system pipes of high quality had a significant probability of gross rupture, and especially of essentially instantaneous rupture, has been controversial. Although some simulated tests involving large pre-existing cracks had led to complete piping failure, and there was some history from the past of the gross failure of large pipes in non-nuclear systems, there existed a school of thought that the sudden "double-ended guillotine break" (involving complete offset of the two pipe ends to form two openings, each having full pipe cross-sectional area) was an inappropriate basis for designing the ECCS and the other safety features associated with this postulated break. In connection with the review of the Browns Ferry 1 and 2 BWRs, which represented a large increase in power over the previously reviewed Dresden and Quad City reactors, the double-ended pipe break controversy arose within the ACRS. After a long, diffi-

---

*The pipe which carries hot water (590° F) away from the reactor pressure vessel is called the hot leg. After giving up heat in the steam generator to the fluid in the secondary system, which is separated from it by a steel tubing, the cooler (530° F) water is pumped back to the vessel through a pipe which is called the cold leg. In the event of a large rupture in one of the cold leg pipes, the relatively hot water in the reactor pressure vessel would rush out the break. Cool water stored in the accumulators would be forced by gas pressure into the other cold leg pipes and thence to the reactor vessel into a large annular (plenum) region. The water was supposed to fall to the bottom of the vessel and then flow up into the reactor core. However, it might flow around the annular plenum and escape through the broken pipe without cooling the fuel and cladding in the core.

cult review in which many issues, new and old, were discussed at great length, the ACRS report of March 14, 1967, on the Browns Ferry reactor included a dissent by member Hanauer. He was dissatisfied with the proposed emergency alternating current power supply which used large diesels to run large pumps in the ECCS, which was needed to cope with rupture of the largest pipe. On the other hand, ACRS member Bush indicated that he also had additional remarks to the effect that the use of the double-ended pipe break was inappropriate and was leading to less safety. Bush did not attach his remarks to the final letter, however, when the ACRS agreed to initiate generic study of the double-ended pipe break question. This matter was pursued by a special subcommittee over the next year, with the conclusion that the double-ended break of the largest pipe should remain among the spectrum of pipe breaks to be analyzed for ECCS performance. Actually, the "dominating" break in ECCS analysis was not always the largest break. And, it is interesting to note that during the subcommittee review, representatives of industry generally agreed that the double-ended pipe break was applicable to the design of ECCS for LWRs (although there were differences of opinion concerning the time for actual rupture, which influences blowdown forces in the reactor vessel for a PWR). To a considerable extent, the report of the task force on emergency core cooling, which is discussed in Chapter 11, "China Syndrome—Part 2," stated the nuclear industry position of that time with regard to LOCA-ECCS. It was relatively optimistic that methods and knowledge were in hand for design and analysis of ECCS to cope with the full spectrum of sizes of pipe break, and that suitable criteria existed.

However, with the passing months and years a variety of matters not anticipated in the task force report arose. One of the first was the observations by R. Ivins at the Argonne National Laboratory that Zircaloy clad exposed to LOCA-like conditions with peak temperatures in the vicinity of 2,500° F (well below the Zircaloy melting point of 3,310° F) embrittled and ruptured, or even shattered on cooling down. This threatened the integrity of the core geometry and hence its continued ability to be cooled. Therefore, instead of the criterion of no (or very little) clad melt, which had been proposed by the vendors and had been accepted for some months, a much lower limit on the highest acceptable clad temperature during a LOCA was indicated, somewhere around 2,200–2,500° F.

Following the Dresden 3 and Indian Point 2 cases in 1966 and the ACRS safety research letter of October 12, 1966, the AEC began to institute a strong safety research program on LOCA–ECCS experiment and analysis, and a beginning was made on forming a LOCA analysis group within the regulatory staff. The vendors all instituted substantial efforts on the development of LOCA–ECCS codes for their reactors, as well as limited experiments.

When the Indian Point 2 reactor was reviewed for an operating license by the ACRS in September 1970, fairly detailed discussions of the anticipated ECCS behavior were held. The reactor vendor and the regulatory staff each expressed confidence in the acceptability of ECCS performance and their understanding of ECCS function. Following blowdown of the primary system from a postulated large pipe break, core-reflooding rates of five to ten inches per second were predicted, and peak clad temperatures of *about* 2,000° F were calculated.

Unfortunately, the physical modeling of ECCS performance for both PWRs and BWRs in the mid- and late 1960s was deficient and even wrong in many significant areas of the analysis. Fortunately, there appears to have been enough margin in other areas of the analysis to roughly compensate for the deficiencies.

In early 1971, some tests in the Semi-Scale Facility at the Idaho National Reactor Testing Station showed that the modeling of water flow from the accumulators in a PWR ECCS had been deficient, and that much of this water would leave the reactor vessel during blowdown following a cold leg break, rather than reflood the core immediately. At about the same time, it also became clear that previous analyses of core reflooding in PWRs following the loss of coolant had been far too optimistic in their estimates of reflooding rates. It was now found that the cool ECCS water supplied to the reactor vessel, first from the accumulator and then by pumps, would be slowed down in its flow through the bottom region of the reactor vessel up into the hot core by the buildup of a high steam back pressure above the core. This high back pressure was built up by the flow resistance to the steam as it tried to escape down a long path to the pipe rupture. The steam back-pressure phenomenon had a very large effect on the rate of core reflooding, which dropped from a predicted five or ten inches per second to the vicinity of one inch per second (or less). For medium and large breaks, this resulted in a much longer time during which the fuel elements were uncovered (and poorly cooled) and during which fission product decay heat raised the temperature of both fuel and clad more than had been previously allowed for.

Similarly, a very different concept of how the core-spray system in a BWR actually behaves, compared to the traditionally proposed picture, evolved; and there were other significant deficiencies uncovered in the physical modeling of LOCA–ECCS for both the BWR and PWR.

In March 1971 Mr. Price established a regulatory staff task force under Dr. Hanauer (who joined the staff in 1970) to review the status of LOCA–ECCS.

Great pressure was being exerted on the regulatory staff with regard to the adequacy of ECCS at the ASLB hearings on the Palisades, Shoreham, Dresden 3, and Midland reactors, among others. Mr. Price decided that he needed to develop some position which could be adopted as a matter of policy by the AEC commissioners and would provide the basis for judging ECCS acceptability at all the ASLB hearings, rather than having to treat each case in an ad hoc fashion. The commissioners exerted pressure on the regulatory staff and the ACRS to have some agreed upon position in time for the hearings on ECCS to be held by the Joint Committee on Atomic Energy on June 22, 1971. After much back and forth discussion, including presentations of an alternate proposed approach by a minority group from the regulatory staff, the ACRS gave limited concurrence to the proposed staff interim acceptance criteria at a special meeting, June 16–17, 1971.

At the one hundred fortieth meeting, December 9–11, 1971, the ACRS heard extensive presentations on proposed amendments to the interim acceptance criteria, and prepared a final report on this matter for use by the commission in the upcoming rule-making hearing on acceptance criteria for ECCS. The letter to AEC Chairman Honorable James R. Schlesinger was never signed by ACRS Chairman Spencer H. Bush, and was hand delivered at a meeting December 15, 1971, between ACRS representatives and the commissioners. It is reproduced below.

ACRS Report to the AEC on Interim Acceptance Criteria for ECCS for Light-Water Power Reactors

At its 140th meeting, December 9–11, 1971, the Advisory Committee on Reactor Safeguards completed a review of proposed amendments to the Interim Acceptance Criteria for Emergency Core Cooling Systems for Light-Water Power Reactors, published as an AEC Interim Policy Statement on June 29, 1971. The proposal was considered at a Subcommittee meeting held on December 4, 1971, in Washington, D.C.. The proposed amendments add new evaluation models by Combustion Engineering, Inc. and the Babcock and Wilcox Company to the existing list of acceptable evaluation models.

The Committee believes that the amendments are acceptable, on the same basis as the original Policy Statement Evaluation models. The Committee also believes it desirable to comment briefly concerning this basis.

At the time adoption of the Interim Acceptance Criteria was under consideration, the Committee had opportunity to discuss the draft Criteria with the AEC Regulatory Staff and the Commission, and to offer comments. The Committee comments, made orally to the Commission on June 17, 1971, included the following:

"The ACRS believes that the draft, 'Interim Acceptance Criteria for ECCS for Light-Water Power Reactors,' dated June 17, 1971, is satisfactory for release subject to some clarification of the language."

"The Committee concurs with the suggested approach of publishing the criteria in the Federal Register as an interim policy statement of the Commission. It agrees that Alternate A setting January 1, 1968, as a cut-off date is reasonable. It agrees with the relaxation in the criteria in Section IVB as promoting greater flexibility without freezing design and codes."

"The Committee believes that the document represents an interim solution only, and that more work is required in code development, safety research oriented to LOCA–ECCS, and work on improved ECCS. The Committee is prepared to work closely with Staff and vendors through appropriate Sub-committees. We believe this approach may lead to an orderly solution of the outstanding problems."

"With regard to the course of events of the last few months, the Committee now feels that there has been an over-all gain in that the level of work on ECCS evaluation has increased and more specific goals and objectives have been established. However, the Committee believes that more experimental work is needed to supplement the further analytical development of the evaluation models."

The Committee wishes to emphasize again that because the Interim Acceptance Criteria are properly regarded as temporary criteria, additional research on code development and on improved ECCS design is required to achieve a satisfactory long-term situation with regard to ECCS performance.

Additional remarks by Dr. Herbert S. Isbin and by Dr. David Okrent are attached.

Additional Remarks by Dr. H. S. Isbin

In my opinion, a major usefulness of the Interim Acceptance Criteria has been the highlighting of several problem areas in the evaluation of emergency core cooling systems. The Committee has taken a strong position in encouraging more work in code development, safety research oriented to LOCA–ECCS, and work on improved ECCS including evaluation of possible modifications to current systems. It is my opinion that at the construction permit stage, the applicant should be committed to research and development in these areas, including possible design modifications which would clearly circumvent potential problems such as steam binding and reflooding. At the operating license stage, I believe that the interim solution should be a corresponding reduction in power rather than in peaking factors to meet the Acceptance Criteria.

Additional Remarks by Dr. David Okrent

I believe that because of the low probability of a major loss of primary system integrity, because of the limited number of reactors involved, and because there is a reasonable probability that emergency core cooling systems in plants currently going into operation will be able to maintain core integrity during postulated loss-of-coolant accidents, operation of such plants during an interim period does not represent an undue risk to the health and safety of the public in this regard.

However, I believe that the increase in our still imperfect knowledge of events during postulated loss-of-coolant accidents tells us that the considerable margins in cooling capability calculated to exist a few years ago may be much decreased. Also, the accepted evaluation models include both some requirements that add conservatism and some features which are or may be non-conservative so that the net conservatism is not amenable to quantitative evaluation.

Therefore, I believe that it is timely to begin implementation of the statement in the Commission's Policy Statement of June 29, 1971, that "in connection with the water power reactors yet to be designed and constructed the possibility of accomplishing by changes in design further improvements in the capability of emergency core cooling systems should be considered."

With regard to the specific criteria and models, I should like to note that for the GE BWR, the Interim Criteria include an evaluation model for the core spray system but not for the equally important low pressure flooding system which is designed to provide an alternate, diverse means of protection against postulated large pipe breaks. Also, there exists a probable inconsistency between permitting a 1% average clad—H_2O reaction and the important requirement that the local metal water reaction not be sufficient to produce clad embrittlement.

This letter was held up and never formally released by the ACRS because the commissioners requested its reconsideration. The commissioners suggested that it would be helpful if the ACRS report could be revised to include a more positive statement from the committee that reactors designed in accordance with the interim acceptance criteria would provide adequate protection of the public health and safety. The commission also exerted pressure by noting that the minority comments attached to the ACRS report might subject the minority commenters to subpoena during the rule-making hearing.

The ACRS reconsidered the letter at its one hundred forty-first meeting, January 6–8, 1972. The committee arrived at a new consensus which incorporated the bulk of the minority opinions which had previously been appended to the December letter. The new letter of January 7, 1972, was less than satisfactory to the commission, in that it indicated the interim criteria, per se, were not sufficient determinants of the adequacy of the safeguards to protect the public health and safety. It also urged design improvements for ECCS systems even if the criteria were met. The ACRS report dated January 7, 1972, was sent to the AEC Chairman Schlesinger and signed by ACRS Chairman C. P. Siess. It is reproduced below.

*ACRS Report to the AEC on Interim Acceptance Criteria
for Emergency Core Cooling Systems for Light-Water Power Reactors*

At its 140th meeting, December 9–11, 1971, the Advisory Committee on Reactor Safeguards completed a review of proposed amendments to the Interim Acceptance Criteria for Emergency Core Cooling Systems for Light-Water Power Reactors, published as an AEC Interim Policy Statement on June 29, 1971. The proposal was considered at a Subcommittee meeting held on December 4, 1971, in Washington, D.C. Further consideration was given to the interim criteria at the Committee's 141st meeting, January 6–8, 1972. The proposed amendments add new evaluation models by Combustion Engineering, Inc. and the Babcock and Wilcox Company to the existing list of acceptable evaluation models.

The Committee concluded that the proposed amendments were acceptable, on the same basis as the original Policy Statement evaluation models. Both the original and the amended criteria involve a number of provisions which are clearly conservative as well as some which may not be conservative, but on balance reflect adequate conservatism for interim use with plants similar in design to those which have been reviewed for construction permits. The Committee believes that each plant should be reviewed on a case by case basis to determine the extent to which these criteria are satisfied or modifications to the design or operation of the plant are required.

At the time adoption of the Interim Acceptance Criteria was under consideration, the Committee had opportunity to discuss the draft Criteria with the AEC Regulatory Staff and the Commission, and to offer comments. The Committee comments, made orally to the Commission on June 17, 1971 included the following:

"The ACRS believes that the draft, 'Interim Acceptance Criteria for ECCS for Light-Water Power Reactors', dated June 17, 1971 is satisfactory for release subject to some clarification of the language.

"The Committee concurs with the suggested approach of publishing the criteria in the Federal Register as an interim policy statement by the Commission. It agrees that Alternate A setting January 1, 1968 as a cut-off date is reasonable. It agrees with the relaxation in the criteria in Section IV B as promoting greater flexibility without freezing design and codes.

"The Committee believes that the document represents an interim solution only, and that more work is required in code development, safety research oriented to LOCA–ECCS, and work on improved ECCS. The Committee is prepared to work closely with Staff and vendors through appropriate Subcomittees. We believe this approach may lead to an orderly solution of the outstanding problems.

"With regard to the course of events of the last few months, the Committee now feels that there has been an over-all gain in that the level of work on ECCS evaluation has increased and more specific goals and objectives have been established. However, the Committee believes that more experimental work is needed to supplement the further analytical development of the evaluation models.

"The Committee reemphasizes that the Commission should indicate publicly the need for continuing work on new ECC systems or significant modifications to current designs. It should be indicated that increases in power or power density will require larger margins."

Some of the restrictive assumptions imposed by the Interim Criteria have been introduced because of uncertainties in the behavior of the ECCS. The Committee believes that restrictions should be removed, as possible, either by design changes, or by demonstration that they are unnecessary. The Committee believes that it is timely for further implementation of the statement in the Commission's Policy Statement of June 29, 1971, that "in connection with the water power reactors yet to be designed and constructed the possiblity of accomplishing by changes in design further improvements in the capability of emergency core cooling systems should be considered." Although the Interim Acceptance Criteria should be useful and helpful in the licensing process, the evaluation models prescribed in these criteria are recognized to have only limited usefulness as design tools for improving emergency core cooling systems. The nuclear industry should respond in a more direct fashion with realistic design methods, based upon additional scaled experiments and analytical studies. Design changes which would clearly eliminate any potential steam binding problem during a loss-of-coolant accident represent an example of the type of improvement considered to be of particular importance. The Committee recommends that design changes to improve ECCS capability should be sought and, to the extent practical, employed in plants for which construction permit applications are received in the future, irrespective of whether the plant design without such changes appears to meet the provisions of the interim Acceptance Criteria.

The Committee believes that there is reasonable assurance that, with appropriate use of the Interim Acceptance Criteria and other applicable design and evaluation criteria, water reactors of current design can be operated without undue risk to the health and safety of the public.

A long and controversial rule-making hearing on acceptance criteria for ECCS was held during 1972–1973. In August 1973 the Atomic Energy Commissioners requested the ACRS to comment on the matter, and at its one hundred sixty-first meeting, the committee prepared a response, dated September 10, 1973, from which a few salient excerpts are given below:

The ACRS believes that the Concluding Statement of Position of the Regulatory Staff's Public Rulemaking Hearing on Acceptance Criteria for Emergency Core Cooling Systems for Light-Water-Cooled Nuclear Power Reactors (April 16, 1973) represents an improvement over the Interim Acceptance Criteria.

In general, the Position retains previous restrictive assumptions, and several others have been added. Although there is evidence to support less restrictive evaluations, the Committee believes that a more substantial demonstration in terms of analyses and experimental data is required before relaxation of the restrictions can be affected. In achieving these goals, the development of the needed technical bases would be considerably enhanced if the proprietary aspects could be minimized.

The Committee reaffirms its position that, in the future, design changes to improve ECCS capability should be sought and, to the extent practical, employed, irrespective of whether the plant design without such changes appears to meet the provisions of the Interim Acceptance Criteria and the proposed changes in these criteria. In its approach to nuclear safety, the Committee has sought to make allowances for the state-of-the-art knowledge on issues, to encourage further acquisition of background knowledge, to determine whether added safety improvements or margins are achievable, and then to encourage incorporation of practical improvements. It is the Committee's judgment, that for an expanding nuclear industry, the cumulative effects of the added improvements represent prudent goals.

The Committee views the following as examples of measures which contribute to design improvements: 1) improved reliability of the ECCS and system components, including approaches intended to minimize the potential for common failure modes; 2) reactor core designs and operating modes which reduce the potential for high temperatures, clad swelling or perforation in postulated LOCA's; 3) ECCS whose proper functioning is relatively insensitive to reactor or ECCS design parameters and to proper functioning of other components such as steam generators or reactor containment; 4) ECCS having redundancy, diversity and abundance of flow such that its adequacy is subject to evaluation without undue requirement for complex evaluation techniques; 5) other measures which further reduce the probabilities and consequences of a LOCA.

The committee generally supported the new position taken by the regulatory staff in their "concluding statement of position" (April 16, 1973). However, the ACRS reaffirmed its recommendation for design changes to improve ECCS capability.

In their decision on the rulemaking hearing about the acceptance criteria for ECCS for LWRs, the commissioners made relatively modest modifications in the final staff position. In an accompanying statement, the commisioners included a recommendation for work on improved ECCS.

However, the reactor designs proposed for construction in the period 1973–1977 show rather little change in this regard, except for a general trend toward smaller diameter fuel rods, which permitted the same or higher core power densities with lower fuel temperatures at normal operation conditions, and hence somewhat lower calculated peak clad temperatures in a LOCA.

Despite a rather considerable experimental and analytical effort in the ensuing five-year period, the basis by which performance of an ECCS is determined has remained imperfectly defined: For the PWR, for example, it has remained a controversial matter whether the possible effects of water carryover to the steam generator, with a resultant increase in steam binding, might not lead to a worsening of performance, if both cold and hot leg injection of ECCS water were used, as in the German PWRs, rather than only cold leg injection as in the United States PWRs.

In addition, the NRC safety research program refrained from any aggressive pursuit of possibly improved ECCS features, taking the position that the law required that the NRC do only "confirmatory" research. The Electric Power Research Institute, the research arm of the utilities, also chose not to pursue such possible improvements.

It remains to be seen what changes in ECCS, if any, will be produced by congressional action in 1977 requiring the NRC to develop a safety research program on new and improved safety concepts and features. The Three Mile Island accident is likely to produce major changes in the previous priorities for safety research, but the new directions will probably not emphasize large loss-of-coolant accidents.

LWR Safety Research

Reactor safety research, in a general sense, began essentially as soon as nuclear reactors began. Safety studies were done by many of the groups working on reactor development. At the third meeting of the statutory ACRS in December 1957, there was a major presentation on a new, fairly comprehensive program on the safety of fast reactors. And prior to this time, there had been major experiments performed on the behavior of light water reactors during very large, rapid power increases (called reactivity transients); for example, the experiment in which the core of the BORAX 1 reactor in Idaho was deliberately destroyed, and the subsequent series of similar experiments on the SPERT reactors in Idaho.

In 1958 the total dollar amount of the experimental reactor safety program was $7 million; the program covered three areas, namely, reactor kinetics and control, chemical reactions, and containment.

In 1961, the growing importance of safety research was reflected by the establishment of the post of Assistant Director for Reactor Safety Research within the AEC Division of Reactor Development. At its forty-fifth meeting, December 13–15, 1962, the ACRS sent a letter report dated December 31, 1962, to AEC General Manager A. R. Luedecke concerning aspects of the AEC safety research program. The letter recommended "that destructive tests be performed as soon as possible on low enrichment oxide cores, since such cores are used in the majority of power reactors." Large destructive reactivity transients were not a "design basis" or "maximum credible accident" for the light water power reactors then under regulatory review. However, a damaging reactivity transient had occurred the previous year in the SL1, a small, experimental boiling water power reactor in Idaho. Of course, reactivity transients had probably been the major safety issue during the 1950s.

At its forty-eighth meeting the ACRS completed a second report on reactor safety research, dated August 1, 1963, and sent it to Luedecke. In this letter the ACRS stressed "the dependence placed on engineering safeguards to reduce the effects of credible reactor accidents." Most of the letter was devoted to research programs on fission product release from the fuel, and migration and possible holdup of fission products in the primary system or containment building. Industry was anxious to find ways of arguing that the fraction of the postulated release available for leakage from an intact containment building following a postulated core meltdown was much less than that assumed in the reactor site criteria, 10 CFR Part 100.

The ACRS letter mentioned the proposed Loss-of-fluid test (LOFT) in which a 50 MWt reactor was to be deliberately subjected to a LOCA, the core allowed to melt, and the pathways of the fission products inside and outside the containment building followed in detail. The ACRS expressed only mild enthusiasm for this proposed LOFT experiment, since it would represent only a single empirical data point.

At its fifty-second meeting, January 9–10, 1964, the ACRS heard encouraging results concerning the ability of the low-enrichment uranium oxide fuel in the SPERT reactor to withstand a severe reactivity transient with essentially no damage. As expected, the transient was terminated by the negative Doppler effect on reactivity, which made the reactor sub-critical (unable to sustain a chain reaction) before it got too hot. The experience in SPERT was directly applicable to the PWRs and BWRs under consideration in the regulatory process.

At its seventy-eighth meeting, October, 1966, the ACRS completed a very significant letter report to AEC Chairman Seaborg on reactor safety research. This letter, which was prepared shortly after "resolution" of the China Syndrome issue for the Dresden 3 and Indian Point 2 reactors, and ACRS agreement to the establishment of a task force by the AEC, placed major emphasis on research into core melt pheonomena, LOCA–ECCS, and improved primary system integrity. The letter was signed by ACRS Chairman Okrent and is reproduced below.

ACRS Report to the AEC on Reactor Safety Research Program

For the past several years the Advisory Committee on Reactor Safeguards, at the request of the Division of Reactor Development and Technology, has reviewed and commented on the Commission's reactor safety research program. This activity gives the Committee an opportunity to keep currently informed of the content and status of the research program and to suggest modifications and additions on the basis of recent Committee experience in the course of reactor licensing reviews.

The following comments, therefore, are intended to emphasize the Committee's interest in the present program, and to suggest certain areas in which the Committee believes more research results are now or will soon be needed.

    A. The Committee attaches special importance to the following areas.

        1.   A vigorous research program should be initiated promptly on the poten-tial modes of interaction between sizeable masses of molten mixtures of fuel, clad and other materials with water and steam, particularly with

respect to steam explosions, hydrogen generation, and possible explosive atmospheres. Work should be directed toward understanding the mechanisms of heat transfer connected with such molten masses of material, the kinds of layers formed at cooled surfaces, the nature and consequences of any boiling of the fuel, and the manner and forms in which fission products escape from bulk molten fuel mixtures. Further, studies should be initiated by industry to develop nuclear reactor design concepts with additional inherent safety features or new safeguards to deal with low-probability accidents involving primary system rupture followed by a functional failure of the emergency core cooling system.

2. Because of the importance of emergency core cooling as an engineered safeguard, studies on core cooling processes already underway within the AEC and industry should receive continued attention. Coolant distribution and heat transfer phenomena which could influence emergency cooling significantly should be examined to remove existing uncertainties, including those related to an assumed course of events where cooling is marginal or inadequate in sections of the core. Tests of actual spray cooling and core flooding systems under accident conditions warrant careful consideration.

3. Development of practical, effective methods for extensive periodic inspection of pressure vessels is of great importance. The current program in AEC and industry should be augmented, as necessary, to assure this. One or more practical systems for such inspection should be developed as soon as possible.

4. A strong program on the properties, homogeneity, and behavior of thick steel pressure vessel sections, including research areas described in the recently proposed program of the Pressure Vessel Research Committee, should be implemented by industry and the AEC. The work on thick-walled vessels should include a thorough study of potential failure modes under pneumatic loading for various flaw sizes and types, and the significance of the reduction in the energy absorption shelf as a function of neutron irradiation.

5. Because the Commission may be called upon to consider proposals to construct reactors utilizing prestressed concrete pressure vessels, the nuclear industry and the AEC should promptly institute a very active safety research program into such vessels, including their design for seismic effects. This program should include research into anomalous failure modes of such vessels, particularly under pneumatic loading. This work should encompass effects of potential structural defects or overloads and problems associated with closures, penetrations, and anchors.

6. The further development of advanced methods of calculating destructive reactivity transients in water-cooled reactors, including predictions of damage to the primary system, is recommended. Important phenomena, such as the mode, time-sequence, and effect of fuel element failure, should be identified and studied so that the phenomena are dealt with adequately

in the over-all analyses. Also, the role of space-dependent kinetic effects should be fully identified.

When it reaches the operational stage, the planned PBF (Power Burst Facility) program should play an important part in identifying fuel failure modes. The large transient experiments in the SPERT program have already been very useful. Further experiments with low-enrichment-fuel water reactors should be considered as a means of providing additional calibration points with which to test improved theoretical methods of predicting large reactivity transients. Such experiments may also uncover new or overlooked phenomena.

B. The following areas are, in the Committee's view, also of current and increasing significance.

  1. The trend toward higher reactor powers may require larger pressure vessels, thus favoring steels stronger than modified A-302-B. Preliminary AEC and industrial irradiation effects programs on some higher strength steels have revealed a limited NDT shift at relatively high neutron exposures. Work on potential new pressure vessel steels should be suggested so that adequate information is available on radiation effects, fabrication and joining processes, fatigue effects, etc., prior to their application to reactor pressure vessels.

  2. The complex interrelationships between flaw size, shape and orientation, flaw growth, and neutron exposure in pressure vessels are still incompletely established. The Committee suggests a continuation and expansion of existing programs in fracture mechanics and irradiation damage, including their relation to fabrication, welding, and heat treatment parameters. Potential for propagation into the base metal of cracks that initiate in cladding should be further examined.

  3. A review and evaluation of nonnuclear industrial experience on pressure vessels, steam generators, valves, and other components of reactor primary systems, similar to the recent study of piping, should contribute to the safety of power reactors by providing recommendations for changes in their current design, fabrication, inspection, or operational practice. A thorough review should be instituted and information obtained from insurance companies, state regulatory groups, and industry associations.

  4. In connection with the review of power reactors for construction permits and operating licenses by regulatory bodies, standard methods of calculating various important events, ranging from reactivity excursions to reactor blowdown following a postulated coolant system rupture, usually do not exist. Applicants use a variety of methods employing a range of parameters. It would be helpful if a series of calculational methods could be developed and placed into use, against which the methods and parameters used by individual applicants could be compared.

Alternatively, a series of reference problems (and solutions) might be established which the applicant could calculate by his particular methods, and the results studied to help judge their degree of conservatism.

Some computational methods are already being developed in connection with various safety research programs. Others could be developed. In expanding this program, the Division of Reactor Development and Technology should work closely with the Regulatory Staff to establish an appropriate series of standard methods or reference problems.

5. In view of the large amounts of recycled plutonium fuel that will probably be used in thermal reactors in the future, potential safety problems arising from the use of such fuel should be identified and appropriate information developed in timely fashion.

6. Since early detection of small leaks in primary coolant systems of reactors can provide considerable protection against more serious difficulties, existing leak detection methods should be evaluated from the safety standpoint and new techniques developed, if appropriate.

7. The dilution, dispersion, and transport of liquid radioactive wastes in surface waters (rivers, lakes, estuaries, bays and open ocean) are important factors in the siting of nuclear reactors. In addition to these phenomena, attention freqently needs to be directed toward biological concentration of radionuclides in aquatic life. It may be desirable to review previous work on this subject, including related research on discharge of municipal and industrial liquid wastes. Preparation of a state of the art review of current knowledge, and delineation of areas where further research is needed, would be useful. A special evaluation of the impact of siting many reactors on the shores of the Great Lakes, in relation to retention and flushing characteristics and to accumulation of radionuclides in aquatic organisms, may also be desirable.

It should be noted that information developed in connection with several items listed above would not only help to enhance public safety, but would also contribute toward a more expeditious review of the large number of reactor projects anticipated for the future.

Following this letter, the LOFT program was entirely changed from one involving melting of the core so that fission product retention and dispersion could be measured, to one emphasizing studies of LOCA–ECCS; and a considerable growth in analytical and small-scale experimental studies on LOCA–ECCS resulted. However, essentially no effort was initiated by the AEC on phenomena related to core melt. The ACRS in a letter dated April 14, 1967, to Milton Shaw, Director of the Division of Reactor Development, called attention to the lack of effort on core melt, but this letter had no effect.

An ACRS–AEC–Industry meeting on safety research for LWRs was held on February 27–28, 1968. The following excerpt from the meeting minutes reflects the industry attitude of that time:

The representatives of the reactor designers asked what the AEC wanted in the way of safety margin. They expressed their opinion that they have sufficient information for current designs and plan no major changes in the immediate future. They expressed their opinions that their design are adequately safe without further major R&D. They said that if the AEC did not agree, then the AEC should provide information as to what it considered to be adequate safety.

In January 1969 the AEC issued a detailed water reactor safety research program plan which included priorities for the various potential tasks. In a letter dated March 20, 1969, the ACRS commented as follows:

The Committee wishes to take this opportunity to emphasize the continued great importance to the health and safety of the public of the Commission's Nuclear Safety Research Programs on water reactors and on other power reactor types, and urges continued vigorous support of this work.

The ACRS agrees with much of the general emphasis placed by the Plan on various facets of water reactor safety research. However, the Committee believes that, in view of overall funding limitations, the considerable expenditures projected for the LOFT facility and its nuclear experimental program warrant re-evaluation with the benefit of a careful review of the specific objectives and anticipated accomplishments of this portion of the program. The Committee also believes that more effort should be devoted to gaining an understanding of modes and mechanisms of fuel failure, possible propagation of fuel failure, and generation of locally high pressures if hot fuel and coolant are mixed, and that effort should commence on gaining an understanding of the various mechanisms of potential importance in describing the course of events following partial or large scale core melting, either at power or in the unlikely event of a loss-of-coolant accident.

The Committee urges that increasingly strong direct support be given the Regulatory Staff by capable, experienced personnel working in the AEC's safety research program. Quantitative evaluation of safety questions arising in construction permit applications will not only help ease the current workload problem of the Regulatory Staff but should make for more meaningful research efforts.

The last paragraph reflected an effort to make the AEC safety research program, which was tightly controlled by Milton Shaw, the Director of the Division of Reactor Development, more useful and responsive to the needs of the regulatory staff.

Several months later, in a letter dated November 12, 1969, to AEC Chairman Seaborg, the ACRS expressed considerable concern about a reduction in reactor safety funding for fiscal years 1970 and 1971. During the ensuing months there was continued discussion within the ACRS of possible recommendations for changes in the administration of LWR safety research in order to make it more responsive to regulatory needs.

The minutes of the one hundred thirty-ninth ACRS meeting, November 11-13, 1971, record that ACRS member Lombard Squires had informed the Director of Regulation, Mr. Muntzing, that the committee had previously recommended (to the commissioners) that the regulatory staff have its own budget for independent work in water reactor safety research. In response, Mr. Muntzing stated that he felt the recommendation had considerable merit, and that it would be particularly desirable for the regulatory staff to have such funds, particularly if separation of regulatory and promotional activities of the AEC should become a reality.

The minutes of the one hundred forty-sixth meeting, June 8–10, 1972, show that the ACRS orally advised the commissioners: "The Committee believes that it is timely to consider providing for the safety research intended to meet specific regulatory needs within a separate organizational branch responsible to the Director of Regulation."

Inadequate funding of water reactor safety research remained a continuing problem in mid-1971. The AEC Division of Reactor Development favored a decrease in funding of water reactor safety research and an increase in fast breeder studies. The Bureau of the Budget saw that a large water reactor industry existed and questioned the need for the AEC to provide funds for water reactor safety research. And the Congress did not respond favorably to an AEC request for increased funding.

However, late in 1971, in large part as a result of the controversy over ECCS and the interim acceptance criteria adopted by the regulatory staff, there occurred a reversal in the trend toward decreased LWR safety research in the AEC. This was reflected in the ACRS letter of February 10, 1972, which said:

The Committee finds that a desirable increase in ECCS-related research by reactor vendors and the AEC has occurred during the past years. Also, it appears that the electric utility industry may in the future take an active role in funding, directing, and applying such research.

However, the Committee finds that the relative roles and responsibilities of the utilities, the reactor vendors, and the AEC with regard to safety research have not been clearly defined. Further, the Committee finds that, while a Water Reactor Safety Program Plan document has been published by the AEC, and there are activities in process or planned for modification and augmentation of the plan, there has not yet been formulated a sufficiently specific definition of the national safety research needs for water reactors, including the means and schedules to be used in resolving problems.

The AEC has frequently stated that the ultimate responsibility for safety rests in the hands of the owner, the utility. However, certain safety questions are generic, applying to a class of light-water reactors, or to all light-water reactors. The ACRS believes that responsibility for the solution of each generic question should be assumed by the nuclear industry. The Committee recommends that the Regulatory Staff assure itself that the overall industrial program is well delineated, is funded and implemented, and is adequate to provide proper assurance of public safety. The

ACRS believes also that the AEC should continue to support at a high funding level a continuing light-water reactor safety research program designed to provide independent confirmation of the adequacy of solutions of identified problems and improved engineered safety features. This program would also provide a valuable and important source of expert consultants to the Regulatory groups.

The ACRS recommends that, in the future, the AEC safety research program should reflect more directly in extent and in detail the recommendations and needs of the Regulatory Staff and the ACRS.

An independent comment on the AEC's water reactor safety program during the late 1960s and early 1970s was given by Ralph Lapp in testimony on January 27, 1972, to the AEC rule-making hearing on interim acceptance criteria. A few excerpts follow:

It is difficult for me to reconcile the fact that much AEC safety research is in the future tense, whereas power reactors are in operation. It seems to me that this situation places the AEC's Regulatory Staff in an awkward position when it is called upon to approve new plant construction and operation. The position will be very much more awkward when utilities come in with reactor designs employing higher power densities. I believe that the Atomic Energy Commission has allowed reactor safety research to lag so that its Regulatory Staff is called upon to judge reactor applications without an adequate experimental base which verifies the evaluation models and checks out the calculational codes of the safety statements submitted by the utilities. Furthermore, it is my opinion that new mechanisms are required to provide independent checks and balances for the protection of the public health and safety in areas where high power reactors are sited.

It appears to me that part of the explanation for the faltering AEC safety program in ECCS may be ascribed to the undefined role of the nuclear industry in this area. There was apparently a belief within the Atomic Energy Commission that it had fulfilled its promotional aspects of reactor development during the late 1960s and that it was up to industry to assume responsibility for the reactors which were being marketed.

It is my own impression that interest within the AEC shifted from safety research on water reactors to programs oriented toward the power-breeder and that this also accounts, in part, for deficiencies in the present water reactor safety program.

A very considerable controversy swirled around the management of LWR safety research under Milton Shaw during the early 1970s. Some of this was aired in a series of articles by Robert Gillette in *Science* (Gillette 1972).

Among other things, Gillette reported allegations that:

(1) Between 1963 and 1971, the Division of Reactor Development and Technology bootlegged money from water reactor safety to accelerate the breeder, and in the process killed or cut back a number of key research

projects that had begun to raise questions about nuclear plants coming up for licensing.

(2) Shaw had shown considerable indifference toward urgen needs of the regulatory branch for technical help during this period, and for several years forbade direct contact between safety researchers and AEC's regulatory staff.

In 1973, Dixy Lee Ray, the new chairman of the AEC, reorganized the safety research program, taking the LWR safety program away from Shaw and placing it in the hands of a new Division Director, Herbert Kouts, a former ACRS member. There was also a major expansion in the funding level for LWR safety. The ACRS letter of November 20, 1974, to Chairman Ray reflected satisfaction with this change, which led to an LWR safety research program which was much more responsive to the requests of the regulatory staff. However, the ACRS noted that:

the Committee finds that the relative roles and responsibilities of the utilities, the reactor vendors, and the AEC with regard to safety research have not yet been sufficiently clearly defined. More specifically, it is not clear that industry, as an entity, has developed a comprehensive program of scope and schedule commensurate with the need. The role of architect-engineers in safety research warrants examination. Clarification of the roles and responsibilities should be pursued to insure that needed programs will have the appropriate attention and funding priorities.

When the Nuclear Regulatory Commission was formed by the Energy Reorganization Act of 1974, all the LWR safety work was sent to the NRC, and Kouts was named Director of the Office of Regulatory Research. A continued growth of the program occurred, with great emphasis placed on a better solution of the ECCS controversy. The congressional committee report on the law establishing the NRC stated that the NRC safety research work should be confirmatory in nature; and the NRC chose to interpret the wording in such a way that it placed little emphasis on advanced concepts or other improvements in safety, or on exploratory research. In 1977, the Congress required the NRC to develop a program aimed at new or improved safety systems for nuclear power plants (NRC 1978b). However, the actual program was not initiated until April 1979, after the Three Mile Island accident, and then on a small scale. In its first phase, the program included a modest effort on a containment concept intended to reduce large atmospheric releases of radioactivity from most postulated core melt accidents, but it still refrained from performing conceptual design studies on systems intended to retain a molten core. However, by the end of 1979, largely as a result of pressures arising from the Three Mile Island accident, there were signs of a considerable increase in NRC emphasis on safety research related to mitigation of core melt accidents. Quite naturally, however, most of the new research is aimed at small LOCAs and complex transients, as exemplified by the Three Mile Island accident.

19

The WASH-1400 Report
and the Three Mile Island Accident

Some Effects and Implications of the WASH-1400 Report

The reactor safety study, WASH-1400 (NRC 1975a), may have had part of its genesis in a letter from Senator John Pastore to AEC Chairman Schlesinger dated October 7, 1971, which requested a comprehensive assessment of reactor safety.

It seems that the first formal report (AEC 1973b) produced by the AEC in response to this letter was entitled *The Safety of Nuclear Power Reactors (Light Water-Cooled) and Related Facilities*, WASH-1250 (Final Draft), dated July 1973. An earlier draft version had been circulated for comment in late 1972. However, this report did not provide a quantitative assessment of risk in a probabilistic fashion as discussed in Senator Pastore's letter.

In the summer of 1972, the AEC initiated a major study on LWR risk assessment with Professor Norman C. Rasmussen of Massachusetts Institute of Technology serving (half-time) as the study director. Mr. Saul Levine, an AEC employee and former member of the regulatory staff, served as full-time staff director of a group within the AEC that performed the study with the aid of many contractors and consultants.

The first draft WASH-1400 report (AEC 1974a) was issued by the AEC for comment in August 1974; the final report, WASH-1400 (NRC 1975a), was issued by the Nuclear Regulatory Commission in October 1975 and was comprised of a main report and 11 appendices.

The WASH-1400 report (NRC 1975a) represents the most-detailed study performed for light water nuclear power reactors of the potential effects of accidents on the public health and safety. Nothing of this depth exists for other nuclear power types, and much less, if anything exists concerning the potential for accidents from hydroelectric sources. Of course, all energy sources pose risks to the public health and safety, as do most other facets of society.

The WASH-1400 report (NRC 1975a) represents an evaluation of the potential for a serious accident for two specific reactors, averaged over many reactor sites to obtain consequences representative of 100 reactors as a group. The principal results in the report are reproduced in Tables A2 and A3 of Appendix I, "A Brief Discussion of Serious Accidents in Light Water Reactors," and are estimated to have an uncertainty of about a factor of five, both in frequency of occurrence of accidents and in their consequences.

The principal architects of the study, Messrs. Rasmussen and Levine, sometimes summarized the results by noting that, while the probability of core melt was estimated to be higher than many people had expected,* the magnitude of the potential consequences of core melt has a wide range of values, and there was a high probability that the consequences would be modest compared to other risks, and small in absolute value.

When the draft WASH-1400, reactor safety study (AEC 1974a), was issued by the AEC for comment in August of 1974, it drew a very substantial response from many quarters of industry, government and the public. Appendix XI of the final version of the WASH-1400 report (NRC 1975a) was devoted to a discussion of what the authors of the report considered to be the principal comments on the draft report, and their response to these comments.

For assessments of low-probability events, whether one is considering reactor core meltdown, failure of a specific large dam, release of large quantities of liquified natural gas in port, or possibly catastrophic, unanticipated effects from biological research or some new vaccine or drug, large uncertainties will exist. Many questions have been raised concerning the

---

*For example, E. G. Case of the AEC regulatory staff, in testimony at the Public Hearing for Operating Licenses for Prairie Island Units 1 and 2, and in an article in *Nuclear Safety* (volume 15, no. 3, May–June, 1974) presented that staff conclusion that the "likelihood of a sudden major LOCA accompanied by failure of the ECCS to cool the core to the degree necessary to prevent breach of the containment is so extremely small—i.e., less than one chance in ten million per reactor year—that the environmental risk of such an accident can be considered to be negligible."

The probability referred to by Case appears to be the product of an estimate of 10^{-3} to 10^{-5} per reactor-year for a sudden major LOCA, and 10^{-3} to 10^{-4} for the likelihood of ECCS failure severe enough to lead to containment breach.

In the WASH-1400 report, on the other hand, core melt, followed by containment breach, was estimated to have a much higher probability, namely, about 1 in 20,000 per reactor-year.

results in the WASH–1400 report (NRC 1975a), and one important one is: Are there undiscovered accident sequences having large probabilities? More specific questions concern the treatment in the report of common-mode failures, reactor aging, human errors, fires, earthquakes, and sabotage, among others, in ascertaining the probability of core meltdown. The use of median rather than mean values in obtaining best estimate results has also been questioned. And there is a considerable body of opinion that the stated uncertainty range of plus or minus a factor of about 5 in the core melt estimates has not been validated and is much too small.

Perhaps the most important questions raised with regard to the accident consequences presented in the report are the following: Is the modeling of containment failure modes and probability satisfactory, given a core melt? Is the estimate of health effects due to low-level radiation sufficiently conservative? Has the efficiency of evacuation and decontamination procedures been over-estimated? Should not the risk be given for each individual site rather than some weighted average? Are the potential effects on drinking water and other long-term radiation effects treated adequately?

The ACRS sent a relatively concise letter, dated April 8, 1975, on the WASH–1400 report (NRC 1975a) to NRC Chairman William A. Anders. On July 14, 1976, and on December 16, 1976, the ACRS sent letters to Congressman Morris K. Udall, Chairman of the House Subcommittee on Energy and Environment, Committee on Interior and Insular Affairs, in reponse to his request for comment on 11 issues. The letters to Anders and Udall are reproduced in Appendices A, B, and C of this chapter.

One ACRS opinion was that the uncertainty in the results was larger than that assigned in the WASH–1400 report. This was a frequent comment by those critiquing the report, including the Risk Assessment Review Group (or Lewis committee) (NRC 1978a), which was established by the Nuclear Regulatory Commission in response to a request by Congressman Udall for a reevaluation of the executive summary to the WASH–1400 report (NRC 1975a). This review committee found that the error bounds on the absolute probabilities of accident sequences were, in general, greatly understated, in part because of an inadequate data base in many cases, in part because of an inability to quantify common cause failures, and in part because of some questionable methodological and statistical procedures. As to the probabilities of accidents, the Lewis committee was unable to determine whether they were high or low, but cited factors working in both directions. The Lewis committee also criticized the executive summary of the WASH–1400 report as not adequately representing the full report and found that the peer review process had been defective in many ways. The Lewis committee also noted that the WASH–1400 report was a conscientious and honest effort to apply advanced risk assessment methods to reactor accidents, and that, despite its shortcomings, it provided the most complete single picture of accident probabilities associated with nuclear reactors available at that time. The Nuclear Regulatory Commission accepted the findings of the review committee.

Some personal comments on the WASH–1400 report (NRC 1975a) follow:

(1) If the WASH–1400 report was correct in its assessment that transients and small LOCAs are the dominant contributors to the probability of core melt in current LWRs, the emphasis placed on pressure vessel and primary system integrity and on improved ECCS since 1966 was effective to the extent that it reduced the probability of a large LOCA and provided an ECCS to cope with large pipe ruptures. However, the WASH–1400 report suggested and the Three Mile Island accident showed that insufficient attention had been given to complex transients, to small LOCAs, and to reliable decay heat removal in design, in analysis, and in operations.

(2) The WASH–1400 report did not adequately flag the safety problem that can arise when failure of one system causes failure of an entirely different system. Similarly, it did not call adequate attention to problems that might be initiated by the failure of control systems, which are not usually reviewed by the regulatory staff. And while the WASH–1400 report treated human error, it did not give adequate attention to correlated or sequential operator errors. The report failed to account for design errors in any meaningful way.

The study of the way fires contribute to risk, a very difficult subject to analyze, is inadequate in the WASH–1400 report. As is noted in Chapter 17 on seismic safety, the WASH–1400 report grossly underestimated the potential of earthquakes to risk from LWRs. And it may have missed out on floods.*

An important defect in the report was a failure to allow adequately for the possibility that the two reactors analyzed were not sufficiently representative of the reactors then in or near operation. Application of the methodology used in the report to other reactors during 1979 and 1980 has demonstrated that large variations exist among LWRs both in the reliability of systems important to safety, and in the probability of a serious accident arising from specific failures.

(3) The WASH–1400 report illustrated clearly that there are many different individual accident paths having the potential to cause core melt,

*The WASH–1400 report did not deal with the influence of sabotage on risk from LWRs, and a separate section on sabotage considerations has not been included in this manuscript for various reasons. In particular, it is inappropriate to discuss in such a public document any specific avenues that a prospective saboteur might take. Although sabotage was pointed out as a potentially important safety consideration as early as 1950 in the WASH–3 report (AEC 1950), it only began to receive continuing attention and emphasis in the regulatory process in the early 1970s. Since then it has been a rapidly developing aspect of reactor safety. However, it remains a matter which is very difficult to quantify for purposes of overall risk evaluation.

only some of which are peculiar to the design of a specific reactor. Hence, unless a few serious accident scenarios have significantly larger probabilities than any predicted in the report, it will be difficult to reduce the risk by a large factor, say 100, by trying to reduce the probability of those few events more likely to cause core melt for a particular reactor. If the few leading accident initiators are made far less probable by design changes, a large number of other initiators become candidates for importance. Also, because of the uncertainties arising from failures due to common or related causes, it is frequently difficult to accomplish major reductions in accident probability with high confidence, once the probability is already low.

(4) The WASH-1400 report demonstrated fairly unequivocally that accident risks from reactors arise from class 9 accidents, even though the AEC and NRC environmental assessment of radiological risks included only accidents up to class 8. The AEC presumably chose to do this because it thought the probability of class 9 accidents is sufficiently low to render their effect negligible and despite the fact that class 9 events are, with rare exception, not treated in safety analysis reports or staff safety evaluations. (Appendix D to this chapter explains the classification of nuclear accidents.)

(5) The effect of routine releases of radioactive liquid waste products or of small accidental spills on drinking water supplies or other aquatic resources has been reviewed routinely. However, only on rare occasions prior to 1974 was the potential effect of a core melt on a large water resource discussed even informally as part of a power reactor licensing review. One such occasion was in connection with the proposed siting of an LWR on the edge of a relatively small, poorly flushed, important lake; for other reasons this review was never completed and the issue did not receive great emphasis.

Presumably, the prevailing thinking was that to the extent that a risk existed from core melt, it was the airborne radioactivity and not the liquid pathway which was dominant.

In commenting on the WASH-1400 report, the Department of Interior raised questions concerning the potential effect on water resources of the solidified residue of a molten core which has penetrated the containment foundation after the unlikely event of core melt. This question has not been posed as part of the licensing process for land-based reactors but has become part of the regulatory review of the floating nuclear plant, on the recommendation of the ACRS. A limited study of the question was performed for the WASH-1400 report, and more extensive studies have since been made. In 1979 a regulatory staff task force recommended that new siting criteria include hydrological considerations of serious accidents.

(6) The WASH-1400 report suggested a low-average risk for all United States sites. However, it did not provide specific evaluation of the risk for

the most-populated sites and for the very remote sites. Such a comparison would have provided additional input into the judgment of site acceptability for future reactors, as well as more insight into the existing risk.

(7) Criticism persists of some of the methods used in the WASH-1400 report for calculating the health effects of accidents involving a large release of radioactive material. In addition, recent studies of the course of some core melt scenarios in an LWR have suggested an increased potential for failure of the containment building due to over-pressurization, with a larger release of radioactivity than was crudely estimated in the WASH-1400 report. All in all, the average consequences, given an accident, may be more or less severe than the best estimates given in the report, and it is difficult at this time to judge whether such a shift lies outside the estimated error band of a factor of five. It seems likely that the estimated costs of property damage from severe accidents are too low in the WASH-1400 report.

(8) The WASH-1400 report reinforced, and the Three Mile Island accident confirmed, a point of view previously held by many ACRS members that the single-failure criterion is useful but is not necessarily adequate. The report suggested several specific systems for which the single-failure criterion may have provided less reliability than desirable. In 1979, the regulatory staff agreed that some modification of the single-failure criterion was necessary. The revision may well appear in the form of a requirement for supplementary quantitative reliability analysis, plus an evaluation of the consequences of system failure. Such additional information could provide a basis for judgment of the acceptability of the single-failure criterion, system by system.

(9) The ACRS in its report of April 8, 1975, to NRC Chairman Anders and its letter dated July 14, 1976, to Congressman Udall, recommended that the methodology of the WASH-1400 report be applied to improve the safety of reactors in operation and under construction. The risk assessment review group repeated the recommendation (NRC 1978a). However, it took the Three Mile Island accident before this advice was adopted by the regulatory staff. The first results obtained by such studies confirm that there are substantial gains in safety to be made. They also provide considerable confirmation to those, including me, who believed that the actual probability per reactor-year of core melt for existing reactors is significantly larger than the "best estimate" provided by the WASH-1400 report.

(10) The studies in the WASH-1400 report into failure of containment buildings, coupled with later studies, showed that different containment designs (e.g., the large dry PWR, the ice condenser, and the Mark I BWR pressure suppression types) may seem to lead to similar off-site doses for the standard design basis accidents, yet be quite different in their protective capability againt core melt. The WASH-1400 report led to some effort to

study containment modifications which could provide a reduction in risk from core melt. However, it took the Three Mile Island accident before serious attention was given by the NRC to the possibility of mitigation of the consequences of core melt. In my opinion, priority should be given to developing and applying to existing and future LWRs practical changes in containment design which provide a significant improvement in the capability of the containment to reduce off-site consequences from an accident involving major core damage or core melt.

(11) The WASH-1400 report vividly showed the inconsistencies in the methodology of the 10 CFR Part 100 site criteria. I believe that a major change in the licensing process is appropriate, one in which the full spectrum of possible accidents is evaluated probabilistically, and quantitative risk criteria are employed judiciously. In a letter dated May 16, 1979, to NRC Chairman Joseph N. Hendrie, the ACRS recommended the development of quantitative risk acceptance criteria for light water reactors. This was at a time when the regulatory staff was still shying away from the quantitative use of probabilistic methodology, following disavowal by the NRC commissioners in early 1979 of any previous NRC endorsement or use of the WASH-1400 results. Since the ACRS recommendation, there has been increased interest by the NRC in the development of such quantitative criteria. While I strongly favor such efforts, it is relevant to point out the potentially large uncertainties which are inherent in most estimates of low-probability hazards. This will require prudence both in the performance and use of such calculations and will make it difficult to devise workable and defensible quantitative risk acceptance criteria whose application, in practice, do not require large elements of judgment. The development of quantitative risk criteria faces many other difficult questions, not the least of which is whether to count one improbable large accident, say involving 1,000 deaths, as equivalent to, or worse than, 1,000 accidents, each involving 1 death.

Despite these and other difficulties, I favor the development and use of quantitative risk criteria as an important input into the decision-making process regarding acceptability for all technologies. Such criteria may well vary in form and accepted risk level among technologies. This should be a societal decision approved explicitly or implicitly by the United States Congress. For each technology the level of accepted risk and the uncertainties inherent in the estimates of such risk should be fully displayed.

Some Legalistic Aspects of Siting

The authors of the WASH-1400 report (NRC 1975a) concluded that such risk as may ensue from accidents in light water reactors clearly arises from "worse than Part 100"* accidents in which core melt and a loss of containment integrity are involved. Review of the WASH–1400 results by Cave and Ilberg (1977) led them to conclude that the containment building as currently designed in LWRs reduces the overall risk, when averaged over all accidents, by only about a factor of 10 from that of the uncontained reactor (on the assumption that the accident probabilities and consequences in the WASH–1400 report are correct). The actual factor may, of course, be substantially different from 10. At Three Mile Island the presence of a containment building reduced the off-site dose by several factors of 10. Nevertheless, there is reason to doubt that the requirement for very small containment leak rates and the safety features which are added to make a reactor fit 10 CFR Part 100 at sites having a small exclusion area and/or small distance to the outer edge of the low-population zone have a major effect on the average risk.

The continued existence of 10 CFR Part 100 and the legal interpretations of its words by ASLB hearing boards and other judicial bodies have resulted in situations which are seemingly anomalous from the point of view of public safety. The matter is well illustrated by reviewing the decision of April 7, 1977, of the Atomic Safety and Licensing Appeal Board in the matter of New England Power Company (New England Power Units 1 and 2) and Public Service Company of New Hampshire (Seabrook station, Units 1 and 2).

The appeal board interpreted 10 CFR Part 100 in such a way as to rule out any requirement for emergency preparedness measures outside the low-population zone, despite recommendations to the contrary by the ACRS and the regulatory staff, and despite the fact that the authors of the WASH–1400 report (NRC 1975a) assume that both evacuation and interdiction measures are effective beyond the low-population zone, and thereby calculate a significantly reduced risk. In the words of the ASLB appeal board, "The regulations require a showing of the possibility of evacuation only from the low population zone."

The other anomaly arises in the way both the regulatory staff and the licensing board have reacted to a change in population distribution around a reactor (e.g., an increase due to new housing development, as at St. Lucie Unit 1 in Florida). The staff found that there was a reduced population center distance and low-population zone, and required those additional measures needed to meet 10 CFR Part 100, assuming the usual recipe for large fission product release to an intact containment building. How much real reduction in risk was accomplished by these changed requirements was not ascertained.

---

*Accidents leading to an off-site radioactive dose exceeding the guidelines limits in 10 CFR Part 100.

The hearing boards have reinforced this approach by the staff. If the population growth is large, they "maintain public safety" by requiring the use of a reduced distance to the outer edge of the low-population zone in calculating the radioactive doses in accordance with the recipe in 10 CFR Part 100.

Much of this problem appears to have arisen from an unwillingness, and perhaps a legal inability under the existing rules and regulations, of the regulatory staff to treat class 9 accidents beyond making an evaluation that such accidents are of sufficiently low probability to be acceptable, i.e., to be neglected, in meeting 10 CFR Part 100.

The approach which involved neglecting class 9 accidents was officially adopted by the AEC in its standard guidelines for evaluating environmental impact of reactor accidents (AEC 1971b). In a press release of November 29, 1971, the AEC position was "Class nine and some class eight accidents would not have to be evaluated because the probability of such an accident occurring is so small that the environmental risk is quite low."

Further insight into the regulatory staff approach to the role of class 9 events in reactor licensing comes out of the record of the ASLB hearing in 1977 on the Black Fox reactors in Oklahoma. During a pre-hearing conference, a member of the ASLB raised a question as to the extent to which class 9 accidents can be dealt with in a hearing. The counsel for the NRC staff filed a memorandum which took the position that class 9 accidents are not generally excluded. However, the staff recommended and the ASLB agreed that class 9 accidents need not be considered if they are not "credible." "Credibility," however, was ill defined, if at all, in the NRC staff memorandum. The ASLB supported the staff with highly qualitative reasoning.

The AEC commissioners also appeared to pursue a qualitative approach, similar to that of the regulatory staff, in a ruling on Indian Point 2 in 1972. The commissioners ruled that compliance with the standards was not automatically sufficient to foreclose further inquiry into the matter of vessel failure in the course of a licensing proceeding. However, they supported the staff position that protection against the consequences of vessel failure need not be required for a particular facility "unless it has been determined that for such facility there are special considerations that make it necessary that potential vessel failure be considered." No basis for deciding whether the matter need be considered for a particular facility was given in the decision.

Historically speaking, the statement of considerations which accompanied publication of a proposed version of 10 CFR Part 100 on February 11, 1961 (see appendix A to Chapter 4), indicated that one of the basic objectives was to assure that "even if a more serious accident (not normally considered credible) should occur, the number of people killed should not be catastrophic." This stated objective was modified in the statement of considerations which accompanied publication of the effective 10 CFR part 100 on April 12, 1962 (see appendix B to Chapter 4). It was there stated that an underlying objective was to assure that "the cumulative exposure dose to large numbers of people as a consequence of any nuclear accident should be low. . . . " Furthermore, the population center distance criterion was included in 10 CFR Part 100.11 (a)(3) "to provide for protection against excessive exposure doses to people in large centers, where effective protective measures might not be feasible" in recognition that "accidents of greater potential hazard than those commonly postulated as representing an upper limit are conceivable, although highly improbable." Thus, an accident seems to be regarded as "credible" for some purposes and "non-credible" for others. Part 100 contains no definition of the term "credible." The commission itself has never specified the probability of occurrence which separates "credible" accidents from "non-credible" accidents, but the commission's staff has taken the position that the figure 10^{-7} per reactor-year should generally be used for this purpose (see the WASH-1270 report, *Anticipated Transients Without Scram for Water-Cooled Power Reactors*, AEC 1973a).

One of the conclusions of the WASH-1400 report (NRC 1975a) was that the probability of core melt was 1 in 20,000 per reactor-year, a number considerably larger than the probability used by the regulatory staff in the past. Perhaps because the staff had believed the probability was much lower, core melt accidents followed by containment failure had not been regarded in the past as "credible" accidents within the meaning of Part 100. However, the WASH-1400 report raised a question as to whether these accidents should continue to be regarded as outside the plant design basis and "non-credible."

Thus it can be argued that staff implementation of Part 100 has had certain deficiencies—deficiencies associated with failure to consider "non-credible" accident consequences where very large cities are involved, and to evaluate routinely certain core melt accidents.

In 1977 and 1978, the question of a potentially explicit role for class 9 accidents in reactor siting began to receive increasing emphasis within various sectors of the NRC, including the licensing staff, the ACRS, and the NRC commissioners themselves. A trend toward some explicit use of class 9 events was indicated when the staff recommended measures to delay or prevent a molten core from getting into the ocean waters by penetration of the base of the floating nuclear plant. This trend was greatly reinforced by the Three Mile Island accident. The NRC staff recommended rulemaking by the NRC on possible requirements to mitigate accidents involving highly degraded or molten cores. The ACRS recommended that the NRC require studies by each reactor owner on the possible use of containment features to mitigate core melt accidents. The NRC required such studies of Indian Point and Zion, the two most densely populated reactor sites in the United States.

As of the middle of 1980, the NRC commissioners were beginning to face the task of formulating policy on this major issue.

A Few Comments on the Three Mile Island Accident

The accident at Three Mile Island occurred in a 900 MWe pressurized water reactor on March 28, 1979. The accident was initiated by a loss of normal feedwater flow to the steam generators, an event which is expected to occur once or twice a year. The resulting increase in primary system pressure led to opening of a power-operated relief valve and to a reactor scram. The emergency feedwater pumps started but could not immediately supply make-up water to the secondary side of the steam generator, because, contrary to the reactor technical specifications, a pair of valves were closed. This led to dryout of the secondary side of the steam generator; that situation was recognized and remedied after about eight minutes. More importantly, the power-operated relief valve, which is located on top of the pressurizer in the primary system, stuck open, creating a small leak, even though the actuating mechanism which permits valve closure operated and signaled to the control room that the valve was closed. During the next 60 to 90 minutes, the reactor operators limited the flow which would have normally been automatically supplied by the high-pressure emergency core-cooling system to make up for the water lost out of the open valve. In part, the operators' action appears to have been influenced by their not having recognized that, for such a stuck-open relief valve, the reactor pressurizer could be nearly full of fluid while the level of fluid was dropping dangerously within the core. Had they been able to interpret the available symptoms correctly, they could have quickly closed another valve, ending the entire event.

As the inventory of water in the primary system was depleted, the primary system pumps began to vibrate. At 75 minutes into the accident the operators turned off one set of primary system pumps and at 100 minutes they turned off the other set of primary system pumps. At this point cooling of the core fuel elements rapidly decreased, and considerable overheating of the cladding resulted. A chemical reaction between steam and much of the Zircaloy cladding led to the generation of substantial amounts of hydrogen, some of which escaped into the containment building and some of which stayed in the primary system.

The operators took various measures to restore cooling water to the core and at sixteen hours into the accident this was accomplished and successfully maintained thereafter.

A large fraction of the gaseous fission products in the fuel escaped into the reactor containment building, and some of the more volatile fission products escaped from the fuel and got into the reactor coolant. The bulk of the radioactivity remained in the containment building, although a relatively small fraction escaped into the auxiliary building, together with some primary system water.

Some of the hydrogen which escaped into the containment building burned rapidly, generating a pressure pulse well below the containment design pressure at about 10 hours into the accident. The hydrogen remaining in the primary system during the first several days introduced an unnecessary concern with regard to the possibility of its rapid combustion within the primary system. The hydrogen did represent a potentially important means of disrupting continued core cooling, however, and considerable care was exercised to remove it, thus enabling much greater assurance of abundant cooling of the damaged core.

The Three Mile Island (TMI) accident was unusual, surprising, and anomalous in many ways. It involved a considerable number of matters in which human performance was less than perfect, including design, maintenance, operation, and regulatory review. The core was damaged far more severely than should occur for any of the so-called design basis accidents, and far more fission products escaped from the primary system than would be expected from the "realistic" estimate given in the environmental impact statement prepared by the regulatory staff.

Although some radioactive water and gas was unintentionally released from the containment building to the less well-guarded auxiliary building, the total radioactivity release and the off-site doses were very small, much less than those in the 10 CFR Part 100 site criteria (by a factor of perhaps 250), and much less than would have occurred with no containment.

The Three Mile Island accident, while not statistically a determining factor, provides considerable support to those who believe that the frequency of an accident causing very severe damage or core melt is higher than the median value of 1 in 20,000 per reactor-year presented in the WASH–1400 report (NRC 1975a). There is little doubt that the Three Mile Island accident will act as a catalyst to the implementation of many improvements in reactor safety, particularly those arising from lessons easily drawn directly from this experience. Some of the more significant improvements in LWR safety may arise from the great emphasis placed, since the accident, on the use of probabilistic methodology to look for weak points in the design of all reactors in operation or under construction.

However, there are many possible paths to an accident which could severely damage or melt the core of a light water reactor, and the residual probability of such an accident may not be drastically reduced by these improvements, at least not to a frequency less than the estimates given in the WASH–1400 report (NRC 1975a).

Hence, if further improvements in LWR safety are to be sought, it will probably be necessary to apply new design features to mitigate the consequences of accidents involving core melt.

In the second half of 1979, the regulatory staff recommended that the NRC institute rule making on such possible requirements for all LWRs. And, of still greater historical interest is the fact that in the fall of 1979, 13 years after the AEC turned aside an ACRS recommendation for such mitigation measures, the NRC commissioners and staff initiated steps to require a special safety approach for the most highly populated approved sites, such as Indian Point and Zion, with considerable emphasis on possible design measures to mitigate core melt accidents. In the eyes of the NRC all approved reactor sites were no longer equal, except as they posed requirements arising from 10 CFR part 100 reactor site criteria. The NRC had finally arrived at a long-standing ACRS position.

Risks from Nuclear and Other Energy Sources

The principal conclusions on reactor safety presented in the Ford Foundation—Mitre Corporation study, "Nuclear Power Issues and Choices," provide an interesting perspective on the issue of LWR safety (Ford-Mitre 1977). In brief, the conclusions of this study, which was published in 1977, were as follows:

(1) Neither the excellent safety record of nuclear power reactors to date nor the history of "abnormal occurrences" yields statistical data for predictions covering the rest of the century.

(2) Although it is a valuable resource for study of reactor safety, the WASH–1400 report (NRC 1975a), should not be used as a definitive guide for policy since it understates the uncertainties and has serious methodological deficiencies.

(3) On an average rate-of-loss basis, nuclear power compares favorably with coal even when the possibility of accidents is included.

(4) The adverse health and property consequences of even an extremely serious nuclear reactor accident would not be out of line with other major catastrophes that our society has been able to handle without major long-term impact.

(5) The reasonable upper bound on the possibility of an extremely serious nuclear accident is not in itself unacceptable.

With regard to the last point, the Ford-Mitre study (1977) estimated an upper bound of about a 25% chance of an extremely serious accident occurring in 5,000 reactor-years. The report stated that the upper bound is extremely unlikely, probably at least by a factor of 10 or 100.

I thought that these conclusions of the Ford-Mitre study were reasonable in 1977, and the TMI accident has not changed my opinion. While the safety of nuclear power can be improved, all of the information at my disposal strongly supports the conclusion that nuclear power imposes risks to the most affected individual or to society which are comparable to or less than those from coal or oil as an electricity source, and it imposes far less risk than most other technologies or large-scale industrial or agricultural activities today. This conclusion is supportable when one examines the potential for low-probability–high-consequence events, such as failure of a dam or a large chemical storage tank. It is also supportable in a comparison of nuclear with most, if not all, proposed future energy sources, if the risk accounting includes allowance for those risks incurred in the acquisition of materials for construction and their fabrication and for backup energy sources, and if a realistic design is used for the evaluation.

Society is not risk-free, and it is not practical to eliminate all risks from energy sources or from other aspects of society, ranging from the production of food and industrial goods to the practice of medicine. At some level of risk from energy sources, other risks in society will not only be greater, but also will be more amenable to reduction for a similar expenditure of our limited resources. The storage of hazardous chemicals, the risk from earthquakes in our cities, and the potential carcinogenic and mutagenic hazard from pollutants released from industry and agriculture afford a few examples of the many risks in society.

Concluding Remarks

In writing this history, I hoped that by examining several aspects of light water reactor safety in varying detail I could illustrate, by example, the way in which safety evolved and continues to evolve. Such a treatment inevitably leads to a discussion of differences of opinion, of changes in specific policy, of errors and omissions in technical judgment, and of imperfections in the reactor licensing process. It must be recognized that such phenomena are a normal part of any regulatory process for a complex system. It is the broad record of accomplishment and the overall integrity of the process which should provide the basis for evaluation of its success.

In retrospect, the conscious policy of trying to make nuclear power reactors safer than other industrial or technological enterprises, a policy which was adopted in the 1950s and which was spelled out in the ACRS letter to AEC Chairman McCone of October 22, 1960, was particularly vital and long-lasting. Over the years the detailed approach toward implementation of this policy varied among the regulatory groups and even within a single entity such as the ACRS. However, this continuing policy provided sustained general guidance, and I believe it is equally applicable today.

The complex history of reactor siting records several major changes in technical approach by the AEC. Nevertheless, despite several efforts in the 1960s by the nuclear industry to introduce urban and even truly metropolitan siting in the United States, the Indian Point site, which was approved in 1955–1956, remains the most densely populated site approved for a large power reactor. The reasons for failure of the industry to gain approval of more densely populated sites varied from case to case. Although prior to 1966 containment was viewed as a independent protective bulwark against most accidents involving core melt, the various regulatory groups were unwilling to approve metropolitan siting at that time, at least for the reactors proposed.

With recognition of the interrelation between core melt and containment failure in 1966, a revolution in LWR safety and licensing occurred. Although the 10 CFR Part 100 site criteria still remained as an AEC regulation to be satisfied, the major emphasis shifted from containment and its associated engineered safeguards to reducing the probability of occurrence of potentially severe initiating events, and to preventing core melt should an event occur. Some interest in metropolitan siting continued to exist within the nuclear industry, but not as much as before. And, with recognition of the problems associated with core melt, the AEC position tended to move away from favorable consideration of such sites.

Within the AEC, some differences remained between the philosophic approaches of the ACRS and the regulatory staff on whether increased safety measures were appropriate for the most densely populated sites receiving approval (beyond the somewhat artificial requirements imposed by Part 100). The staff tended to treat all acceptable sites as equal, while the ACRS tended to try to balance the increased numbers of people at risk with additional safety features. The staff's "black or white" approach may have arisen, in large part, from legal constraints, namely, that all reactors that were approved had to meet the existing rules and regulations which differentiated among sites only through Part 100.

In retrospect, the results of consequence studies like those in the WASH–1400 report (NRC 1975a), show large relative differences in the risk between sites like Indian Point and Zion and the more remote sites. Of course, if the risk is acceptably low at the most densely populated site which has been approved, then it is still lower at other sites; and one can argue that the staff approach was appropriate. However, some philosophic questions then arise concerning the basis by which still more densely populated sites were rejected.

One trend that emerges from the historical review is the general reversal in relative conservatism between the ACRS and the regulatory staff with regard to seismic safety requirements. In the late 1950s conventional industrial seismic design practice was applied to the eastern reactors. In the early 1960s the utilities and their seismic consultants first proposed seismic design bases for reactors in California similar to those previously used for fossil-fueled electricity generating stations in the same area. Then, during the review of the Bodega Bay, Malibu, and San Onofre sites, the applicant's originally proposed seismic design basis was made more conservative. This seemed to occur more because of the initiative of the regulatory staff and the advice of its consultants than because of the ACRS. Of course, the staff ended up opposing the Bodega Bay site while the ACRS wrote a report favorable to its construction. (In retrospect, both the staff and the ACRS were probably non-conservative in their decision on acceptable vibratory motion.)

Similarly, for the Connecticutt Yankee reactor in 1965, the ACRS seemed to feel that the seismic requirements recommended by the staff and its advisors might be excessive.

After 1966, when differences on seismic safety design bases arose between the staff and the ACRS, the ACRS was usually though not always, on the more conservative side. Why this change occurred is not obvious. It might be due to changes in personnel both on the staff and on the ACRS. It might be due to the fact that the staff formed an in-house capability and called less often on the United States Geological Survey for advice. Or it may be that before 1966, containment was receiving ACRS emphasis while after 1966 the emphasis changed to reducing the probability of an event which could lead to an accident involving core melt.

The various generic safety issues have been resolved in very different ways. The steam-line break issue, ATWS, and the matter of fires provide three illustrative examples. Fire had been recognized as a potential safety concern of considerable importance for at least a decade before occurrence of the Browns Ferry fire in 1975. There were large differences in opinion concerning the magnitude of the threat to safety and in the measures needed to make the probability of a serious accident from this cause acceptably low. Significant fires did occur at the San Onofre and Indian Point reactors, and these led to some changes in specific requirements. In addition, evolutionary improvements were being made in the LWRs under construction, and an industry standard (later shown to be inadequate) was developed. In retrospect, an early clean-cut decision to take design measures with regard to limiting fire damage would have been preferable to the actual course of action. Fortunately, no harm to the public health and safety resulted from the semi-empirical process of learning which was followed. This culminated in the Browns Ferry fire, which resulted in changed requirements.

The steam-line-break issue represents almost the opposite extreme in regulatory action. When it was recognized in 1972 that the existing design basis for high-energy steam and other process lines outside containment buildings did not include gross rupture of the largest pipes, the regulatory staff promptly initiated a program which required appropriate changes in existing plants and established new design requirements for those to be constructed. No such accident (that is, rupture of a large steam line) had occurred in a reactor. However, it was judged and judged quickly that all plants had to be protected against such an eventuality.

For obscure reasons, ATWS has had a very different history. Not long after the matter was identified, the staff judged that scram unreliability was unacceptably high. A quick solution was indicated as needed at least for BWRs where the safety concerns seemed to be relatively well defined. Not too long afterwards, at least some of the PWR designs seemed to have unacceptably high primary system pressures for ATWS events.

In the period 1970–1971, the ACRS exercised some delaying effect on regulatory action while more detailed technical information was being developed on both the effects of an ATWS event and possible remedial measures. Since 1972, the regulatory staff has been struggling to try to reach a firm position, while more and more reactors have gone into operation and many more have been designed and constructed without including measures to accomodate ATWS, or even the flexibility to incorporate such measures readily.

Fires, steam line breaks, and ATWS all represent potential safety concerns whose probability is and was very difficult to assess, let alone quantify with confidence. It is possible that, in the future, essentially all the reactors will be judged to have been adequately designed without additional provisions for any of these matters. Or, it may be determined in the future that still more provisions would have been preferable.

The absence of a quantified risk acceptance criterion, as well as the lack of ability to quantify the risk from such events with high confidence, leaves engineering judgment as the means for resolution of such matters.

That a major effort has been made to identify, review, and judge such safety problems in advance of the occurrence of an accident having severe effects on the public health and safety, is relatively unique in the regulatory field. Most technological ventures have approached safety empirically, with corrections made after the occurrence of one or more bad accidents leading to many fatalities. So, if the process had been imperfect, at least it has existed.

Chapter 19: Appendix A:
ACRS Report to the AEC on Reactor Safety Study, WASH-1400

Since the release of the draft Reactor Safety Study, WASH-1400 (RSS) in August 1974, the Advisory Committee on Reactor Safeguards has been reviewing the considerable body of information presented in the report, its appendices, and the comments received on it, giving primary attention to the potential implications of the draft report on the reactor licensing process. In its review, the Committee has had the benefit of Subcommittee meetings held on October 9, November 22, and December 20, 1974, and March 5, 1975, and of full Committee meetings held on October 10-12, October 31-November 2, November 14-16, December 5-7, 1974, and January 9-11, February 6-8, March 6-8, April 3-5, 1975.

The ACRS believes that the RSS represents a valuable contribution to the understanding of light water reactor safety in its categorization of hypothetical accidents, identification of potential weak links for the two reactors studied, and its efforts to develop comparative and quantitative risk assessments for accident sequences examined. The Committee believes that a continuing effort and better data will be required to evaluate the validity of the quantitative results in absolute terms. Special emphasis should be given to quantification of the initiators, probabilities, and consequences of core melting.

The Committee believes that the methodology of the RSS should be applied to other types and designs of reactors, other site conditions and other accident initiators and sequences, and that the current efforts to compile, categorize, and evaluate nuclear experience should be extended in breadth and depth to improve the data base for future studies of this type.

The Committee believes, further, that the RSS can serve as a model for similar studies of the failure probabilities, consequences, and resulting risks of other hazards (both nuclear and non-nuclear) to the health and safety of the public.

The Committee believes that many of the techniques used in the RSS can and should be used by reactor designers to improve safety and by the NRC Staff as a supplement to safety assessment.

The Committee's review of the RSS has not caused the Committee to alter its judgment that reactors now under construction or in operation do not represent undue risks to the health and safety of the public.

The Committee will continue to review the RSS and will comment further on it in the future.

Chapter 19: Appendix B:
ACRS Letter (July 14, 1976) to the Ho..orable Morris K. Udall

At its 195th meeting on July 8-10, 1976, the Advisory Committee on Reactor Safeguards (ACRS) considered the points raised in your June 14, 1976, letter on the

Reactor Safety Study (RSS, WASH–1400, NUREG 75/014). The ACRS reviewed the draft version of the Reactor Safety Study in late 1974 and early 1975 and submitted a report to the Nuclear Regulatory Commission on April 8, 1975. A copy of the ACRS report is attached.

Your letter identified eleven issues on which you requested comment and the Committee is pleased to respond to issues 1, 3, 4, 6, 8, 9 and 10. However, extensive time and effort would be required by the ACRS to respond adequately to the other topics and the needed effort would have to be factored into overall considerations of other ACRS functions, including mandatory review of applications for construction permits and operating licenses for commercial nuclear power plants.

The Committee's reponses follow:

1. "The extent that the NUREG 75/014 fault-tree analysis adds to understanding of the likelihood of major nuclear reactor accidents."

The ACRS believes that the fault-tree methodology used in the Reactor Safety Study to develop comparative and quantitative risk assessments for postulated accident sequences represents a valuable contribution to the understanding of the likelihood of major nuclear reactors accidents.

3. "Adequacy of data base for NUREG 75/014 type fault-tree analysis."

As noted in our report of April 8, 1975, the ACRS believes that a better data base will be required to evaluate the vaility of the RSS's quantitative estimates of the likelihood of low probability high consequence events, and recommends that current efforts to compile, categorize and evaluate nuclear and other applicable industrial experience be extended in breadth and depth to improve the data base for further studies of this type.

4. "Sensitivity of NUREG 75/014 conclusions to differences in reactor design, in site characteristics, in local meteorological conditions and in population distributions."

All of the factors noted above will some effect on the probability or consequences of a serious accident. The Committee has recommended that the methodology of the Study be applied to other types and designs of reactors, other site conditions and other accident initiators and sequences. If this is done, it will provide greater insight into the sensitivity of differing reactor designs and safety features.

6. "Adequacy of NUREG 75/014 methodology to take account of gradual degradation of plant safety over plant lifetime."

The Committee believes the methodology is capable of taking into account wear out of components and degradation of equipment over the lifetime of the plant but an appropriate data base needs to be developed.

8. "Need for periodic updating of NUREG 75/014 to take account of new data."

The Committee believes that a continuing effort is desirable in the application of the methodology developed by the Reactor Safety Study not only to factor in new data but also to consider design variations and new concepts.

9. "Need for continuing analysis of NUREG 75/014 for purposes of delineating areas of research and data collection."

The Committee believes that the NUREG 75/014 methodology should be used to aid in delineating areas for further research. Special emphasis should be given to quantification of the initiators, probabilities, and consequences of core melting.

10. "The extent to which NUREG 75/014 can be used to aid development of regulatory policies concerning design, construction, and operations."

The Committee has recommended to the NRC that many of the techniques used in the Study can and should be used by the reactor designers to improve safety and by the NRC Staff as a supplement to their safety assessment.

Chapter 19: Appendix C:
ACRS Letter (December 16, 1976) to the Honorable Morris K. Udall

At its 200th meeting, December 9–11, 1976, the Advisory Committee on Reactor Safeguards (ACRS) continued its consideration of the points raised in your June 14, 1976, letter on the Reactor Safety Study (RSS, WASH–1400, NUREG 75/014). The ACRS had previously considered these matters at its 196th and 199th meetings and had responded to issues, 1, 3, 4, 6, 8, 9 and 10 in its letter to you dated July 14, 1976. In its further consideration of the remaining four issues, the Committee had the benefit of meetings of its Reactor Safety Study Working Group with the Nuclear Regulatory Commission Staff in Washington, D.C., on October 12, 1976, and November 10, 1976,

The ACRS is continuing to evaluate the considerable body of information presented in the RSS report, its appendices, and the comments received on it, giving primary attention to the potential implications of the report for the reactor licensing process. This letter provides the Committee on Interior and Insular Affairs a brief resume of current ACRS thought on issues 2, 5, 7 and 11.

"2. Adequacy and appropriateness of analysis used in NUREG 75/014 for purposes of estimating the likelihood of low probability, high consequence events."

The ACRS believes that the methodology of NUREG 75/014 is useful for purposes of identifying important accident sequences and for attempting to develop comparative and quantitative risk assessments for low probability, high-consequences accidents. However, the ACRS believes that considerable effort by more than a single group over an extended period of time will be required to evaluate the validity of the results in NUREG 75/014 in absolute terms. Among the matters which will warrant emphasis in such an evaluation are the following: improved quantification of accident initiators; the identification and evaluation of atypical reactors; the influence of design errors; improved quantification of the role of operator errors; improved quantification of consequence modeling; and the development of improved data for systems, components and instruments under normal and accident-related environmental conditions in a nuclear reactor.

The ACRS believes that NUREG 75/014 represents a very considerable contribution to the understanding of reactor safety and provides a point of departure for quantitative assessment.

"5. Adequacy of NUREG 75/014 methodology to take account of multiple correlated errors in procedures, design, judgment, and construction such as those leading to the Browns Ferry fire."

The ACRS believes that the methodology of NUREG 75/014 is useful in accounting for that portion of the risk resulting from identifiable potential common mode or dependent failures, and can be used to search out the possibility of multiple correlated errors. However, the methodology cannot guarantee that all major contributors to risk will be identified, and a considerable element of subjective judgment is involved in assigning many of the quantitative input parameters. Both for nuclear and non-nuclear applications, for complex systems, where multiple, correlated failures or common cause failures may be significant, the record shows that investigators working independently will frequently make estimates of system unreliability which differ from one another by a large factor. At this stage of its review, the ACRS believes that a substantial effort may be required to develop and apply dependable methods for quantitatively accounting for the very large number of multiple correlated or dependent failure paths and to obtain the necessary failure rate data bases.

Whether multiple, correlated errors will dominate the overall risk, however, is subject to question, particularly if simpler postulated accident sequences are generally the dominant contributors to the likelihood of system failure.

"7. Extent to which the final version of NUREG 75/014 takes into account comments on the draft version."

The ACRS is in the process of reviewing the disposition of selected comments received by the Reactor Safety Study Group, particularly as they have implications for short or long-term improvements in reactor safety. The ACRS plans to continue this type of activity; however, it is beyond the scope or available working time of the ACRS to review in detail the extent to which the final version of NUREG 75/014 takes into account the comments received.

"11. Validity of NUREG 75/014 conclusions regarding accident consequences."

As stated in its report to you of July 14, 1976 and as indicated in its reponse to other questions in this group, the ACRS believes that considerably more effort on the part of the various contributors is needed to evaluate the quantitative validity of NUREG 75/014 conclusions regarding accident consequences. Based on information currently available, the ACRS would assign a greater uncertainty to the results than that given in NUREG 75/014.

The ACRS believes that the past and current practice of trying both to make accidents very improbable and to provide means to cope with or ameliorate the effects of accidents has been the correct approach to nuclear reactor safety.

The ACRS review of the Reactor Safety Study has not caused the ACRS to alter its judgment that operation of reactors now under construction or in operation does not represent an undue risk to the health and safety of the public. The ACRS believes that NUREG 75/014 has suggested many fruitful areas for study and evaluation for potential improvements in light water power reactor safety. The ACRS also believes that the extension of such risk assessment methodology to the total spectrum of activities involved in the production of nuclear power and in the production of electric power by other means, as well as to other technological aspects of society, could add significantly to our overall understanding of risk.

Chapter 19: Appendix D
Classification of Nuclear Accidents

In response to the requirements of the National Environmental Protection Act (NEPA) the Atomic Energy Commission in 1971 published 10 CFR Part 51

Licensing and Environmental Policy and Procedures for Environmental Protection (AEC 1971b). A major part of this document relates to the identification of nine classes of accidents. These ranged from trivial accidents, falling in class 1, through accidents of moderate frequency and modest consequences, say class 5, up to accident intiation events considered in the design basis evaluation in the regulatory staff safety analysis report (class 8). Hypothetical sequences of failures more severe than those postulated in class 8, sequences which could lead to a release of radioactivity having consequences exceeding the guideline doses in 10 CFR Part 100, were designated class 9. The types of accidents falling in the various classes are illustrated in table 3.

Another important aspect of 10 CFR Part 51 was the use of realistic rather than conservative fission product releases in assessing the environmental effects. Thus, in a large LOCA, rather than assuming the fission product release to the containment specified in 10 CFR Part 100, only a small fraction of this quantity is estimated to escape from the fuel into the containment building.

In a policy paper dated November 15, 1971, to the commissioners proposing an AEC approach to the preparation of environmental reports, the regulatory staff estimated that for the accident in which an irradiated fuel bundle is dropped during its removal from the reactor (provided as a typical example), the realistic assumptions to be used in environmental reports would lead to an off-site dose 100,000 times smaller than that estimated in the safety evaluation report. Of more interest is the fact that the staff estimated the probability of hypothetical class 9 accidents leading to substantial core meltdown as 10^{-8} per reactor-year.

TABLE 3

Classification of Postulated Accidents and Occurrences

No. of Class	Descriptions	Example(s)
1	Trivial incidents	Small spills Small leaks inside containment
2	Miscellaneous small release outside containment	Spills Leaks and pipe breaks
3	Radwaste system failures	Equipment failure Serious malfunction or human error
4	Events that release radioactivity into the primary system	Fuel failures during normal operation Transients outside expected range variables
5	Events that release radioactivity into secondary system	Heat exchanger leak
6	Refueling accidents inside containment	Drop fuel element Drop heavy object onto fuel Mechanical malfunction or loss of cooling in transfer tube
7	Accidents to spent fuel outside containment	Drop fuel element Drop heavy object onto fuel Drop shielding cask— loss of cooling to cask Transportation incident on site
8	Accident initation events considered in design basis evaluation in the safety analysis report	Reactivity transient Rupture of primary piping Flow decrease—steam-line break
9	Hypothetical sequences of failures more severe than class 8	Successive failures of multiple barriers normally provided and maintained

Appendixes

References

Index

Appendix I:
A Brief Discussion of
Serious Accidents in
Light Water Reactors

In a boiling water reactor or BWR, steam is generated directly in the reactor core as the cooling water flows by the thousands of fuel elements, each of which consists of 10- to 12-foot long tubes of Zircaloy, about ½ inch in diameter, stacked with pellets of uranium dioxide. (The pellets resemble large aspirin pills.) The steam, which has a temperature of about 550° F and a pressure of 1,000 psi, flows to a turbine where it gives up about one-third of its energy driving the electric generator which is coupled to the turbine. The steam then goes to a condenser where it gives up the remaining two-thirds of its heat to a waste heat sink, such as river or ocean water. After the steam is condensed, it is pumped back into the reactor pressure vessel as feedwater to begin the process again. Most of the water going through the core does not boil. It is recirculated through the core, together with the feedwater, by large recirculating pumps. In a boiling water reactor, a decrease in pressure or a decrease in flow rate through the core tends to increase the amount of steam being formed, and this in turn tends to reduce the power. Conversely, an increase in pressure, such as might occur if the isolation valves in the steam lines were all to close due to some signal such as excess radiation in the steam, tends to increase the power, requiring that the control rods be rapidly inserted.

In a pressurized water reactor or PWR, the temperature of the cooling water flowing through the core is about the same as in a BWR. However, the pressure is maintained at a much higher level, about 2,200 psi, to prevent the water from boiling as it is heated in the core. The heated water flows into another large pressure vessel, called the steam generator, where the hot water from the core flows through many steel tubes which are surrounded by water from the secondary system. The secondary system water takes heat from the primary system water and boils in the process. The secondary system steam then goes to a turbine followed by a condenser, where it gives up its energy and is pumped back into the secondary side of the steam generator as feedwater. Meanwhile, the hot primary system water, which was cooled about 75° F in going through the steam generator, is pumped back into the reactor pressure vessel to begin the cycle again. Depending on the design, a PWR will have two to four steam generators connected in parallel to the reactor pressure vessel. The PWR also has a special component in the primary system called the pressurizer, which has the function of controlling primary system pressure during normal operation.

The BWR has only a primary system, and the steam which goes to the turbine is very slightly radioactive. Both BWR and PWR have special valves on the primary system, called safety or relief valves, which are provided to open and prevent the system pressure from getting too high during a transient in flow or power.

Very large amounts of radioactivity are generated by the fission process in a nuclear power plant. However, the bulk of this radioactivity (about 98%) remains in the fuel pellets as long as they are adequately cooled. For large amounts of radioactivity to be released from the fuel, it must be grossly overheated or melt. If the chain reaction is shut down in the reactor and no more heat is being generated due to fission, the fuel is still subject to heating from the decay of radioactive materials. Immediately following the shutdown of a reactor that has operated for about one month or longer, the decay heat amounts to about 7% of the prior operating power level. While the afterheat initially decreases rapidly after reactor shutdown, it constitutes a substantial heat source for some time, and continued cooling of the fuel is required.

Overheating of fuel could occur if the heat being generated in the fuel exceeds the rate at which it is being removed. This heat imbalance in the fuel in the reactor core could occur in the following ways:

(1) A loss of coolant event might allow the fuel to overheat (due to decay heat) unless emergency cooling water is supplied to the core.

(2) Overheating of fuel can result from transient events that result in a mismatch between the reactor power level and the heat removal rate by cooling systems.

Loss-of-Coolant Accident (LOCA)

A LOCA would result if the reactor coolant system experienced a break or opening large enough so that the coolant inventory in the system could not be maintained by the normally operating, low capacity makeup system. Nuclear plants include many engineered safety features that are provided to mitigate the consequences of such an event. A brief description of the LOCA sequence, assuming that all safety features operate as designed, is as follows:

(1) A break in the coolant system occurs and the high-pressure, high-temperature coolant water is rapidly discharged into the containment building, with most of the water flashing to steam.

(2) The emergency core-cooling system (ECCS) operates to keep the core adequately cool.

(3) Any radioactivity released from the core is largely retained in the low-leakage containment building.

(4) Natural deposition processes and and radioactivity-removal systems remove the bulk of the released radioactivity from the containment atmosphere.

(5) Heat-removal systems reduce the containment pressure, thereby reducing leakage of radioactivity to the environment.

If the safety features operate as designed, the reactor core would be adequately cooled and no significant off-site consequencs would result. However, the potential consequences could be much larger if safety system failures were to result in overheating of the reactor core.

There are a number of ways in which a LOCA may be initiated. The most commonly considered initiating event would be a break in the reactor coolant system. Piping breaks that could cause a LOCA range in size from about the equivalent of a one-inch diameter hole up to the complete severance of one of the main coolant loop pipes (about three feet in diameter). Pressure vessel failure is another possible cause, albeit of a much lower probability. Large disruptive reactor vessel failures could prevent adequate cooling of the core and could cause failure of the containment building.

Reactor Transients

In general, the term "reactor transient" applies to any significant deviation from the normal operating value of any of the key reactor operating parameters. More specifically, transient events can be assumed to include all those situations (except for LOCA, which is treated separately) which could lead to fuel heat imbalances. When viewed in this way, transients cover the reactor in its shutdown condition as well as in its various operating conditions. The shutdown condition is important because many transient conditions result in shutdown of the reactor, and decay heat-removal systems are needed to prevent fuel heat imbalances due to fission product decay heat.

Transients may occur as a consequence of human error or the malfunction or failure of equipment. Many transients are handled by the reactor control system, which returns the reactor to its normal operating condition. Other are beyond the capability of the control system and require shutdown by the reactor protection system in order to avoid damage to the fuel.

In safety analyses, the principal areas of interest are increases in reactor core power (heat generation), decreases in coolent flow (heat removal), and reactor coolant system pressure increases. Any of these could result from a malfunction or failure, and they represent a potential for damage to the reactor core or the pressure boundary of the coolant system, that is, the reactor pressure vessel, the pumps, the valves, the piping, and any other components which make up the primary system.

It should be noted that the kinetics of the fussion process in LWRs are relatively sluggish. The cores are designed to have less than the optimum amount of water with regard to making the chain reaction proceed; hence any further reduction in average water density (as from a coolant temperature rise) is negative in reactivity effect (it tends to decrease the power). Furthermore, fuel heating causes a negative Doppler effect on reactivity. These factors all tend to reduce the vulnerability of LWRs to power excursions-type transients, and thus to make overpower conditions less important than inadequate heat removal as a possible cause of fuel melting in LWRs.

Finally, it is noted that a light water reactor core cannot explode in the usual sense attributed to an atomic bomb.

Engineered Safety Features

The basic purpose of the engineered safety features is the same for pressurized water reactor (PWR) and boiling water reactor (BWR) plants, the two types of light water reactor plants used in the United States. However, the nature and functions of the engineered safety features differ somewhat between PWRs and BWRs because of differences in the plant designs. A number of the safety features are included in a group termed the emergency core-cooling system (ECCS), whose function is to provide adequate cooling of the reactor core in the event of a LOCA. Other engineered safety features provide rapid reactor shutdown and reduce the containment radioactivity and pressure levels that could result from escape of the reactor coolant in the event of a LOCA. The following functional descriptions of engineered safety features apply to current designs of BWR and PWR plants:

(1) Reactor trip: to stop the fission process and terminate core power generation.

(2) Emergency core cooling: to cool the core, thereby keeping the release of radioactivity from the fuel into the containment building at low levels.

(3) Post-accident radioactivity removal: to remove radioactivity from the containment atmosphere.

(4) Post-accident heat removal: to remove decay heat from within the containment building, thereby preventing overpressurization of the containment building.

(5) Containment integrity: to prevent radioactivity within the containment building from escaping into the environment.

The course of events following a LOCA initiating event is strongly influenced by the degree of successful operation of the various engineered safety features. The ways in which failures of the above functions influence the outcome of LOCAs are discussed briefly below.

Reactor trip is accomplished by rapid insertion of the reactor control rods. The action is initiated automatically by electrical signals generated if any of a number of key operating variables reaches a present level. The way in which failure of the trip function might affect a LOCA is complicated by a number of factors. For example, in the PWR, failure of the trip function in a large LOCA is of no immediate significance since the reactor is rapidly shut down by the loss of water from the core and the injected emergency cooling water contains boron to prevent a return to power. However, there are circumstances in which trip is required.

Emergency core cooling involves a number of systems that deliver emergency coolant to the reactor core. Both PWR and BWR plants include high-pressure systems primarily for coping with small LOCAs and low-pressure systems primarily for large LOCAs. Together, these systems can fulfill the emergency core-cooling requirement over a wide range of small to large pipe breaks. Since the high- and low-pressure systems each have redundant trains of equipment to supply water, and since the various systems overlap in function, to some degree, a range of equipment failures can be tolerated without losing capability for core cooling.

Post-accident radioactivity removal is accomplished differently in the PWR and in the BWR. In the PWR, this function is performed by systems that spray water into the containment atmosphere. The water spray, which includes a chemical additive for enhancing iodine removal, washes radioactivity out of the containment atmosphere. In the BWR, this function is performed by a vapor suppression pool in the containment building through which the flashing discharge from a LOCA is passed, and by a filtering system associated with the reactor building which surrounds the containment building. The vapor suppression pool removes some of the radioactivity that would be released from the core. The filtering system removes radioactivity that might leak from the containment building into the reactor building. Failure of the radioactivity-removal systems would increase the radioactive material inventory available for leakage from the containment building.

Post-accident heat removal is performed by systems that transfer heat from hot water within the containment building to cold water outside it. The containment water that flows through the primary side of the heat exchanger is taken from the reactor building sump in the PWR and from the pressure suppression pool in the BWR. This is a particularly important function, since failure to perform this function can lead to overpressure failure of the containment building and related failure of emergency cooling systems.

Containment integrity is provided by the containment features that serve to isolate the containment atmosphere from the outside environment. The BWR primary containment building completely encloses the reactor system and is provided with a pressure suppression pool which condenses the steam and prevents overpressurization of the containment by the initial steam release to the containment in the event of a LOCA. A secondary confinement barrier is provided by the reactor building and assures that radioactivity leaking from the primary containment is discharged to the environment through filters, usually at an elevated level via a stack. In a PWR, the reactor primary system is also enclosed in a containment building. A system to quench the steam released in a PWR loss-of-coolant accident is not needed to prevent initial overpressurization because of the large volume within the containment building. However, systems are provided to remove heat and reduce the pressure in the containment building and to retain radioactivity that may be released from the core. Both PWR and BWR containments are designed to have low-leakage rates in order to inhibit the release of radioactivity to the environment.

The present designs of engineered safety features have evolved over recent years as LWRs have increased in power capacity. In early power reactors the power level was about one-tenth that of today's large reactors. It was thought that core melting in those low-power reactors would not lead to melt-through of the containment building. Further, since the decay heat was low enough to be readily transferred through the steel containment walls to the outside atmosphere, the containment building could not overpressurize and fail from this cause. Thus, if a LOCA were to occur, and even if the core were to melt, the low-leakage containments that were provided would have permitted the release of only a small amount of radioactivity.

However, as reactors grew larger, several new considerations became apparent. The decay heat levels were now so high that the heat could not be dissipated through the containment walls. Further, in the event of accidents, concrete shielding was required around the outside of the containment building to prevent overexposure of persons in the vicinity of the plant. However, the concrete also insulated the containment. Finally, it was realized that a hot core mass, once molten, could probably melt through the thick concrete containment base into the ground. Thus, a new set of requirements came into being.

Emergency core-cooling systems were needed to prevent core melting, since core melting could, in turn, cause failure of all barriers to the release of radioactivity. Systems were needed to transfer the core decay heat from the containment building to the outside environment in order to prevent the heat from producing internal pressures high enough to rupture the containment building. Systems were also needed to remove radioactivity from the containment atmosphere in order to reduce the amount that could leak from the containment into the environment.

The major goal behind these changes was to provide safety features designed so that the failure of any single barrier would not be likely to cause the failure of any of the other barriers. For example, if the reactor coolant system were to rupture, emergency core-cooling systems were installed to prevent the fuel from melting and thereby protect the integrity of the containment. Other features were added to aid this positive objective. For example, piping restraints and protective shields were required to lessen the likelihood of damage to engineered safety features that could result from pipe whip following a large pipe break. Awareness that large natural forces such as earthquakes and tornadoes could cause multiple failures led to design requirements to reduce the likelihood of dependent failures from such causes.

The net result of the addition of engineered safety features in current large reactors is to reduce the likelihood of accidents that could have significant public impact. However, it is apparent that there are important interrelationships between the possible failures of various safety features and either the need for, or the ability of, other safety features to perform their functions. For instance, in the event of a LOCA, if all safety features operate as designed, the consequences to the public would be negligible because very little radioactivity would be released from the core and much of that would be removed by containment systems provided for that purpose. However, if all of the multiple electric power sources were to fail, the likelihood of reactor trip and containment isolation would be enhanced, but the emergency core cooling, radioactivity removal, and post-accident heat removal functions could not be performed. Thus, the core would be likely to melt, removal of radioactivity from the containment atmosphere would occur only by natural deposition processes, and the containment building might fail due to overpressure.

Containment failure modes, their timing, and the potential radioactive release depend strongly on the operability of the various safety features. In the event of core melting, the WASH–1400 report (NRC 1975a) predicted that the core mass could melt through the bottom of the reactor vessel and the thick, lower concrete structure of the containment in one-half to one day after the accident. This time could provide considerable time for radioactive decay, for plate out of radioactive particles on the containment walls, or for washout of the radioactivity from the containment atmosphere by water sprays, thereby reducing the level of radioactivity available to escape, assuming the containment integrity was not breached at the beginning of the accident. Furthermore, if most of the gaseous and particulate radioactivity that might be released were to be discharged into the ground, the soil acts as an efficient filter and significantly reduces the radioactivity released above ground. Accidents that would follow this path and lead to containment failure in a downward direction into the ground below are thus characterized by lesser releases to the atmosphere and reduced consequences to health and safety. In plants that have containment buildings with relatively large volumes, the melt-through path described represents the most likely course of the accident, according to the WASH–1400 report (NRC 1975a). More recent studies are suggesting that some change is appropriate in the failure mode analyses made in the WASH–1400 report.

Following such a melt-through, there would be the possibility of ground water contamination through a long-term process of leaching of radioactivity from the solidifying mass of fuel and soil. The leaching and contamination processes should occur over an extended period of time (several to many years, depending on the particular radioactive species) and the potential contamination levels might not be substantially larger than the maximum permissible concentrations.

Containment buildings may also fail from overpressure resulting from a combination of steam, the burning of hydrogen, and the accumulation of various noncondensible gases, such as hydrogen and carbon dioxide, released within the containment building as a result of core melting. For small containment buildings, the pressure due to these gases would represent the most likely path to containment failure. Such failures would most likely occur in the above-ground portion of the containment, and several hours from the time of core melting.

At two key stages in the course of a core meltdown there would be conditions which may have the potential to result in a steam explosion that could rupture the reactor vessel or the containment building. The term "steam explosion" refers to a phenomenon in which the fuel is in finely divided form and intimately mixed with water so that its thermal energy is efficiently and rapidly deposited in the water, creating a large amount of steam in a short time. These conditions might occur if molten fuel fell from the core region into water at the bottom of the reactor vessel, or if the core melted through the bottom of the reactor vessel and fell into water in the bottom of the containment building. These modes of containment failure have usually been predicted to have lower probabilities of occurrence than overpressurization or melt-through of the foundation. Recent studies have suggested that the probability of containment failure due to steam explosion is very small.

The only published detailed study of LWR accident possibilities and consequences, for accidents going beyond the plant design basis and including core melting, was carried out in 1973–1975 and is reported in the WASH–1400 report (NRC 1975a).* The WASH–1400 study was performed on two actual plants, one PWR and one BWR. For each reactor, possible accident sequences were collected into radioactivity-release categories and the overall probability and consequences of each release category was calculated.

The probability side of the study was based on the fault tree-event tree methodology.† The authors of the study believed that an understanding of what

*In 1979 the Federal Rupublic of Germany reported in summary form the results of a similar study made for a specific pressurized water reactor in Germany.

†Fault tree and event tree methodology is a systematic way of looking at the structure of a complicated system graphically and examining which combinations of component failures can lead to failure of the entire system and which sets of systems are important in any accident scenario.

makes accidents hazardous and a knowledge of the engineering design of the reactor made it practical for them to handle the large number of theoretically possible accident sequences and to cut this array down to a number that was manageable. Further, the authors stated that it was not essential to include all accident sequences, but only to identify enough of the higher-probability sequences so that the overall probability of a given level of release became insensitive to the addition of more low-probability sequences. In the WASH–1400 report (NRC 1975a), the overall probabilities of the release categories were usually determined by a rather small number (five to ten in each category) of dominant sequences (i.e., high-probability sequences) and many of these are single-failure events. A summary of the results obtained for release probabilities is given in table A1.

On the consequence side, the first step is to characterize the nature and amount of radioactivity release for each release category. Table A1 also lists the essential release information. The most severe categories for both PWRs and BWRs involve release of about half of the total core inventory.

The next steps are transport of radioactivity through the atmosphere with associated plume spread, meander, and depletion processes, and the calculation of inhalation and whole body doses, health effects, and land contamination effects.

The consequences calculated in the WASH–1400 report (NRC 1975a) are presented in tables A2 and A3 for a "synthesized" reactor whose population distribution and meteorology is supposed to be an average of 100 reactors. The uncertainty ranges in the probability and consequences estimates are stated in the WASH–1400 report to be approximately a factor of 5. The consequences are calculated using an evacuation model based on some limited previous experience with evacuation for non-nuclear accidents; it is also assumed that people will not return to land having an excessive radiation level until cleanup measures have been taken.

According to the calculations in the WASH–1400 report, if one hypothesizes a major uncontrolled release of radioactivity due to rupture of the containment building above ground, under the more adverse meteorological conditions (low wind speed, little dispersion of radioactivity), the radiation dose rate in a relatively narrow plume could be lethal out to more than 15 miles. However, such a plume would be less difficult to handle via evacuation measures. In, the WASH–1400 report it is estimated that even if such a large radioactivity release occurred, the probability that a person living 2 miles from the plant would receive a lethal dose is less than 1 in 100; for a person 5 miles from the plant, the probability was estimated to be less than 1 in 1,000.

From such a large release, a considerable number of latent cancer deaths might occur, but the probability for individuals living more than 4 miles away is estimated in the WASH–1400 report (NRC 1975 a) to be less than 1 in 10,000.

Many questions have been raised concerning the quantification of the uncertainty in the best estimate results, and there is a considerable body of opinion that the stated uncertainty range of plus or minus a factor of about 5 in the core melt estimates has not been validated and is too small. More specifically, questions have been raised concerning the WASH–1400 treatment of common-mode failures, reactor aging, human errors, fires, earthquakes, and sabotage, among others, in

TABLE A1
Summary of Accident Involving Core Melting (NRC 1975a)

Release Category	Probability per Reactor-Year	Time of Release (hr.)	Duration of Release (hr.)	Warning Time for Evacuation (hr.)	Elevation of Release (Meters)	Containment Energy Release (10^6 Btu/hr.)
PWR 1	9×10^{-7}	2.5	0.5	1.0	25	520[a]
PWR 2	8×10^{-6}	2.5	0.5	1.0	0	170
PWR 3	4×10^{-6}	5.0	1.5	2.0	0	6
PWR 4	5×10^{-7}	2.0	3.0	2.0	0	1
PWR 5	7×10^{-7}	2.0	4.0	1.0	0	0.3
PWR 6	6×10^{-6}	12.0	10.0	1.0	0	N/A
PWR 7	4×10^{-5}	10.0	10.0	1.0	0	N/A
PWR 8	4×10^{-5}	0.5	0.5	N/A	0	N/A
PWR 9	4×10^{-4}	0.5	0.5	N/A	0	N/A
BWR 1	1×10^{-6}	2.0	2.0	1.5	25	130
BWR 2	6×10^{-6}	30.0	3.0	2.0	0	30
BWR 3	2×10^{-5}	30.0	3.0	2.0	25	20
BWR 4	2×10^{-6}	5.0	2.0	2.0	25	N/A
BWR 5	1×10^{-4}	3.5	5.0	N/A	150	N/A

Fraction of Core Inventory Released							
Xe–Kr	Organic I	I	Cs–Rb	Te–Sb	Ba–Sr	Pu[b]	La[c]
0.9	6×10^{-3}	0.7	0.4	0.4	0.05	0.4	3×10^{-3}
0.9	7×10^{-3}	0.7	0.5	0.3	0.06	0.02	4×10^{-3}
0.8	6×10^{-3}	0.2	0.2	0.3	0.02	0.03	3×10^{-3}
0.6	2×10^{-3}	0.09	0.04	0.03	5×10^{-3}	3×10^{-3}	4×10^{-4}
0.3	2×10^{-3}	0.03	9×10^{-3}	5×10^{-3}	1×10^{-3}	6×10^{-4}	7×10^{-5}
0.3	2×10^{-3}	8×10^{-4}	8×10^{-4}	1×10^{-3}	9×10^{-5}	7×10^{-5}	1×10^{-5}
6×10^{-3}	2×10^{-5}	2×10^{-5}	1×10^{-5}	2×10^{-5}	1×10^{-6}	1×10^{-6}	2×10^{-7}
2×10^{-3}	5×10^{-6}	1×10^{-4}	5×10^{-4}	1×10^{-6}	1×10^{-8}	0	0
3×10^{-6}	7×10^{-9}	1×10^{-7}	6×10^{-7}	1×10^{-9}	1×10^{-11}	0	0
1.0	7×10^{-3}	0.40	0.40	0.70	0.05	0.5	5×10^{-3}
1.0	7×10^{-3}	0.90	0.50	0.30	0.10	0.03	4×10^{-3}
1.0	7×10^{-3}	0.10	0.10	0.30	0.01	0.02	3×10^{-3}
0.6	7×10^{-4}	8×10^{-4}	5×10^{-3}	4×10^{-3}	6×10^{-4}	6×10^{-4}	1×10^{-4}
5×10^{-4}	2×10^{-9}	6×10^{-11}	4×10^{-9}	8×10^{-12}	8×10^{-14}	0	0

[a] A lower energy release rate than this value applies to part of the period over which the radioactivity is being released.

[b] Includes Mo, Rh, Tc, Co.

[c] Includes Nd, Y, Ce, Pr, La, Nb, Am, Cm, Pu, Np, Zr.

ascertaining the probability of core meltdown. And the question remains: Are there undiscovered accident sequences with relatively large probabilities?

Among the principal questions raised with regard to the conquences presented in the WASH–1400 report (NRC 1975a) are the following: Is the estimate of health effects due to low-level radiation sufficiently conservative? Has the efficiency of evacuation and decontamination procedures been over-estimated? Should not the risk be given for each individual site rather than some weighted average? Are the potential effects on drinking water and other long-term dose commitment effects treated adequately?

TABLE A2

Near-Term consequences of reactor accidents for various probabilities for one reactor (NRC 1975a)

	Consequence				
Chance per Reactor-Year	Early Fatalities	Early Illness	Total Property Damage (10^9)	Decontamination Area (Square Miles)	Relocation Area (Square Miles)
One in 20,000[a]	<1.0	<1.0	<0.1	<0.1	<0.1
One in 1,000,000	<1.0	300	0.9	2000	130
One in 10,000,000	110	3000	3	3200	250
One in 100,000,000	900	14,000	8	–	290
One in 1,000,000,000	3300	45,000	14	–	–

[a] This is the predicted chance of core melt per reactor-year.

TABLE A3

Delayed consequences of reactor accidents for various probabilities for one reactor (USNRC 1975)

	Consequences		
Chance per Reactor-Year	Latent Cancer[b] Fatalities (per year)	Thyroid Nodules[b] (per year)	Genetic Effects[c] (per year)
One in 20,000[a]	<1.0	<1.0	<1.0
One in 1,000,000	170	1400	25
One in 10,000,000	460	3500	60
One in 100,000,000	860	6000	110
One in 1,000,000,000	1500	8000	170
Normal Incidence	17,000	8000	8000

[a] This is the predicted chance of core melt per reactor-year.
[b] This rate would occur approximately in the 10 to 40 year period following a potential accident. Multiply by 30 to get total latent cancer fatalities or thyroid nodules.
[c] This rate would apply to the first generation born after a potential accident. Subsequent generations would experience effects at lower rate.

Appendix II:
Licensed United States
Nuclear Power Plants,
November 30, 1979

Facility Name	*State*
Arkansas 1,2	Arkansas
Beaver Valley 1	Pennsylvania
Big Rock Point 1	Michigan
Browns Ferry 1, 2, 3	Alabama
Brunswick 1, 2	North Carolina
Calvert Cliffs, 1, 2	Maryland
Cook 1, 2	Michigan
Cooper Station	Nebraska
Crystal River 3	Florida
Davis-Besse 1	Ohio
Dresden 1, 2, 3	Illinois
Duane Arnold	Iowa
Farley 1	Alabama
Fitzpatrick	New York
Fort Calhoun 1	Nebraska
Fort St. Vrain	Colorado
Ginna	New York
Haddam Neck	Connecticut
Hatch 1, 2	Georgia
Humboldt Bay	California
Indian Point 1, 2, 3	New York
Kewaunee	Wisconsin
La Crosse	Wisconsin
Maine Yankee	Maine
Millstone 1, 2	Connecticut
Monticello	Minnesota
Nine Mile Point 1	New York
North Anna 1	Virginia
Oconee 1, 2, 3	South Carolina

Facility Name	*State*
Oyster Creek 1	New Jersey
Palisades	Michigan
Peach Bottom 2, 3	Pennsylvania
Pilgrim 1	Massachusetts
Point Beach 1, 2	Wisconsin
Prairie Island 1, 2	Minnesota
Quad Cities 1, 2	Illinois
Rancho Seco 1	California
Robinson 2	South Carolina
Salem 1	New Jersey
San Onofre 1	California
St. Lucie 1	Florida
Surry 1, 2	Virginia
Three Mile Island 1, 2	Pennsylvania
Trojan	Oregon
Turkey Point 3, 4	Florida
Vermont Yankee 1	Vermont
Yankee-Rowe 1	Massachusetts
Zion 1, 2	Illinois

Appendix III
United States Nuclear Power Plants
with Construction Permits,
December 31, 1979

Facility Name	*State*
Bailly 1	Indiana
Beaver Valley 2	Pennsylvania
Bellefonte 1, 2	Alabama
Black Fox 1, 2	Oklahoma
Braidwood 1, 2	Illinois
Byron 1, 2	Illinois
Callaway 1, 2	Missouri
Catawba 1, 2	South Carolina
Cherokee 1, 2 3	South Carolina
Clinton 1, 2	Illinois
Comanche Peak 1, 2	Texas
Davis-Besse 2, 3	Ohio
Diablo Canyon 1, 2	California
Enrico Fermi 2	Michigan
Farley 2	Alabama
Forked River	New Jersey
Grand Gulf 1, 2	Mississippi
Harris 1, 2, 3, 4	North Carolina
Hartsville A-1, A-2, B-1, B-2	Tennessee
Hope Creek 1, 2	New Jersey
Jamesport 1, 2	New York
LaSalle 1, 2	Illinois
Limerick 1, 2	Pennsylvania
Marble Hill 1, 2	Indiana
McGuire 1, 2	North Carolina
Midland 1, 2	Michigan
Millstone 3	Connecticut
Nine Mile Point 2	New York
North Anna 2, 3, 4	Virginia

Facility Name	*State*
Palo Verde 1, 2, 3	Arizona
Perry 1, 2	Ohio
Phipps Bend 1, 2	Tennessee
River Bend 1, 2	Louisiana
Salem 2	New Jersey
San Onofre 2, 3	California
Seabrook 1, 2	New Hampshire
Sequoyah 1, 2	Tennessee
Shoreham	New York
South Texas 1, 2	Texas
St. Lucie 2	Florida
Sterling 1	New York
Summer 1	South Carolina
Susquehanna 1, 2	Pennsylvania
Vogtle 1, 2	Georgia
Washington Nuclear 1, 2, 3, 4, 5	Washington
Waterford 3	Louisiana
Watts Bar 1, 2	Tennessee
Wolf Creek 1	Kansas
Yellow Creek 1, 2	Mississippi
Zimmer 1	Ohio

References

ACRS 1974. *Report on integrity of reactor vessels for light water reactors*, U.S. Atomic Energy Commission report WASH-1285.

Adams, C. A., and Stone, C. N. 1967. Safety and siting of nuclear power stations in the United Kingdom. *Containment and siting of nuclear power plants: Proc. Symp. Vienna*, p. 129. IAEA.

AEC 1950. *Summary report of Reactor Safeguards Committee*. U.S. Atomic Energy Commission report WASH-3.

AEC 1957. *Theoretical possibilities and consequences of major accidents in large nuclear power plants*. U.S. Atomic Energy Commision report WASH-740.

AEC 1965a. Record of the AEC regulatory staff symposium held on April 29, 1965 at Germantown, Maryland, on possible zirconium-water reactions in water reactors. Division of Safety Standards, U.S. Atomic Energy Commission.

AEC 1965b. Proposed general design criteria for construction permits for nuclear power plants. U.S. Atomic Energy Commission. *Federal Register* November 22, 1965.

AEC 1967. Proposed general design criteria for nuclear power plants. U.S. Atomic Energy Commission. *Federal Register*, July 10, 1967.

AEC 1971a General design criteria for nuclear power plants, Title 10 Code of Federal Regulations, Part 50. U.S. Atomic Engergy Commission. *Federal Register*, February 19, 1971.

AEC 1971b. Licensing and regulatory policy and procedures for environmental protection. Title 10 Code of Federal Regulations Part 51, U.S. Atomic Energy Commission, *Federal Register*, December 1, 1971

AEC 1973a. *Anticipated transients without scram for water-cooled power reactors*. U.S. Atomic Energy Commission report WASH-1270.

AEC 1973b. *Safety of nuclear power reactors (light water-cooled) and related facilities*. U.S. Atomic Engery Commission report WASH-1250 (final draft).

AEC 1974a. *Draft WASH*-1400, Reactor Safety Study. An assessment of accident risks in U.S. commercial nuclear power plants. U.S. Atomic Energy Commission.

AEC 1974b. *Technical report on analysis of pressure vessel statistics from fossil-fueled power plant service and assessment of reactor vessel reliability in nuclear power plant service*. U.S. Atomic Energy Commission report WASH-1318.

Beck, C. K. 1959 Safety factors to be considered in reactor siting. *Proc. VI Rassegna Internazionale Elettronica e Nucleare, Rome*, p. 21.

Cave, L., and Ilberg, D. 1977. *Relative hazard potential—the basis for definition of safety criteria for fast reactors*. University of California, Los Angeles report UCLA-ENG-7692.

Charlesworth, F. R., and Gronow, W. S. 1967. A summary of experience in the practical application of siting policy in the United Kingdom. *Containment and siting of nuclear power plants: Proc. Symp. Vienna*, p. 168. IAEA.

Clinch River 1977. CRBRP Safety Study. An Assessment of Accident Risks in the CRBRP. Clinch River Breeder Reactor Project office, unpublished report CRBRP 1.

Cornell, C. A., and Newmark, N. M. 1978. On the seismic reliability of nuclear power plants. *Proc. Conf. on probabilistic analysis of nuclear reactor safety, Newport Beach*, vol. 3, p. XIV 1-1. American Nuclear Soc.

DiNunno, J. J., Anderson, F. D., Baker, R. E., and Waterfield, R. L. 1962. *Calculation of distance factors for power and test reactor sites*. U.S. Atomic Energy Commission Technical Information Document TID-14844.

EPRI 1976a. *ATWS: A reappraisal part I, an examination and analysis of WASH-1270, technical report on ATWS for water-cooled power reactors*. Electric Power Research Institute report EPRI NP-251.

EPRI 1976b. *ATWS: A reappraisal part II, evaluation of societal risks due to reactor protection system failure*. Electrtic Power Research Institute report EPRI-NP-265, vol. 1, 2 and 3.

EPRI 1977. *ATWS: A reppraisal part II, evaluation of societal risks due to reactor protection system failure*. Electric Power Research Institute report EPRI NP-265, vol. 4.

EPRI 1978. *ATWS: A reappraisal, part III, frequency of anticipated transients*. Electric Power Research Institute report EPRI NP-801.

Farmer, F. R. 1964. Reactor safety analysis as related to reactor siting. *Proc. 3rd United Nations Conf. on peaceful uses of atomic energy, Geneva*, paper P/182. United Nations.

Farmer, F. R., 1967. Siting criteria—a new approach. *Containment and siting of nuclear power plants: Proc. Symp. Vienna*, p. 303. IAEA.

Ford-Mitre 1977. *Nuclear power issues and choices*. Ballinger Publishing Co., Cambridge, Mass.

Gillette, R., 1972. Nuclear safety. *Science* 177: 771, 867, 970 and 1080.

Hewlett, R. G. 1974. The evolving role of the advisory committee on reactor safeguards. U.S. Atomic Energy Commission unpublished report.

Hsieh, T. and Okrent, D., 1977. On design errors and system degradation in seismic safety, *Proc. 4th Int. Conf. on structural mechanics in reactor technology (SMIRT), San Francisco*, vol. K, p. 9/4. International Association for Structural Mechanics in Reactor Technology.

IAEA 1962. *Safety evaluation of nuclear power plants: Proc. Symp. Vienna*. IAEA.

IAEA 1963. *Siting of reactors in nuclear research centers: Proc. Symp. Bombay*. IAEA.

IAEA 1967. *Containment and siting of nuclear power plants: Proc. Symp. Vienna*. IAEA.

IAEA 1973. *Principles and standards of reactor safety: Proc. Symp. Julich*. IAEA.

IEEE 1967. Proposed standard for nuclear power plant protection systems. Institute of Electrical and Electronics Engineers.

Irvine, W. H., Quirk, A., and Bevitt, E. 1964. Fast fracture of pressure vessels: An appraisal of theoretical aspects and application to operational safety. *J. Brit. Nuclear Energy Soc.* 3: 31.

Johnson, W. E. 1962. Principles and practices in consequences limiting safeguards in facility design. *Reactor safety and hazard evaluation techniques: Proc. Symp. Vienna*, p. 349. IAEA.

Kellerman, O., and Seipel, H. G. 1967. Analysis of the improvement in safety obtained by a containment and by other safety devices for water-cooled reactors, *Containment and siting of nuclear power plants: Proc. Symp. Vienna*, p. 403. IAEA.

Levy, S. 1967. A systems approach to containment design in nuclear power plants. *Containment and siting of nuclear power plants: Proc. Symp. Vienna*, p. 227. IAEA.

Lewis, H. W., Chairman 1975. Report to the American Physical Society by the study group on light water reactor safety. *Rev. Mod. Phys.* 47: Supp 1, S1.

Nichols, R. W., Irvine, W. H., Quirk, A., and Bevitt, E. 1965. *A limit approach to the prevention of pressure vessel failure.* United Kingdom Atomic Energy Authority report TRG 1004(C).

NRC 1975a. *Reactor Safety Study: An assessment of accident risks in U.S. commercial nuclear power plants.* U.S. Nuclear Regulatory Commission report WASH–1400 (NUREG–75/014).

NRC 1975b. *Standard Review Plan.* U.S. Nuclear Regulatory Commission report NUREG 75/087.

NRC 1978a. *Report of risk assessment review group.* H. Lewis, Chairman. U.S. Nuclear Regulatory Commission report NUREG/CR–0400.

NRC 1978b. *Report to Congress by NRC on plan for research to improve the safety of light water nuclear power plants.* U.S. Nuclear Regulatory Commission report NUREG–0438.

NRC 1978c. *Anticipated transients without scram for light water reactors.* U.S. Nuclear Regulatory Commission report. vols. 1 and 2.

Okrent, D. 1975. A survey of expert opinion on low probability earthquakes. *Ann. Nuclear Energy*, 2: 601.

Okrent, D. 1979. On the history of the evolution of light water reactor safety in the United States. University of California, Los Angeles, unpublished.

Shibata, H., 1970. *Aseismic design of nuclear power plants—developments of these ten years.* Japan Atomic Energy Research Institute report NSG–Tr 161.

United Nations 1955. *Proc. 1st United Nations Conference on peaceful uses of atomic energy, Geneva.*

United Nations 1958. *Proc. 2nd United Nations Conference on peaceful uses of atomic energy, Geneva.*

United Nations 1964. *Proc. 3rd United Nations Conference on peaceful uses of atomic energy, Geneva.*

United Nations 1971. *Proc. 4th United Nations Conference on peaceful uses of atomic energy, Geneva.*

Vinck, W. F., and Maurer, H. 1967. Some examples of the relationship between containment and other engineered safeguard requirements, accident analysis, and site conditions. *Containment and siting of nuclear power plants: Proc. Symp. Vienna*, p. 383. IAEA.

Index

Acceptable risk. *See* Risk, acceptable

Accident effects: WASH-740 estimates, 8; public discussion, 94; ACRS estimates, 98; WASH-740 revision, 99, 101, 108–10; Wash-1400, 319–22, 346–49

Accidents: Windscale reactor, 37; SL-1 reactor, 306; classification, 336–38; discussion, 341–50. *See also* Accident effects; Class 9 accidents; Maximum credible accident; Three Mile Island 2 reactor

ACRS. *See* Advisory Committee on Reactor Safeguards

Adams, C.A., 182

Advisory Committee on Reactor Safeguards (ACRS): formation, 4; becomes statutory committee, 7; role in licensing process, 8–11; first dissent, 25–26; called too conservative, 27; criticized for pressure vessel letter, 89; independence attacked by Clifford Beck, 122; causes major changes in LWR safety and regulation, 124; supported by Congressman Aspinall, 183; attacked by Congressman Hosmer, 184; non-mandatory review, 186

—ACRS reports: Indian Point 1, 20; Plum Brook, 25; Elk River, 26; Piqua, 26–27; site comparisons, 27; California sites, 34; Point Loma, 36; Jamestown, 36; reactor siting (October 22, 1960), 40; Los Angeles sites, 60; San Onofre 1, 64; Connecticut Yankee, 64; engineered safeguards, 67; Malibu, 67; metropolitan siting (draft), 79; pressure vessels, 88; Dresden 3, 127; Indian Point 2, 129; problems arising from primary system rupture, 131; testimony to JCAE, 141; oral report on Burlington, 144; Browns Ferry, 147; Zion, 155; safety research (October 12, 1966), 165, 307; safety research (April 14, 1967), 171; ECCS Task Force report, 174; comments

to Milton Shaw on a core retention system, 176; oral report on Trap Rock, 191; location of power reactors at sites of population density greater than Indian Point-Zion (draft), 198; Newbold Island site, 208; Newbold Island, 209; pressure vessel integrity, 224; separation of protection and control, 237, 238, 243; Bodega Bay (April 18, 1963), 264; Bodega Bay (May, 1964 draft), 267; Bodega Bay (October 20, 1964), 269; memo on seismic criteria (June 11, 1973), 282; memo to Harold Price on generic issues, 291; ECCS interim acceptance criteria (draft), 355; ECCS interim acceptance criteria, 302; AEC concluding statement on ECCS, 304; safety research (March 20, 1969), 311; safety research (February 10, 1972), 312; safety research (November 20, 1974), 314; WASH-1400, 333; to Congressman Morris Udall, 333, 335

Ahearne, John, 260

American Nuclear Society, 183, 185

American Physical Society, 180

American Society of Mechanical Engineers (ASME), 221–22, 222*n*

Anders, William A., 257, 318, 321

Anticipated transients without scram (ATWS): definition, 203; and pressure vessel integrity, 225; history, 236–60; Epler letter, 243; 1970 staff position, 247; mitigation measures, 247; 1972 staff position, 252; staff issues WASH-1270, 254–55; staff change position, 255–60; NUREG-0460 issued, 258–59; EPRI role, 258; Browns Ferry failure to scram, 260; as generic item, 291, 294, 332

Argonne National Laboratory, 108, 116, 239, 298

Aspinall, Wayne, 183–84

359

Vented-filtered containment, 168, 170, 180
Vermont Yankee reactor, 290

WASH-740 report: issuance in 1957, 8;
reevaluation, 98–102, 107–11
WASH-1270 report. *See* Anticipated transients without scram
WASH-1285 report. *See* Pressure vessels
WASH-1400. *See* Reactor Safety Study
Weinberg, Alvin, 244
Wensch, Glen, 101, 111
West, J.M., 195–97
Western, F., 108
Westinghouse Electric Company: design of
PWR, 6; Ravenswood reactor, 71; Brookwood reactor, 94–95; pressure vessel
failure, 96; metal-water reactions, 106; and
China Syndrome, 111, 112; core catcher,
112, 115, 117, 118, 152; on metropolitan
siting, 146, 195; Zion reactors, 155; ECCS,
233; separation of protection and control
236–39, 242, 250; and ATWS, 247
West Milton, New York, Submarine Intermediate Reactor, 19
Wexler, Harry, 4
Whelchel, Cornelius C., 267
Whitman, Robert, 285
Wiesemann, Robert A., 74, 106, 111
Williams, Charles R., 44n, 279
Williamson, Robert A., 265, 279

Wilson, James T., 204
Wilson, R.E., 231, 232
Windscale, Great Britain reactor accident,
37
Wolman, Abel, 4, 25, 44n

Yankee reactor, 25

Zabel, Carroll: ACRS meeting attendance,
88, 123n, 144n; China Syndrome, 121;
ACRS relationship with AEC, 123; possible dissent on Browns Ferry, 137; and
Zion letter, 160; Bolsa Island review, 190;
on separation of protection and control,
242
Zion reactors: pressure vessels, 97; metropolitan siting, 135, 146, 189, 201; population density, 136, 188; generally, 150–62;
construction permit application, 150;
comparison with Indian Point, 153;
ACRS report, 155; dissent by Okrent,
157–59; Hendrie letter to Okrent, 159–61;
and Trap Rock site, 192; post TMI 2
review, 328
Zion, sites worse than, 193–202 *passim*
Zircaloy-steam reaction: for Connecticut
Yankee, 65; description, 105; AEC Symposium, 106; study group report, 107. *See
also* Cladding, Zircaloy

JACKET DESIGNED BY MIKE JAYNES
COMPOSED BY DUARTE COMPANY, LEWISTON, MAINE
MANUFACTURED BY INTER-COLLEGIATE PRESS, INC.
SHAWNEE MISSION, KANSAS
TEXT AND DISPLAY LINES ARE SET IN TIMES ROMAN

Library of Congress Cataloging in Publication Data
Okrent, David.
Nuclear reactor safety.
Bibliography: pp. 355-358
Includes index.
1. Nuclear reactors—United States—Safety
measures—History. 2. Nuclear reactors—Safety
regulations—United States—History. 3. United
States. Atomic Energy Commission. Advisory Committee
on Atomic Safeguards—History. I. Title.
TK9152.035   621.48'35   80-53958
ISBN 0-299-08350-0   AACR2